Applied Panel Data Analysis for Economic and Social Surveys

Hans-Jürgen Andreß · Katrin Golsch ·
Alexander W. Schmidt

Applied Panel Data Analysis for Economic and Social Surveys

Hans-Jürgen Andreß
Lehrstuhl für Empirische Sozial-
und Wirtschaftsforschung
Universität Köln
Cologne, Germany

Alexander W. Schmidt
Lehrstuhl für Empirische Sozial-
und Wirtschaftsforschung
Universität Köln
Cologne, Germany

Katrin Golsch
Fakultät für Soziologie
Universität Bielefeld
Bielefeld, Germany

All necessary data sets and Stata syntax files on http://eswf.uni-koeln.de/panel

ISBN 978-3-642-43417-4   ISBN 978-3-642-32914-2 (eBook)
DOI 10.1007/978-3-642-32914-2
Springer Heidelberg New York Dordrecht London

© Springer-Verlag Berlin Heidelberg 2013
Softcover re-print of the Hardcover 1st edition 2013
This work is subject to copyright. All rights are reserved by the Publisher, whether the whole or part of the material is concerned, specifically the rights of translation, reprinting, reuse of illustrations, recitation, broadcasting, reproduction on microfilms or in any other physical way, and transmission or information storage and retrieval, electronic adaptation, computer software, or by similar or dissimilar methodology now known or hereafter developed. Exempted from this legal reservation are brief excerpts in connection with reviews or scholarly analysis or material supplied specifically for the purpose of being entered and executed on a computer system, for exclusive use by the purchaser of the work. Duplication of this publication or parts thereof is permitted only under the provisions of the Copyright Law of the Publisher's location, in its current version, and permission for use must always be obtained from Springer. Permissions for use may be obtained through RightsLink at the Copyright Clearance Center. Violations are liable to prosecution under the respective Copyright Law.
The use of general descriptive names, registered names, trademarks, service marks, etc. in this publication does not imply, even in the absence of a specific statement, that such names are exempt from the relevant protective laws and regulations and therefore free for general use.
While the advice and information in this book are believed to be true and accurate at the date of publication, neither the authors nor the editors nor the publisher can accept any legal responsibility for any errors or omissions that may be made. The publisher makes no warranty, express or implied, with respect to the material contained herein.

Printed on acid-free paper

Springer is part of Springer Science+Business Media (www.springer.com)

Contents

1 Introduction 1
   1.1 Benefits and Challenges of the Panel Design 2
       1.1.1 Benefits 2
       1.1.2 Challenges 7
   1.2 Outline of the Book 12
   1.3 Audience and Prerequisites 12

2 Managing Panel Data 15
   2.1 The Nature of Panel Data 16
   2.2 The Basics of Panel Data Management 19
       2.2.1 Merging and Appending Data 20
       2.2.2 Basic Append and Merge Commands 23
       2.2.3 Building a Working Data Set with Append and Merge Commands 25
       2.2.4 Wide and Long Format 27
       2.2.5 Some General Remarks 30
   2.3 Three Case Studies on Poverty in Germany 31
       2.3.1 Case Study 1: How Many German Citizens Were Poor in 2004? A Cross-Sectional Analysis 33
       2.3.2 Case Study 2: Did Poverty Increase in Germany After 2004? An Analysis of Pooled Cross-Sections 40
       2.3.3 Case Study 3: How Large Is the Risk of Becoming Poor in Germany? A Panel Analysis 42
   2.4 How to Represent a Population with Panel Data? 49
       2.4.1 Weighted and Unweighted Analysis of Cross-Sections ... 50
       2.4.2 Weighting in Balanced and Unbalanced Panels 52
       2.4.3 When to Use Weights? 55
   2.5 Conclusion and Further Reading 58

3 Describing and Modeling Panel Data 61
   3.1 Some Basic Terminology 62
   3.2 Measurements over Time Are Not Independent 65
   3.3 Describing the Dependent Variable 70
   3.4 Explaining the Dependent Variable over Time: Typical Explanatory Variables 79

		3.4.1 Time-Constant and Time-Varying Variables 79
		3.4.2 Serially Correlated Observations 83
	3.5	Modeling Panel Data 86
		3.5.1 Modeling the Level of the Dependent Variable 87
		3.5.2 Modeling Change of the Dependent Variable 89
		3.5.3 Additional Models 92
	3.6	Estimating Models for Panel Data 98
		3.6.1 Omitted Variable Bias (Unobserved Heterogeneity) 99
		3.6.2 Serially Correlated and Heteroscedastic Errors 107
		3.6.3 Measurement Error Bias 109
		3.6.4 A Formal Summary of the Main Estimation Assumptions . 114
	3.7	Overview of Subsequent Chapters 116

4 Panel Analysis of Continuous Variables 119
    4.1 Modeling the Level of Y 120
        4.1.1 Ignoring the Panel Structure 121
        4.1.2 Modeling the Panel Structure 126
        4.1.3 Extensions 174
    4.2 Modeling the Change of Y 180
        4.2.1 Analysis of Change Using Change Scores 180
        4.2.2 Analysis of Change Using Impact Functions 185
        4.2.3 Analysis of Trends 192
    4.3 Conclusion and Further Reading 201

5 Panel Analysis of Categorical Variables 203
    5.1 Modeling the Level of Y: Discrete Response Models
        for Panel Data 204
        5.1.1 Ignoring the Panel Structure 206
        5.1.2 Modeling the Panel Structure 227
        5.1.3 Extensions 247
    5.2 Modeling the Change of Y: Discrete-Time Event History Models
        for Panel Data 248
        5.2.1 Basic Terminology 251
        5.2.2 How to Estimate a Discrete-Time Hazard Model 254
        5.2.3 Applying the Discrete-Time Event History Model 262
        5.2.4 Extensions 281
    5.3 Conclusion and Further Reading 284

6 How to Do Your Own Panel Analysis 287

7 Useful Background Information 293
    7.1 Functions of Random Variables 293
    7.2 Estimation and Testing 295
        7.2.1 Ordinary Least Squares 295
        7.2.2 Maximum Likelihood 307
    7.3 Web Site of the Textbook 312

References 313
Index 319
Author Index 325

List of Figures

Fig. 1.1	Successful interviews with persons and households (SOEP: Samples A and B)	8
Fig. 2.1	Appending data	21
Fig. 2.2	Merging data with one key variable	21
Fig. 2.3	Merging data with two key variables	22
Fig. 2.4	Many-to-one merge	22
Fig. 2.5	Unweighted German poverty rates (2004–2006)	42
Fig. 2.6	Weighted and unweighted German poverty rates (2004–2006)	52
Fig. 3.1	Survival probability by income level	73
Fig. 3.2	Income trajectories of 19 individuals with very high and very low wages	76
Fig. 3.3	Explaining panel data	80
Fig. 3.4	Impact functions	90
Fig. 3.5	Log hourly wages by labor force experience and education	101
Fig. 3.6	Path diagram of hourly wages	112
Fig. 4.1	Residuals (connected by *lines*) for each of ten individuals	125
Fig. 4.2	Government spending in Ireland and Sweden (1961–1993)	137
Fig. 4.3	Explaining between and within variance (uncorrelated u_i)	161
Fig. 4.4	Explaining between and within variance (correlated u_i)	161
Fig. 4.5	Demeaning parameter θ with respect to number of measurements T	172
Fig. 4.6	Path diagram of a RE (*left*) and a FE model (*right panel*)	179
Fig. 4.7	Mean life satisfaction before and after separation	186
Fig. 5.1	Scatter plots of a continuous and dichotomous Y	208
Fig. 5.2	Observed responses and predicted probability of secondary job holding (pooled LPM)	211
Fig. 5.3	The logistic regression function	215
Fig. 5.4	Conditional effects plot using the estimates of the pooled logistic model	223
Fig. 5.5	Linear probability, logistic and probit regression model	226
Fig. 5.6	Different types of censoring	252
Fig. 5.7	Predicted hazard rate of female retirement (using estimates of model 1)	266
Fig. 5.8	True and observed hazard rate of dying for the control and treatment group	268

Fig. 5.9	Observed hazard rate of dying for the treatment and control group	270
Fig. 5.10	Predicted hazard of retiring	273
Fig. 7.1	Example with two continuous variables X and Y	295
Fig. 7.2	Empirical sampling distribution of β_1	299
Fig. 7.3	How to choose between different estimators	300
Fig. 7.4	Drawing inferences from a sample about an unknown population	302
Fig. 7.5	Log likelihood function (β_1 varied, β_0 fixed at $\beta_0 = -1.4606$)	309

List of Tables

Table 1.1	Family size-adjusted income transition tables for US American families with children (using 40, 50 and 60 percent of median income)	4
Table 2.1	Organizing panel data in a data set	17
Table 2.2	Pooled cross-sections	41
Table 2.3	Response patterns in unbalanced panel data	44
Table 2.4	Poverty dynamics in Germany (2004–2006, unweighted data)	48
Table 2.5	Poverty dynamics in Germany (2004–2006, weighted data)	55
Table 3.1	Log hourly wages 1980–1987	66
Table 3.2	Union membership 1980–1987 (frequencies and percentages)	68
Table 3.3	Sequences of union membership status 1980–1983	71
Table 3.4	Survival probability of union members from 1980	72
Table 3.5	Log hourly wages for selected individuals 1980–1987	75
Table 3.6	Descriptive statistics of selected variables	81
Table 3.7	Log hourly wages 1980–1987 (demeaned data)	84
Table 3.8	Three different prototypical correlation structures	85
Table 3.9	Union membership status 1985 by membership status 1984	105
Table 3.10	Union membership status 1981 by membership status 1980	111
Table 4.1	Determinants of log hourly wages: Pooled OLS (1980–1987)	122
Table 4.2	Serial correlations in pooled OLS model and in raw data	124
Table 4.3	Determinants of public spending: Pooled OLS (1961–1993, 15 OECD countries)	130
Table 4.4	Determinants of public spending: LSDV (1961–1993, 15 OECD countries)	132
Table 4.5	Determinants of public spending: FE (1961–1993, 15 OECD countries)	139
Table 4.6	Determinants of log hourly wages: Pooled OLS and FE (1980–1987)	141
Table 4.7	Correlation of the residuals from the FE model	145
Table 4.8	Determinants of psychological distress: Pooled OLS, FE, and RE (1980–1992)	150
Table 4.9	Correlation of the residuals from the pooled OLS model	151
Table 4.10	Correlation of the residuals from the RE model	156
Table 4.11	Determinants of psychological distress: FE, RE, pooled OLS, and BE (1980–1992)	159

Table 4.12	Determinants of psychological distress: Hybrid models (1980–1992)	166
Table 4.13	Change in overall life satisfaction due to separation (FD estimation)	183
Table 4.14	Life satisfaction before and after separation (RE, FE, and FD estimation)	188
Table 4.15	Psychological distress before and after divorce	192
Table 4.16	Determinants of post-materialism (1984–1996) (FGLS and ML estimates)	200
Table 5.1	Measures used in the analysis of the `heineck-schwarze` data	207
Table 5.2	Transitions between panel waves	209
Table 5.3	Determinants of secondary job holding (pooled LPM)	210
Table 5.4	Correlation of residuals from the pooled LPM	212
Table 5.5	Determinants of secondary job holding (pooled LPM with robust standard errors)	213
Table 5.6	Determinants of secondary job holding (pooled logistic model)	221
Table 5.7	Determinants of secondary job holding (odds ratios for pooled logistic model)	222
Table 5.8	Determinants of secondary job holding (pooled linear probability, logistic, and probit model with cluster robust standard errors)	225
Table 5.9	Determinants of secondary job holding (pooled and FE logistic model)	235
Table 5.10	Determinants of secondary job holding (RE logistic and RE probit model)	241
Table 5.11	Determinants of secondary job holding (RE logistic hybrid model)	246
Table 5.12	Measures used in the analysis of the `hank` data	250
Table 5.13	Hypothetical data set for estimating discrete-time event history data	259
Table 5.14	Determinants of female retirement (discrete-time logistic hazard model)	264
Table 5.15	Determinants of female retirement (results of competing model specifications)	265
Table 5.16	True hazard rates	268
Table 5.17	Determinants of female retirement (logistic and RE logistic hazard model)	272
Table 5.18	Women's retirement by home ownership (in percent)	276
Table 5.19	Home ownership by women's retirement (in percent)	277
Table 5.20	Determinants of female retirement (complementary log–log hazard model)	282
Table 6.1	Characteristics of selected panel studies	288
Table 7.1	Expected values and variances for functions of random variables	294
Table 7.2	Example with two continuous variables X and Y	295

List of Textboxes

Textbox 3.1 Intra-class correlation coefficient 77
Textbox 3.2 Hierarchical linear models: Extended notation 96
Textbox 4.1 OLS assumptions 122
Textbox 4.2 FE assumptions 135
Textbox 4.3 Degrees of freedom for time-demeaned data 137
Textbox 4.4 RE assumptions 152
Textbox 4.5 BE estimation 158
Textbox 4.6 FD assumptions 184
Textbox 4.7 FE versus FD estimation 189
Textbox 4.8 ML estimation of random effects models 197
Textbox 5.1 ML assumptions 219
Textbox 5.2 Latent variable specification for discrete response model 228

List of Examples

Example 2.1 `mypanel` data 19
Example 2.2 `SOEP` data 31
Example 3.1 `wagepan` data 65
Example 3.2 `hetbias` and `nohetbias` data 100
Example 4.1 `garmit` data 128
Example 4.2 `johnson-wu` data 147
Example 4.3 `efficiency` data 160
Example 4.4 `genderdiff` data 181
Example 4.5 `postmat` data 193
Example 5.1 `heineck-schwarze` data 205
Example 5.2 `hank` data 249
Example 5.3 `cancer` data 267
Example 7.1 `sixcases` data 295
Example 7.2 `wpgen` data 297

Introduction 1

A research design that collects information of the same units repeatedly over time is called a *panel*. Traditionally, panel studies use surveys and focus on individuals. But increasingly, this design is also applied to the analysis of firms, nations, and other social entities using all kinds of source (official statistics, process-produced data, etc.).

The collection of panel data in academic research dates back to the 1940s when Paul F. Lazarsfeld (Lazarsfeld and Fiske, 1938; Lazarsfeld, 1940) started to introduce this methodology from market research into the analysis of public opinion. The first classical panel study (also known as the Erie County study) was an analysis of voting behavior during the 1940 presidential campaign and was conducted by the Bureau of Applied Social Research of Columbia University under the direction of Lazarsfeld himself (Lazarsfeld et al., 1944). Ten years later, the ELMIRA study was published that analyzed some of the open questions of the Erie County study using panel data collected during the 1948 presidential campaign (Berelson et al., 1954).

In the present day, numerous panels are available. They can be found in all social and life sciences. Chapter 6 lists some of the most prominent social science examples (see Table 6.1). The classical examples are the US American National Longitudinal Surveys of Labor Market Experience (NLS) and the University of Michigan's Panel Study of Income Dynamics (PSID) that were started in the 1960s. In many respects, both studies have been prototypes for many other household panels. In Europe, various countries have their own national household panel studies, among them the German Socioeconomic Panel Study (SOEP), the British Household Panel Survey (BHPS), and the Swiss Household Panel (SHP). In response to the increasing demand in the European Union for comparable information across Member States, Eurostat has coordinated in the 1990s a European Community Household Panel (ECHP), which later has been replaced by the European Union Statistics of Income and Living Conditions (EU-SILC). Many countries outside the US and Europe have initiated similar panel studies (e.g., Korea Labor Income Panel Study; Household, Income and Labor Dynamics in Australia Survey). A research project at the Department of Policy Analysis and Management at Cornell University has integrated some of these data in a large comparative panel data set, the Cross National Equivalent

File (CNEF), which includes data from Australia, Canada, Germany, Great Britain, Korea, Switzerland, and the US.

All of the previous panel studies focus on individuals (in households), but Table 6.1 mentions also some other examples. For instance, the Organization for Economic Development (OECD) provides a Social Expenditure Database (SOCX) that includes yearly social policy indicators for 34 OECD countries since 1980. In this case the unit of analysis is the country. Another example is the IAB Establishment Panel (IAB-EP) of the Institute for Employment Research (IAB) of the German Federal Employment Agency. It is a yearly repeated survey of German establishments, which began 1993 in West Germany and 1996 in East Germany. Here the unit of analysis is the single establishment.

As in the aforementioned household panels, the IAB-EP is a survey, while OECD's Social Expenditure Database uses official government statistics. However, the establishments in IAB-EP can be matched with data on employees generated in labor administration and social security data processing. Obviously, the method of data collection varies between different panel studies. Therefore, by using the term "panel" we refer to a specific research design (repeated measurements of identical units) and not to a particular method of data collection.

1.1 Benefits and Challenges of the Panel Design

As the increasing number of panel studies in the recent years shows, the panel design has become increasingly attractive in social research. It can answer more research questions in a much more convincing manner than other research designs. However, a panel is a complex research design and presents many new challenges for social science methodology. We start by summarizing some of its benefits, before we briefly mention the most important challenges.

1.1.1 Benefits

1.1.1.1 Measuring Change at the Individual Level

The main motivation for collecting panel data is an interest in the analysis of change; more specifically, an interest in the analysis of change at the (individual) level of units. What is meant by this can be illustrated with a classical example from poverty research.

How to measure poverty and whether it is a social problem public policy should take care of, is a constant controversy in public discourse. The conventional poverty indicator measures the number of individuals having less economic resources than 40, 50 or 60 % of the median income in their home country. For instance, the European Union defines individuals falling below 60 % of the median income at risk of poverty (Atkinson et al., 2002). Of course, the details of this indicator are much more involved (Which incomes to look at? How to compare single persons and individuals living in families?) but for our present purpose it is enough to say that such a measure exists.

1.1 Benefits and Challenges of the Panel Design

According to the data in Duncan et al. (1993, 231) the average at-risk-of-poverty rate in the US between 1980 and 1985 amounted to 27.9 %. In the following six-year-period from 1981 to 1986, the rate was slightly higher (on average 28.6 %). If more than one-quarter of US citizens according to this definition are poor, it looks like the US had a dramatic poverty problem in the early 1980s. Yet, some scholars argue that the 60 % threshold is much too generous; it measures individuals *at the risk* of poverty, they argue, but not those *in* poverty. For that purpose, the 40 % threshold should be used and according to that measure fewer US citizens were estimated as being poor in the early 1980s (on average 13.6 %) (see Duncan et al., 1993, 231). Whether this percentage is a less dramatic number, is a difficult question because a benchmark is missing. However, if it would be significantly larger than the corresponding poverty rate in other countries or if it would increase over time in the US, it would certainly be a matter of concern.

All of these questions can be answered by using cross-sectional (income) surveys in the corresponding countries and years. Likewise, the aforementioned poverty rates could have been computed from cross-sectional surveys in the years 1980–1986. In other words: For estimates of the *level* and *trend* of poverty rates we do not need panel data. However, if someone asks how many of these poor people are also poor in the following year, cross-sectional data would not provide the answer, because more information is needed than the (aggregate) poverty rate in the following year. One must know the poverty status for each individual in the following year, which presupposes a second (repeated) measurement of the same individual's income. This kind of information measures change (and stability) at the individual level and is only available from panel data. Clearly, a situation in which a significant proportion of this year's poor individuals escapes poverty would be less of a concern than a situation in which the poor remain in poverty for a longer time. Furthermore, transient poverty may have other causes and needs other policy measures than permanent poverty.

Note that similar questions about stability and change at the individual level are asked in other fields of social inquiry, among them voting and consumer behavior where, as we have seen, panel designs were used for the first time. For example, party preferences at the aggregate level may be quite stable, but at the individual level only some voters may have stable preferences, while the majority of voters is not committed to a certain party and may change their party vote quite quickly. Obviously, political parties have an interest in strengthening the bonds to their stable electorate and to convince as many of the undecided voters, and it may be necessary to design different campaigns for both groups of voters. Similarly, producers of consumer goods are confronted with the problem of brand loyalty. On the one hand, they are interested in knowing who the loyal clients are and how to strengthen their preferences for the product. On the other hand, they want to increase their market share and for that reason they need to know how to gain new consumers.

But let us turn back to the poverty example and see what can be done with panel data. Table 1.1 shows the results for the US during the early 1980s. According to these data, 71.3 % (=9.7/13.6) of the severely poor (those below the 40 % threshold) remain in poverty in the following year. Note that this is an average of all yearly

Table 1.1 Family size-adjusted income transition tables for US American families with children (using 40, 50 and 60 percent of median income)

	<40	40–50	50–60	≥60	All
	Percent				
<40	9.7	1.8	0.8	1.3	13.6
40–50	2.1	2.0	1.1	1.5	6.7
50–60	1.1	1.4	2.1	3.0	7.5
≥60	1.6	1.5	3.4	65.6	72.2
All	14.5	6.6	7.4	71.4	100.0

Source: Duncan et al. (1993, 231) using PSID data ($n = 17{,}427$)

transitions between 1980 and 1986, but it would be easy to compute the corresponding percentage for a specific year, say 1984. A slightly larger (average) stability rate of 79.2 % is observed for the group of US citizens at the risk of poverty (those below the 60 % threshold). All the statistics in Table 1.1 have been estimated using data from the US American Panel Study of Income Dynamics (PSID). However, as the margins of the table demonstrate, panel data can also be used to estimate (cross-sectional) poverty rates for specific years. The right column (labeled "all") shows that in the period from $t = 1980$ to $t = 1985$ on average 13.6 % have been severely poor and that poverty increased slightly to 14.5 % in the following years $t + 1$ (see the last row labeled "all").

Hence, besides answering questions on *individual change*, panel data can also be used to answer typical cross-sectional questions about *level* and *trend*. In other words: panel data allow us to address all the research questions that we are used to analyze with cross-sectional data and some additional questions that cross-sectional data cannot deal with; among them the question of individual change.[1] Nevertheless, some purists argue that panel data should be used for the analysis of change only, especially so because panel data have their problems too when it comes to the analysis of long-term trends (see the problem of panel attrition below). We agree, however, with the majority of researchers who think that this would be a waste of resources. If this rich data are available, they should also be used for the analysis of levels and trends, especially so if no other longitudinal information is available. Repeated cross-section surveys are not abundant and often do not include the variables of interest.

The distinction between level and change is one of the guiding principles that structures the material presented in this textbook. Furthermore, we differentiate with respect to the type of the dependent variable that is of interest. Poverty status, party preference, and consumption pattern are called *categorical variables*, while income, political interest, and consumption expenditures are *continuous variables*. This text-

---

[1] Of course, it is true that a cross-sectional survey can also ask retrospective questions and in doing so measure what has changed since some former point in time. However, the amount of retrospective information is usually quite limited and always prone to recall bias.

book will show how to analyze the level and change of continuous and categorical panel data.

1.1.1.2 Separating Age and Cohort Effects

When analyzing change, researchers often want to separate generational from maturation effects. While the former relate to the time when the units of interest started to exist (e.g., year of birth in case of individuals or year of foundation in case of business companies), the latter relate to the time that has passed since the starting date (i.e., the age of individuals or business companies). With a cross-sectional design it is impossible to disentangle both effects, because if one knows the age of an individual (company) at the time of measurement (t), one can easily compute its year of birth (foundation). By definition, with only one measurement both variables are perfectly related to each other: $birth = t - age$. With a panel design, on the other hand, each unit is observed repeatedly over time and hence, units belonging to the same generation are measured at different ages. Now it becomes possible to analyze how maturation (age) affects the characteristics of different generations (sometimes also called cohorts).

In principle, this analysis can also be done by combining *several* cross-sections over time (the *pooled cross-sectional design*), given we are not interested in change at the individual level. For example, a cross-sectional survey conducted in the year 2000 will include individuals from different birth cohorts, among them individuals born in 1950. Another cross-sectional survey sampling the same population in 2005 will again include individuals from the 1950 generation, however at a later age (55 instead of 50). Combining (pooling) both surveys provides us with two measures of age for the 1950 and all other birth cohorts, which also allow us to separate maturation (age) from generation (cohort) effects. However, compared to the panel design, individuals from the 1950 generation sampled in 2005 will not be the same individuals that have been sampled in 2000 (except some rare cases that incidentally have been sampled in both years). Therefore, the pooled cross-sectional design provides us only with so-called *synthetic cohorts*. Analyzing differences with respect to age with synthetic cohorts always has to control for possible chance fluctuations in these differences that are due to sampling repeatedly from the corresponding birth cohorts as is done when using several cross-sections. In case of a panel design, on the other hand, we measure the same members of a birth cohort repeatedly over time and hence, with these "true" cohort data we can make a much stronger case for maturation effects.

1.1.1.3 Controlling for Omitted Variable Bias

Another problem that ails all empirical research is the fact that we often do not know all the determinants of our dependent variables and even if we know them theoretically, we often do not have measures of them. Therefore, we always have to be aware that our models may be incomplete and our estimates possibly biased, because we have omitted important explanatory variables from our models. With cross-sectional data, there is not much we can do about omitted variable bias except make simplifying assumptions about the effects of these omitted variables. The situation is less hopeless with panel data.

As we will show in the following chapters, panel data allow us to control for at least part of this unobserved heterogeneity. The fact that we have access to repeated measurements of the same units allows us to control at least for their unknown characteristics that are constant over time. Units are used as their own controls, a technique known from experimental research as the *pre-test post-test design*. The underlying idea is the following: if a variable X influences the variable of interest Y, then a change of X at some time point t should result in a different value of Y at $t + 1$ than the value of Y at $t - 1$. Since this design compares identical units measured at $t - 1$ and $t + 1$, it also controls for all their characteristics that do not change in between.

1.1.1.4 Assessing Causality

Talking about influences and effects instantly leads to the question of causality. This introductory chapter is not a good place to discuss criteria of causality and causality assessment. Nevertheless, to understand the potential of panel data compared to other research designs, an informal definition of causality is sufficient. According to this definition, (i) two variables X and Y should correlate with each other, when they are causally related. (ii) This correlation should not be spurious in the sense that the correlation between X and Y is due to the correlation of both variables with some other (third) variables. (iii) Finally, whether X has a causal effect on Y (and not Y a causal effect on X) should be demonstrated by manipulating X and analyzing the changes of Y. At least, changes of X should precede changes of Y.

These criteria are most easily assessed with an experiment. One can manipulate X under controlled conditions and analyze whether that results in changes of Y. Other determinants of Y are controlled for by the experimental setup and by selecting units randomly into the treatment and control group (randomization).

A cross-section is the most inappropriate design to assess causality. First, it does not allow one to disentangle the time order of X and Y because all variables are measured at the same point in time. Second, in a real-life situation X is possibly correlated with other variables that cannot be controlled for because they are unknown or have not been measured.

As the discussion in the previous section showed, with panel data it is at least possible to control for those unknown or unmeasured determinants of Y that are constant over time. Moreover, since panel data include repeated measures of X and Y it is much easier to assess whether changes of X precede changes of Y or vice versa. This does not mean that all problems of causality assessment are solved with panel data, but the panel design has much more power than many other designs.

1.1.1.5 Obtaining Larger Sample Sizes

In most cases, small sample sizes are not a problem for survey researchers. Given enough financial resources, it is just a matter of time to collect data on a sample of several thousand individuals. However, social scientists interested in the quantitative analysis of macro phenomena (political systems, national economies, and so on) often have to deal with small sample sizes.

For example, scholars interested in social expenditures in modern capitalist welfare states often decide to analyze OECD countries, simply because the OECD provides so many statistics about them. At present, this population includes only 34 units (countries) and given this low number, it does not make sense to draw a sample. Such small data sets are typical for many analyses at the country level, as you find them in political science, macroeconomics and macro sociology. The limited sample size severely limits possible statistical analyses. In this case, many scholars recommend to extend the data in the time dimension and measure each (macro) unit at several points in time (a panel design). However, it is important to keep in mind that a sample of 30 units observed more than 20 times (see, e.g., the SOCX data base in Table 6.1) is not equivalent to a sample of 600 units, because repeated measurements of identical units do not provide totally independent information. Nevertheless, a panel of this size certainly provides more information than a cross-section of only 30 units.

1.1.1.6 Measurement Error

As we all know, social science data are prone to measurement error, which contaminates the statistical associations that we observe in our data to a greater or lesser extent. Therefore, we would like to have measures of the reliability of our data in order to correct our estimates of the statistical associations. One method to assess the reliability of a variable is to compare several measurements of this variable over time (*test–retest reliability*). This is easily done with panel data, while reliability analyses with cross-section data require that we have *parallel* measurements of the same underlying construct, which may be hard to defend in some cases.

Hence, panel data are a perfect tool to examine measurement error. On the other hand, measurement error is also a challenge for panel data. If we want to analyze change, we have to deal with the problem that part of the observed change is due to measurement error. In order to achieve both a measure of reliability and a measure of "true" error-free change, we need more than just two measurements over time. Therefore, extending statistical models to cope with measurement problems is easily done with panel data, but may raise additional questions of identification.

1.1.2 Challenges

As the discussion in the previous sections has shown, a panel allows answering many more research questions than other kinds of research designs. However, it is no panacea! Naturally, a panel is also a much more complex design that leads to many new challenges when putting it into practice.

1.1.2.1 How to Represent the Population over Time?

The most prominent challenge is the issue of sampling and representing the population over time. Of course, if one studies a census of the population (like the SOCX that includes *all* OECD member states), sampling and representation are not an issue. However, most of the aforementioned panel studies use a sample of a well-defined population.

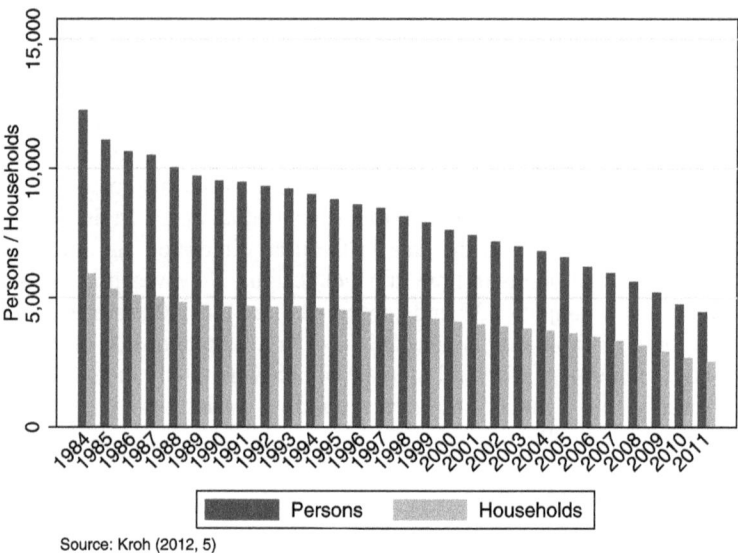

Fig. 1.1 Successful interviews with persons and households (SOEP: Samples A and B)

For example, when the SOEP was started in 1984, the survey institute selected a stratified random sample of all German private households and interviewed all household members aged 16 and older. The initial sample included 5,921 households and 12,245 individuals. According to the panel design, these 12,245 individuals should have been re-interviewed every year starting from 1985. This is not an easy task. Some of them may have moved away, others may not be available in a particular year, some may refuse to continue participating and finally, a few may have died. All of these events result in missing information for some of the original sample members: either temporarily (no interview in a specific year) or permanently (dropout out of the panel). Temporarily missing information is less of a problem, because it can be imputed from the available information in the other years. The main problem arises when sample members permanently drop out of the panel. This process is called *panel attrition* or *panel mortality* and as Fig. 1.1 shows (Kroh, 2012), the number of dropouts is quite significant, especially when re-interviewing respondents for the first time (in the second wave).

What is so problematic about panel attrition? Since the SOEP is supposed to represent the 16+ population living 1984 in (West) Germany, all dropout events that cannot also happen to a member of the population are potentially harmful to the representativeness of the sample. For example, if a sample member dies and for that reason drops out of the panel, this is a personal tragedy, but from a statistical point of view it is unproblematic because it represents an event that also happens in the population. The same applies to a birth of a child, as long as it is included in the sample (as it is in the population). At age 16, the child will be interviewed for the first time. However, if a sample member cannot be contacted or refuses to participate, this is potentially harmful because in principle every member of the living population can

1.1 Benefits and Challenges of the Panel Design

be contacted and no one can refuse to be member of the population. If these and other kinds of non-response are selective, then the available (non-missing) information provides a biased picture of the population.

For example, the former Table 1.1 is based on all PSID members that provided information for both years t and $t + 1$. Hence, it does not include individuals that refused to participate or could not be contacted in $t + 1$ (or in t). If these are the individuals in permanent poverty, the former stability measures will underestimate the true percentage of permanently poor. Furthermore, the level of poverty in both years t and $t + 1$ will be underestimated too. Since an unbiased estimate of these poverty rates (not the stability measures) can be obtained from cross-sectional surveys, panel attrition is a clear disadvantage of the panel design and a strong argument in support of the cross-sectional design. Although the cross-sectional design cannot answer all research questions (e.g., with respect to individual change), it is not negatively affected by selective non-response that is due to repeated measurements.

Quite generally, representing a population over time is a much more complex issue than representing a population at a given point in time. This is due to the fact that the population itself changes over time. Demographers distinguish between natural changes of the population (births, marriages, divorces, and deaths) on the one side and processes of immigration and emigration on the other.

From the very beginning, SOEP tried to represent natural changes of the population by including all "new" SOEP household members into the survey and by following all "existing" SOEP members founding a (new) household of their own. The first group includes SOEP children that are "born into" the interview age (16) and individuals moving into existing SOEP households (e.g., by marriage). The second group includes, e.g., SOEP youngsters leaving their parent households or SOEP adults that have to found a new household due to divorce. However, these inclusion rules fail if significant changes of the population happen outside existing SOEP households, as is the case if there is heavy migration into or out of the country.

Hence, besides the problem of panel attrition the panel design also suffers from significant *changes of the population* itself. To put it differently: Even if SOEP would not suffer from panel attrition and therefore, represent correctly the German population from 1984, the 1984 population is no more representative of Germany today, which now includes also the population of the former Democratic Republic of Germany and which has experienced a massive immigration of native Germans and other nationals after the fall of the Iron Curtain. A recent cross-sectional survey would not have these problems, because it would sample the present population.

In sum, a panel has the ability to answer research questions that the cross-sectional design cannot address. However, it has selectivity problems due to panel attrition and population change. Hence, to exploit the unique features of the panel design, much effort must be invested to minimize these problems of representation.

Counter measures include intensified field work (a tracking concept) to contact as many of the previously selected households as possible and to motivate as many of the former respondents to continue participating in the panel. The remaining non-response has to be either imputed (in case of temporarily missing information) or compensated for by re-weighting the remaining units (in case of panel attrition). However, at a certain point the loss due to panel attrition will be so large that weighting the few remaining units does not make sense anymore. At that point, a *refreshment sample* is necessary. Changing populations, on the other hand, can be dealt with by drawing new samples either at regular points in time (called *rotating panels*) or when necessary (e.g., after a period of massive immigration). The EU-SILC is an example of the first sort, the SOEP immigration sample begun in 1995 is an example of the second sort (Schupp and Wagner, 1995).

1.1.2.2 How to Obtain Valid and Reliable Measurements over Time?

If repeated measurements are the main purpose of the panel design, then every effort has to be undertaken to ensure they are valid and reliable. Some scholars argue that the repetition itself may be harmful to the validity of the measures. However, a closer look at the scholarly literature on the so-called *panel effect* (*panel conditioning*) provides positive and negative views.

On the one side, it is correct that posing identical survey questions over and over again elicits stereotypical and streamlined answers. Respondents and interviewers also "learn" how to avoid difficult and time-consuming questions, e.g., by answering filter questions strategically (Van der Zouwen and Van Tilburg, 2001). On the other side, answering repeatedly the same questions over time induces also positive learning effects and attentiveness. Respondents may become more "knowledgeable", when asked repeatedly over time the same knowledge questions (Das et al., 2011). Complicated questions referring, for example, to the various income sources of the household may be difficult in the first panel wave, but become easier after having answered them several times (Frick et al., 2006; with respect to attitudes see Sturgis et al., 2009). Hence, the panel effect may bias the measures, but also decrease non-response and increase validity. Whether and how it works has to be found out by comparing data from a panel with measurements from independent cross-sections.

Another challenge is to keep the survey instruments equivalent across time. For example, survey questions may need to be changed because their repeated application shows that they have low quality, because they become obsolete during the course of time, because they have to be adapted to the actual historical context, or because survey methods change over time (e.g., changing from face-to-face to telephone interviews). Equally, new questions have to be developed if new aspects attract the attention of researchers. Finally, even if questions are identical across time, their meaning may change over time. Overall, the practice of many panel surveys shows that longitudinal analyses are often hampered by non-equivalent survey instruments over time. All the more reason it is necessary to restrict instrument changes to the absolute minimum and to assess their equivalence at regular intervals.

Traditionally, panel measurements provide information for each unit of analysis at $t = 1, \ldots, T$ *discrete* points in time. For example, Lazarsfeld's Erie county study measured political attitudes at $T = 7$ monthly measurements (May–November) during the 1940 presidential campaign. There was no attempt to measure political attitudes *between* the seven survey dates, assuming that attitude change can be approximated by monthly measurements. Similarly, the large household panels mentioned in Table 6.1 are conducted mostly on a yearly basis. However, researchers have become increasingly interested in what happens in between, also because a yearly interval between the measurements is quite large. For example, measuring income at only one point in time during the course of the year is not very useful, when incomes change rapidly due to changes in employment. In that case, panel researchers try to collect *continuous* employment and income histories either by retrospective questioning or by merging the panel information with process-produced data from other sources. Of course, retrospective questioning is not without risks due to recall bias and seam effects.

For example, the SOEP uses a monthly job calendar, in which respondents can report their employment status for each month in the year before the interview date. If they frequently change their status, they may have problems recalling all transitions with their exact dates. Furthermore, an analysis of the job calendars shows that many job changes happen at the end of the year. This is, however, often a methodological artifact at the seams of the yearly job calendars. Some respondents have forgotten what job they specified for December in the previous job calendar and report a seemingly "new" job for January in the present job calendar.

In sum, although this textbook focuses on managing and analyzing panel data, it should be stressed that the collection of panel data is a methodology of its own. Obviously, the collection of panel data includes many pitfalls that may hamper later statistical analyses. However, it should also be stressed that at the same time repeated measurements are a perfect tool to assess all kinds of measurement errors (see Sect. 1.1.1.6).

1.1.2.3 How Much Does It Cost?

Finally, the question comes up: How much does it cost in terms of money, time, and manpower? And does it not cost too much to make the effort worthwhile? Certainly, a panel is much more costly than a single cross-section. But is it really more costly than a pooled cross-section design that could also answer some of the longitudinal research questions and at various places performed better than the panel design? Both designs need resources for (i) sampling, (ii) data collection, (iii) data management, (iv) weighting, and (v) documentation. Most of these cost factors are more-or-less identical for both designs. In both cases, data have to be put into a data analysis system, weighted and documented. Perhaps data management, weighting, and documenting are a little bit more complicated for panel data, but the differences will not be significant in terms of resources needed.

What is different between both designs is sampling and data collection. While a panel, in the ideal case, only needs a fresh sample at the beginning, the pooled cross-section design needs a new sample for each additional cross-section. Furthermore, resources are needed for collecting data for each panel wave and each cross-section. This is certainly more expensive for the panel design, because a specialized tracking concept is needed to minimize panel attrition. Nevertheless, these additional field work expenses are less costly than selecting new samples for each cross-section. Hence, considering the main cost factors, the panel design does not perform as badly as one might think from the beginning.

1.2 Outline of the Book

As already mentioned, this textbook focuses on methods of managing and analyzing panel data. Hence, most of the chapters will be devoted to statistical methods of panel analysis. The only exception is the following Chap. 2 that shows how to prepare panel data for statistical analysis. In the previous sections, we have shown that panel data are used both for the analysis of level and change. Moreover, statistical models are often differentiated with respect to characteristics of the variables they focus on. Like many other statistical textbooks, we distinguish between continuous and categorical dependent variables and discuss the corresponding panel models in different chapters (Chaps. 4 and 5). Within each chapter we start with models focusing on the level of the dependent variable and then continue with models focusing on change of the dependent variable. All in all, the presented material is quite comprehensive, covering two different types of dependent variable and two modes of analyzing them. Chapter 3 shows how to describe panel data and how to decide between the different models presented in Chaps. 4 and 5. Finally, Chap. 6 concludes with some suggestions on how to do your own panel analysis. It shows you what panel data are available for secondary analysis, but gives you also some references on how to design and collect your own panel data. Moreover, it discusses typical applications in different social science disciplines and mentions other sources that you can read to know more about the specific methods that we only alluded to without discussing them in detail.

1.3 Audience and Prerequisites

There are several excellent textbooks available on panel data analysis (among others Baltagi, 2008; Cameron and Trivedi, 2005; Hsiao, 2003; Wooldridge, 2010), but all of them require a fairly good understanding of matrix algebra and advanced econometric methods (e.g., instrumental variable estimation). At an introductory level, several software and econometric textbooks also treat methods for panel data analysis (e.g., Cameron and Trivedi, 2008; Rabe-Hesketh and Skrondal, 2008; Wooldridge, 2009). However, when things get complicated these sources usually refer to the more advanced literature. Moreover, methods for categorical data are hardly treated in these introductory texts (the textbooks by Cameron and Trivedi (2008) and Hsiao (2003) are exceptions).

This textbook provides an introduction into panel data analysis that does not use matrix algebra and instrumental variables estimation. It does not only focus on linear models and least squares estimation; it also provides an introduction into maximum likelihood estimation, which is a necessary tool when modeling categorical data with non-linear models. The focus is on applications of panel models and less so on the underlying statistical theory. We illustrate all methods with real research examples from scholarly journals from different social science disciplines (sociology, political science, economics).

Readers should be familiar with linear regression and have a good understanding of ordinary least squares estimation. It is also helpful to have some experiences with logistic regression and perhaps maximum likelihood estimation, but as already mentioned these techniques will be introduced in greater detail in this book (see Chap. 5). Naturally, restricting ourselves to such a limited set of mathematical and statistical tools implies that we cannot go into more advanced methods of panel data analysis. We hope, however, that the basic panel regression models are introduced in a way that few questions remain open and readers can go on to the more advanced literature.

The text is written without any specific software in mind to estimate these models, but certainly statistical software is needed to do this job. Stata is a perfect choice, when it comes to regression models for panel data, but other statistical software like SPSS or SAS is equally well suited, at least for the models discussed in this textbook. On the web site of this textbook you find all the data sets used in our examples accompanied by Stata syntax files that replicate our results (see Sect. 7.3).

Leaving matrix algebra and instrumental variables aside does not mean that we can refrain from mathematics. Indeed, rather than simple introductions, we want to make sure that readers understand the mathematics behind the basic panel regression models. Nevertheless, we tried to keep a simple and unified mathematical notation across all chapters. Most of it will be explained in the methodological overview (see Chap. 3) and if necessary in the method-specific chapters (Chaps. 4 and 5).

At this point you only need to know that we distinguish between variables and the values (realizations) that these variables obtain for each unit of analysis. Variables are symbolized with capital letters (Y, X, Z, T, U, E), their realizations (values) with small letters (y, x, z, t, u, e). We distinguish between dependent (Y) and independent (explanatory) variables (X, Z), process time (T), and independent (explanatory) variables that are unobserved (U, E). Realizations of these variables refer to measurements for a specific unit at a given point in time. To denote them as precisely as possible we use the indices i and t (for example, y_{it}). Estimation results are presented in tables and interpreted in the text. Estimates in the text are usually rounded and hence, slight differences between text and tables may happen because of rounding errors.

Acknowledgements This text has a long history and parts of it have served as background literature in lectures, seminars, and workshops both in Cologne and in other places. We thank the participants of these courses for their questions and comments, which helped us to formulate our ideas more precisely and hopefully in a more coherent form. Our special thanks go to our colleagues Josef Brüderl, Kenneth Bollen, Romana Careja, Marco Gießelmann, Achim Goerres, Heiner Meulemann, Henning Lohmann, Luis Maldonado, Ulrich Pötter, Götz Rohwer, and Hawal Shamon who discussed numerous versions of this text with us and contributed valuable improvements. Thorsten Meiser, Ingo Rohlfing, Elmar Schlüter, and Dirk Temme provided helpful literature references from their field of methodological expertise. A special word of thanks goes to all the people that supported our intention to write an applied textbook introducing panel analysis with real research examples from scholarly journals. Helmut Dietl, Bernd Fitzenberger, Geoffrey Garrett, Karsten Hank, Guido Heineck, David Johnson, Markus Klein, Richard Lucas, Pasi Moisio, Stephanie Moller, and John Stephens provided us with their data. Some of them will be used in this textbook, the rest will be provided on the web site for secondary analysis. We especially like to thank Jan Goebel at the Research Data Center of the SOEP and Heather Laurie at the Institute for Social and Economic Research for the permission to use anonymized versions of SOEP and BHPS data in our textbook. Evelyn Funk, Claudia Ubben, and Ravena Penning helped with typesetting the manuscript and producing tables, figures, and the index. Donatas Akmanavičius at VTeX Book Production did the final editing. Finally, Joscha Dick rewrote all our Stata syntax files to publish them on the book's web site. Martin Spitzenpfeil prepared the data files that are available for secondary analysis on the web site. He also programmed Excel spreadsheets to illustrate examples from Sect. 7.2.

Managing Panel Data 2

In this chapter we will discuss some practical problems and challenges we usually face when we analyze data, in particular, panel data. Our main concern is to show how to generate and manage panel data sets. As panel data are more complex than simple cross-sectional data, the data management tasks can be quite complex as well.

Many of the publicly available panel surveys provide their raw data in a number of single data sets which have to be combined in order to obtain a working data set for the analysis with statistical software. A working data set is built by extracting certain information from a number of raw data sets and combining them into a new data set. This process requires a definition of the *target population*, the *time period* under investigation and the *variables* of interest.

We will show how to prepare such a working data set by using data from the German Socioeconomic Panel Study (SOEP) as an example. Necessarily, this demonstration is somehow technical and we will use Stata syntax for this purpose. However, as you will see in the examples below, you will be able to easily translate the Stata commands into the statistical language of your choice.

We will also address some other issues we usually face when doing empirical research. Most statistical methods, like those presented in this book, are based on the assumption that the observations are a simple random sample of the population. In practice this assumption is often violated. Very often surveys employ multi-stage sampling techniques using stratification and clustering. Such complex survey designs result in different selection probabilities for the units of analysis. The complexities of estimation and testing in such cases are beyond the limits of this textbook and we refer the reader to the specialized literature (see the references in Sect. 2.5). However, even when the sampling design is a simple random sample, data collection may still be selective. Usually, some of the sampled units refuse to participate or cannot be contacted at all (*unit non-response*).

In the case of panel data, we distinguish between *balanced* and *unbalanced* panels. With balanced panel data, each unit is observed in each wave. In unbalanced panels, the number of observations per unit can differ. Unbalanced panels occur when some respondents do not participate in each wave (*temporary unit non-*

response), when respondents drop out of the panel (*panel attrition*), or when respondents enter the panel at a later point in time (*late entry*).

Generally speaking, social science data often have missing values. There are different ways of dealing with these data problems. As a countermeasure we can either use statistical weighting to compensate for the selectivity of participation or we can use imputation techniques to fill in the missing values. In any case, we have to decide how we treat the missing data when we build a working data set. Shall we include weights in our data set? Shall we generate a balanced or an unbalanced data set? How do we have to weight balanced and unbalanced data sets? These decisions might seem to be of practical nature only, but they have significant methodological consequences and it is important to be aware of these consequences. In this chapter we will briefly discuss different ways of dealing with these issues and, where possible, give recommendations on which methods to choose.

In Sect. 2.1 we will discuss some general characteristics of panel data. In Sect. 2.2 we will present the basic operations of data management and show how to apply these operations within Stata. In Sect. 2.3 we will show how to apply the presented data management tools to real survey data. We present three case studies from poverty research, using data from the SOEP. Section 2.3 will also discuss how to define a target population for cross-sectional and for longitudinal analyses. Finally, in Sect. 2.4, we will demonstrate how to use statistical weights and we will discuss how to generate and weight balanced and unbalanced panels. The chapter concludes with some suggestions for further reading (Sect. 2.5).

2.1 The Nature of Panel Data

Panel data contain repeated observations of the same units. In principle, panel data can be seen as a *data cube* with three dimensions: units $i = 1, \ldots, n$, time points $t = 1, \ldots, T$ and variables $v = 1, \ldots, V$. In order to analyze panel data with a statistical computer software we need to rearrange the three-dimensional data cube into a two-dimensional *working data set*. In general, there are two ways of organizing panel data in a working data set:

- In *wide format* each unit occupies one row of the data matrix. All measurements over time are included in each row. The matrix has n rows and $T \cdot V$ columns. The wide format is the traditional way of organizing panel data.
- The *long format*, which originates from time-series analysis, is the "modern" form of organizing panel data. In long format each single measurement occupies one row. Thus, we have T rows per unit and $N = n \cdot T$ rows in total. The number of columns equals the number of variables V.

Which format you use for data input is not of importance since all statistical software packages include commands for transforming data in wide format to data in long format and vice versa. Usually, the format you use for data input will be defined by the raw data as they are provided by the institution collecting the data. Table 2.1 exemplifies the two forms of organizing panel data with an excerpt from the data

2.1 The Nature of Panel Data

Table 2.1 Organizing panel data in a data set

(a) Wide format

Persnr	income2004	income2005	income2006
21	1,245	1,245	814
41	480	502	524

(b) Long format

Persnr	Year	income
21	2004	1,245
21	2005	1,245
21	2006	814
41	2004	480
41	2005	502
41	2006	524

Source: SOEP data (see Example 2.2)

we will use in the example below. The table shows two individuals (`Persnr` = 21 and `Persnr` = 41) from the SOEP data (an explanation of these data follows below in Example 2.2).

Organizing data in wide format means to integrate the time dimension of the data cube into the columns of the two-dimensional working data set. Therefore, variable names have to include characters indicating the time point at which the information was recorded. In long format the time dimension is included in the rows of the working data set. Both tables include exactly the same information but a striking difference is that data in long format need to include *key variables* indicating the structure of the data. In wide format each cell is uniquely identified with regard to the unit and the time point to which it pertains. This is because each column represents a variable measured at a certain point in time and each row represents one unit. In long format, we need one key variable indicating to which unit a row in the data pertains (here `Persnr`) and a second key variable indicating the time point t to which a row in the data set pertains (here `Year`). Both key variables together identify each observation uniquely.

For the methods presented in this book you will have to use data in long format because, for regression-type analyses, statistical software packages prefer data in long format. Furthermore, the long format is the more efficient way of organizing data. If the number of variables and time points increases, a data matrix in wide format becomes excessively large.

Imagine a data set with 1000 units ($n = 1,000$) and 100 variables ($V = 100$), which have been measured yearly for a period of ten years ($T = 10$). Now think about the distinction between balanced and unbalanced panels. Assume that we find a response rate of $r = 90\%$ (=900 units) for each wave (10 % temporary unit non-response). In long format, the data set will have nine thousand rows ($n \cdot r \cdot T = 1,000 \cdot 0.9 \cdot 10 = 9,000$) and 100 columns ($V$). This makes a total of 900,000 cells. In wide format, the data set will have $V \cdot T = 1,000$ columns and

$n = 1,000$ rows. In total, this data set includes one million cells. The difference in the number of cells between wide and long format occurs because of temporary unit non-response.

In long format, a temporary unit non-response decreases the size of the table by one row. In wide format, the size of the data set is unaffected by temporary unit non-response. In our example, each temporary unit non-response will produce 100 empty cells if the data is in wide format.

The wide format, on the other hand, has its advantages, too. From a technical point of view, it is much easier to analyze statistical associations between the repeated measurements because, in wide format, each new measurement is a variable on its own. From a didactic point of view, it also makes clear that you have only n independent units of analysis. Panel data provide information about the *same* units at different points in time. As already indicated in Sect. 1.1.1.5, this does not increase the sample size. In wide format, each repeated measurement adds a new column to the data matrix.

For instance, in Table 2.1(a) the income of 2007 would add an additional column to the table (income2007). Obviously, this information pertains to individuals we have already observed before. In long format, each new measurement adds an additional row to the data matrix. Some scholars argue that the number of cases and, thus, the statistical power increases with these repeated observations. This pitfall occurs because in a simple cross-sectional data set the number of rows equals the number of units, i.e., the data matrix has n rows.

Compared to this, a panel data set in long format with a data matrix of $N = n \cdot T$ rows looks like a huge amount of information. However, the repeated observations are not independent of each other and do not provide as much information as their sheer number indicates. Therefore, when we analyze panel data we need to use statistical tools which take into account that the number of units n is smaller than the number of rows N and that the observations within one unit are not statistically independent. In Sect. 3.2, we use data in long format to demonstrate these statistical dependencies and to discuss how they influence our statistical analyses.

Panel data share the feature of dependent observations with other *nested* or *hierarchical* data. In the case of panel data, each unit is observed T times, which means that the single measurements at the lower level (level 1) are nested within units i at the higher level (level 2). In contrast to other hierarchical data, such as pupils in schools or individuals in countries, panel data have an inherent order at the lowest level (time points). The chronological order of these time points should be visible in the data, i.e., we have to sort the single observations for each unit according to the time points at which they were recorded. In cross-sectional hierarchical data a similar natural order is often absent. Individuals might be nested within countries, households, or schools, but the order of these individuals within each higher-level unit is of no importance.

The *key variables* in panel data fulfill three different functions. First, we need the key variables in order to organize the data in a way that reflects the nested structure and the chronological order of the measurements. Second, if we analyze panel data with software, the information included in the key variables is used to account for the statistical dependencies in the data. Therefore, you will usually have to tell the computer which variables it should use as key variables. Finally, the key variables allow us to generate a working data set from the raw data sets, because the key variables are included in each raw data set in order to identify the data that belong together.

2.2 The Basics of Panel Data Management

Raw data from large economic and social surveys are mostly provided in a form that requires some data management before we can start a statistical analysis. In the previous section we saw that panel data need to be organized in hierarchical form with time points nested in units. Unfortunately, the data structure that results from the process of data collection usually is organized the other way around. Panel data is collected in regular intervals. Each additional wave adds a new data set to the panel. Hence, the data structure resulting from the data collection is units nested in time points. In order to achieve a working data set, with time points nested in units, we have to extract the required information from the single waves and combine it into a panel data set.[1] Furthermore, the wave-specific data may be distributed across different data files, because each survey wave may consist of different parts.

For example, information is gathered about individuals and households or about adults and children. If different survey instruments are used (e.g., different questionnaires for adults and children), the survey institute will often provide the raw data in separate data files. We will have a closer look at real survey data from the SOEP when we present some real life examples in Sect. 2.3.

In this section, we will discuss the general principles of data management. Therefore, we will use a set of very small example data sets. The purpose of these examples is to visualize the basic operations that we use to manage data, in general, and panel data, in particular.

Example 2.1 (mypanel data) Imagine you are doing a small panel survey, in which you collect information on adults, children, and their households. The survey institute provides you with some sample files so that your data administrator can test the implementation of the data at your institute. The test data include data from three panel waves. The information on households, adults, and children is stored in different files:

---

[1] Most panel studies mentioned in Table 6.1 provide their raw data in wave-specific files. EU-SILC, NLS, and SOEP additionally provide their raw data also in long format.

adult08				child08				house08		
ID	HHNR	Year	Sex	ID	HHNR	Year	Sex	HHNR	Year	Income
1	10	2008	0	3	10	2008	1	10	2008	3,000
2	11	2008	1	4	11	2008	0	11	2008	4,000

adult09				child09				house09		
ID	HHNR	Year	Sex	ID	HHNR	Year	Sex	HHNR	Year	Income
1	10	2009	0	3	10	2009	1	10	2009	3,500
2	11	2009	1	4	11	2009	0	1	2009	4,500

adult10				child10				house10		
ID	HHNR	Year	Sex	ID	HHNR	Year	Sex	HHNR	Year	Income
1	10	2010	0	3	10	2010	1	10	2010	4,000
2	11	2010	1	4	11	2010	0	11	2010	5,000

The test data include two households (HHNR = 10, 11) and four individuals (ID = 1, ..., 4) living in these households. The two households and the four individuals have been observed for three years from 2008 to 2010. Besides the key variables ID, HHNR, and Year, each data set includes only one variable of interest: the household income (Income) in the household-level files and gender (Sex) in the individual-level files (0 = *woman*, 1 = *man*).

In total, the example consists of three cross-sections that have to be pooled together to obtain a panel data set. Each of these cross-sections consists of three single data sets. In Sect. 2.2.1 we present the basic operations to combine these data. Then, we are going to show how we use statistical software to perform these basic operations (Sect. 2.2.2). Finally, in Sect. 2.2.3 we demonstrate how we can actually combine the nine data sets into a complete panel data set.

2.2.1 Merging and Appending Data

There are two basic operations which bring together information from different data sets. To imagine what is going on here, we can think of them as either putting the data sets next to each other (adding columns) or putting one below the other (adding rows). In both cases, the data sets that ought to be combined have to have partly identical characteristics. Only data sets that are related to each other, either through their columns (variables) or their rows (observations), can be combined.

When we add rows to the data set, we often speak of *appending* data. A data set can be appended to another data set if both data sets include *identical variables* but *different observations*. For example, the data sets adult08 and child08 (see Example 2.1) could be appended, because both of them include the variables ID,

2.2 The Basics of Panel Data Management

adult08					child08			
ID	HHNR	Year	Sex		ID	HHNR	Year	Sex
1	10	2008	0	append	3	10	2008	1
2	11	2008	1		4	11	2008	0

⇓

ID	HHNR	Year	Sex
1	10	2008	0
2	11	2008	1
3	10	2008	1
4	11	2008	0

Source: mypanel data (see Example 2.1)

Fig. 2.1 Appending data

adult08					house08		
ID	HHNR	Year	Sex		HHNR	Year	Income
1	10	2008	0	merge	10	2008	3,000
2	11	2008	1		11	2008	4,000

⇓

ID	HHNR	Year	Sex	Income
1	10	2008	0	3,000
2	11	2008	1	4,000

Source: mypanel data (see Example 2.1)

Fig. 2.2 Merging data with one key variable

HHNR, Year, and Sex. Figure 2.1 shows the new data set including the units ID = 1, 2, ..., 4.

When we put tables next to each other, we usually use the term *merging* data. Two data sets can be merged if the *observations* are *identical* but the *variables* are *different*. For merging it is necessary that both data sets include one or more *key variables* identifying the observations that have to be put together. For instance, the data sets adult08 and house08 (see Example 2.1) could be merged using the key variable HHNR. The new data set contains units 1 and 2 and the variables Sex and (household) Income (see Fig. 2.2). It is important to notice that the units in the data set house08 are households, while the units in the data set adult08 are individuals. However, we can merge the two data sets because the individual-level data set adult08 includes the household identification number HHNR.

Merging data can require using more than one key variable. Consider the two data sets shown in Fig. 2.3. The data set on the left hand side results from appending adult08 and adult09; the data set on the right hand side results from appending house08 and house09. Both data sets have been sorted by the key variables ID,

adult08+adult09			
ID	HHNR	Year	Sex
1	10	2008	0
1	10	2009	0
2	11	2008	1
2	11	2009	1

house08+house09		
HHNR	Year	Income
10	2008	3,000
10	2009	3,500
11	2008	4,000
11	2009	4,500

merge

⇓

ID	HHNR	Year	Sex	Income
1	10	2008	0	3,000
1	10	2009	0	3,500
2	11	2008	1	4,000
2	11	2009	1	4,500

Source: mypanel data (see Example 2.1)

Fig. 2.3 Merging data with two key variables

adult08+child08			
ID	HHNR	Year	Sex
1	10	2008	0
3	10	2008	1
2	11	2008	1
4	11	2008	0

house08		
HHNR	Year	Income
10	2008	3,000
11	2008	4,000

merge

⇓

ID	HHNR	Year	Sex	Income
1	10	2008	0	3,000
3	10	2008	1	3,000
2	11	2008	1	4,000
4	11	2008	0	4,000

Source: mypanel data (see Example 2.1)

Fig. 2.4 Many-to-one merge

HHNR, and Year to transform them into the usual long format of panel data. To merge the two data sets we need to specify two key variables: HHNR and Year. This is because the variable HHNR no longer uniquely identifies the observations within both data sets.

Finally, data can also be merged if the observations within *one* of the data sets are not uniquely identified. This kind of merging is often referred to as a *one-to-many* or *many-to-one* merge, because the observations in one data set have to be assigned to more than one observation in the other data set. Figure 2.4 exemplifies a many-to-one merge. The data set on the left hand side results from appending adult08

2.2 The Basics of Panel Data Management

and `child08` (and sorting the data by `HHNR` and `ID`). Each household income has to be assigned to two observations in the individual-level file, if we want to include the information from the data set `house08` into the data on the left hand side. This merge can be done using the key variable `HHNR`. Quite generally, a many-to-one merge occurs when the units of analysis are nested within some higher-level units. In the example, we merge data observed on the household-level to all individuals within these households. Similar situations occur when we have a panel data set with repeated observations of the same units and merge time-constant information about these units to the data, or when we assign country-level data to individual- or household-level data, and so on.

2.2.2 Basic Append and Merge Commands

To generate a panel data set including all the information from Example (2.1), we have to use a series of append and merge commands. In Sect. 2.2.3 we demonstrate how to do this efficiently. In this section we present the very basic syntax for appending and merging data. For the purpose of this demonstration we will use Stata. However, the general principles demonstrated here work the same way in all statistical program packages.

From the data sets of Example 2.1 we constructed nine Stata data sets named `adult08`, `child08`,..., `house10`. The syntax for appending two tables is quite simple. It consists of three basic elements. First, the command (in Stata `append using`) and, second, the two data sets which ought to be combined. In Stata we have to name only one of the data sets, while the other one has to be loaded into memory before (this data set is called the *master* data set).[2] In the following example we append the two data sets `adult08` and `child08`:

```
. use adult08
. list

     +----------------------------+
     |  ID    HHNR    Year    Sex |
     |----------------------------|
  1. |   1      10    2008      0 |
  2. |   2      11    2008      1 |
     +----------------------------+

. append using child08
. list

     +----------------------------+
     |  ID    HHNR    Year    Sex |
     |----------------------------|
  1. |   1      10    2008      0 |
  2. |   2      11    2008      1 |
  3. |   3      10    2008      1 |
  4. |   4      11    2008      0 |
     +----------------------------+
```

---

[2] In many statistical program packages we can directly name all data sets we want to combine. From Version 11.0 onwards this is also possible in Stata.

The code above is taken from Stata's output. In this output, a line beginning with a dot is a command line. Lines without a dot are output in response to these commands. After we loaded the data set `adult08` (use `adult08`), we have a look at the data using the `list` command. Then, we state that the data set `child08` should be appended to the data in memory (append using `child08`). Finally, the `list` command shows the combined dataset. As in Fig. 2.1, it includes the information from the data sets `adult08` and `child08`. From a technical point of view, it is crucial that the variables in both data sets have exactly identical names. Otherwise, the computer will not be able to match the columns of both data sets correctly. The following Stata output demonstrates this:

```
. use child08a
. list

     +---------------------------+
     | ID   HHNR   Year   Sxe |
     |---------------------------|
  1. |  3     10   2008    1  |
  2. |  4     11   2008    0  |
     +---------------------------+

. use adult08
. append using child08a
. list

     +----------------------------------+
     | ID   HHNR   Year   Sex   Sxe |
     |----------------------------------|
  1. |  1     10   2008    0     .  |
  2. |  2     11   2008    1     .  |
  3. |  3     10   2008    .     1  |
  4. |  4     11   2008    .     0  |
     +----------------------------------+
```

The data set `child08a` is identical to the data set `child08`, but there is a typo in the variable names (`Sxe` instead of `Sex`). Obviously, the resulting data set does not have the form we want it to have. In Sect. 2.2.5 we will present a systematic list of all potential pitfalls that can occur when combining data.

The syntax for merging data is a little more complicated. It consists of four basic elements: the command, the two data sets which ought to be combined and the *key variable(s)*. Furthermore, it is important to sort both data sets by the key variable(s) before merging the data:[3]

```
. use house08
. sort HHNR
. save house08, replace
. use adult08
. sort HHNR
```

---

[3] In this book we use Stata's old merge syntax. Since Stata 11.0 there is a new version of this command. We present the old version because its structure is similar to the structure of merge commands in other statistical packages. However, if you work with Stata you should use the new version of the merge command.

2.2 The Basics of Panel Data Management

```
. merge HHNR using house08
. list
        +--------------------------------+
        | ID   HHNR   Year   Sex   Income |
        |--------------------------------|
     1. | 1    10     2008   0     3000  |
     2. | 2    11     2008   1     4000  |
        +--------------------------------+
```

In the first three lines of the syntax above we load the data set house08, sort it by the variable HHNR and save it in this form. The next two commands load the data set adult08 and sort it by the variable HHNR as well. The merge command, then, merges the information from the data set house08 (using house08) and the data in memory (adult08) and uses the specified variable HHNR as the key variable. From a technical point of view, it is important that the key variable(s) have exactly the same names in both data sets.

Moreover, it is important to sort the observations by the key variable(s) to correctly merge the information of both data sets. However, in some statistical program packages this step is (or can be) integrated in the merge procedure and it is not necessary to do this in a separate step. Knowing the two basic modes of combining different data sets we can now start to think about ways to build a panel data set.

2.2.3 Building a Working Data Set with Append and Merge Commands

Combining the information from all nine data sets into one panel data set can be done in various ways. For instance, we could start to combine the various data sets within each cross-section and go on with pooling the three cross-sections. Logically, the number of solutions increases with the number of raw data sets. When working with real survey data, we will usually have to extract certain units, variables, and time points from the raw data.

In this example, we simply want to combine all information included in the nine data sets. We can find an efficient solution by identifying all data sets which include the same variables. In our example, six data sets include the variable Sex and three include the variable Income. As merging is more complex than appending, it is reasonable to combine as many data sets as possible using the append command. Some statistical software packages can even append several data sets simultaneously. With the corresponding commands we can easily generate a data set on individuals (adults and children) and a data set on households. In Stata the syntax looks like this:

```
. use adult08
. append using child08 adult09 child09 adult10 child10
. sort ID Year
. list, sepby(ID)
```

```
     +------------------------------+
     |  ID    HHNR    Year    Sex  |
     |------------------------------|
 1.  |   1      10    2008     0   |
 2.  |   1      10    2009     0   |
 3.  |   1      10    2010     0   |
     |------------------------------|
 4.  |   2      11    2008     1   |
 5.  |   2      11    2009     1   |
 6.  |   2      11    2010     1   |
     |------------------------------|
 7.  |   3      10    2008     1   |
 8.  |   3      10    2009     1   |
 9.  |   3      10    2010     1   |
     |------------------------------|
10.  |   4      11    2008     0   |
11.  |   4      11    2009     0   |
12.  |   4      11    2010     0   |
     +------------------------------+
. save myfirstpanel
. use house08
. append using house09 house10
. sort HHNR Year
. list, sepby(HHNR)
     +------------------------+
     | HHNR    Year    Income |
     |------------------------|
 1.  |   10    2008     3000  |
 2.  |   10    2009     3500  |
 3.  |   10    2010     4000  |
     |------------------------|
 4.  |   11    2008     4000  |
 5.  |   11    2009     4500  |
 6.  |   11    2010     5000  |
     +------------------------+
. save mysecondpanel
```

The first two commands append all data sets including the variable Sex. Sorting the resulting data set by the variables ID and Year results in a panel data set in long format. After saving this first panel data set (save myfirstpanel), we also append the data sets that include the variable Income. Sorting the resulting data set by the variables HHNR and Year provides us again with a (second) panel data set in long format.

Now, in a final step we need one merge command to combine the two panel data sets. Before we do so, we have to take a closer look at the structure of both data sets. The first data set we generated (myfirstpanel) includes information on individuals (ID = 1, ..., 4). The second data set (mysecondpanel) includes information on households. As in Fig. 2.4 we have to perform a many-to-one merge, which looks like this in Stata:

2.2 The Basics of Panel Data Management

```
. use myfirstpanel
. sort HHNR Year
. merge HHNR Year using mysecondpanel, uniqusing
variables HHNR Year do not uniquely identify observations in
the master data
. sort HHNR ID Year
. list HHNR ID Year Sex Income, sepby(ID)
     +------------------------------------+
     | HHNR   ID   Year   Sex   Income   |
     |------------------------------------|
  1. |  10    1    2008    0     3000    |
  2. |  10    1    2009    0     3500    |
  3. |  10    1    2010    0     4000    |
     |------------------------------------|
  4. |  10    3    2008    1     3000    |
  5. |  10    3    2009    1     3500    |
  6. |  10    3    2010    1     4000    |
     |------------------------------------|
  7. |  11    2    2008    1     4000    |
  8. |  11    2    2009    1     4500    |
  9. |  11    2    2010    1     5000    |
     |------------------------------------|
 10. |  11    4    2008    0     4000    |
 11. |  11    4    2009    0     4500    |
 12. |  11    4    2010    0     5000    |
     +------------------------------------+
```

In order to force Stata to merge the two data sets we have to specify the option `uniqusing`, indicating that the observations are only uniquely identified within the using data set. Without this option Stata would refuse to perform the merging and return an error. It depends on the statistical software that you are using how to deal with this kind of merging.[4] You should notice that we need to specify two key variables in order to merge the two data sets `myfirstpanel` and `mysecondpanel`. The simple key variable HHNR no longer uniquely identifies the observations, neither in the data set `myfirstpanel` nor in the data set `mysecondpanel`. Using HHNR and Year as key variables, in contrast, identifies each observation in the data set `mysecondpanel` uniquely.

2.2.4 Wide and Long Format

Before we apply the presented techniques to real survey data we will have a very short look at panel data formats again. In the example above we generated a data set in long format. To that end, we basically appended the data from three waves and sorted them by HHNR, ID and Year. Alternatively, we could have generated data

---

[4] Stata by default does not allow the user to merge data sets if the observations are not uniquely identified in one of the data sets. If we want to do so, we have to specify an option. In SPSS we need to declare the file with uniquely identified observations as a *table*. In SAS we can directly match the files without any additional step.

in wide format. To that end, we would have to use the merge instead of the append command. Here we give you two small examples in which the data sets house08 and house09 are combined into a (household) panel, first using the long and then the wide format:

Long format

```
. use house08
. append using house09
. sort HHNR Year
. list, sepby(HHNR)
```

```
     +------------------------+
     |  HHNR    Year   Income |
     |------------------------|
  1. |    10    2008     3000 |
  2. |    10    2009     3500 |
     |------------------------|
  3. |    11    2008     4000 |
  4. |    11    2009     4500 |
     +------------------------+
```

```
. save housepanL
```

Wide format

```
. use house09
. rename Income Income09
. sort HHNR
. save house09_, replace
. use house08, clear
. rename Income Income08
. sort HHNR
. merge HHNR using house09_
. order HHNR Income08 Income09
. list HHNR Income08 Income09
```

```
     +----------------------------+
     |  HHNR    Income08  Income09 |
     |----------------------------|
  1. |    10        3000      3500 |
  2. |    11        4000      4500 |
     +----------------------------+
```

```
. save housepanW
```

You should be familiar with the example on the left hand side. On the right hand side we generate a panel data set in wide format. Therefore, we have to rename all variables except the identification variables (HHNR and Year). As already mentioned, in wide format the time dimension is included in the columns (variables) of the data set. Therefore, it is necessary to include the information from the time identifying key variable (Year) into the variable names. In the example above we do so by adding the year to each variable name (rename Income Income09 and rename Income Income08). After renaming the variables in each data set and sorting them by the variable HHNR we can merge the household data from 2008 and 2009.

Ordering the variables according to their content and temporal order produces a data set in wide format. Earlier we mentioned that the format you use for data input will often be defined by the raw data. However, as the example shows, it is always possible to obtain both formats with the same raw data. Nevertheless, as you can easily see, the data are much easier to combine into a data set in long format than into a data set in wide format.

We already said that all statistical program packages provide specialized commands for transforming data from long into wide format and vice versa. To complete this general introduction, we show how panel data can be transformed from long into wide and from wide into long format using the two basic commands append and merge. To that end, we use the data sets housepanL and housepanW from the example above. These data sets contain the units HHNR = 10, 11, the years 2008 and 2009 and the variable Income (compare output above). We can transform these data sets using the following syntax:

2.2 The Basics of Panel Data Management

Long to wide

```
. use housepanL
. keep if Year == 2008
(2 observations deleted)
. rename Income Income08
. sort HHNR
. save house08_
. use housepanL
. keep if Year == 2009
(2 observations deleted)
. rename Income Income09
. sort HHNR
. merge HHNR using house08_
. order HHNR Income08 Income09
. list HHNR Income08 Income09

     +--------------------------------+
     |  HHNR     Income08    Income09 |
     |--------------------------------|
  1. |    10         3000        3500 |
  2. |    11         4000        4500 |
     +--------------------------------+
```

Wide to long

```
. use housepanW
. drop Income09
. generate Year = 2008
. rename Income08 Income
. save house08a
. use housepanW
. drop Income08
. generate Year = 2009
. rename Income09 Income
. append using house08a
. order HHNR Year Income
. sort HHNR Year
. list HHNR Year Income

     +----------------------+
     |  HHNR    Year  Income|
     |----------------------|
  1. |    10    2008    3000|
  2. |    10    2009    3500|
  3. |    11    2008    4000|
  4. |    11    2009    4500|
     +----------------------+
```

To transform data from long to wide format (left hand side), we have to split up the data set by time points and then merge the data from each time point. In other words, we are taking all rows representing one time point and making them new columns in the wide data set. Here we use the `keep if` command to delete all observations which are not from 2008 (`keep if Year == 2008`). Again, we have to rename all variables except the key variables to include the time dimension in the variable names. Then, we save the data set (as `house08_`) and build a second data set which includes only observations from 2009. Again, we rename the variables to include the time dimension in the variable names. After merging both data sets and ordering the variables, we come up with a data set in wide format.

To obtain a data set in long format starting from a data set in wide format (right hand side), we have to do exactly the opposite. We split up the data set by deleting single variables. The resulting data sets have to include only the variables for one time point. In each data set we generate a new variable (`Year`) indicating to which time point the information pertains. Finally, after renaming the variables (from `Income08` to `Income`, from `Income09` to `Income` etc.), we can append the single data sets to come up with a data set in long format.

The purpose of this demonstration was to provide a basic understanding of how to transform data from long to wide format and vice versa. Therefore, we used the two basic commands `merge` and `append`. As already mentioned, all statistical programs provide specialized commands that allow us to specify both transformations with a few terms (in Stata this is the `reshape` command). However, these specialized commands are based on the two basic operations merge and append. Basically, they start programs like those we have just discussed.

2.2.5 Some General Remarks

Both of the basic operations, merging and appending, share an important feature. In both cases, the combination of data from two data sets into one is based on their qualitative similarity. If data is merged together, the data sets have to include observations that fit together. If one data set is appended to another, the data sets have to have the same variables. Abstractly speaking, data sets can be combined if there is an identifying element either in the columns (appending) or in the rows (merging) of the data matrices. From a technical point of view, the variables (observations) have to have the same names (identification numbers) if they are equal and different names (identification numbers) if they are not. This might sound like a trivial technical matter but it is very important to be aware that our computer will do exactly what we tell it and not what we want it to do. In the example above we combined tables that were prepared for a perfect fit. When working with real data, we have to make sure that we understand the structure of all data sets we want to combine. There are four potential pitfalls when combining data:

1. Appending
   (a) Two variables measure the same substantive variable, but do not have the same name.
   (b) Two variables have the same name, but do not measure the same substantive variable.
2. Merging
   (a) Two observations belong to each other, but do not have the same identification number(s).
   (b) Two observations have the same identification number(s), but do not belong to each other.

More problematic are those cases in which two variables (observations) have the same name (identification number) but do not correspond to each other. In these cases the computer will simply put the information together and it can be quite difficult to detect the error. When the same substantive variable has a different name in both data sets, we will certainly find the mistake because the resulting data set will include two variables where we expect only one and, as shown in the example in Sect. 2.2.2, both of these resulting variables will have a lot of missing values. The same is true for merged data sets: If two connected observations have different identification variables, the data set will include two observations where we expect only one.

However, ensuring that all variable names and identification numbers are correct is only the first point on our checklist. Having correct variable names and identification numbers does not guarantee that the combination of two data sets works the way we want it. The second point is to make sure the variables we want to combine are identically coded (scaled). The introduction of the Euro currency in 2002, for instance, resulted in a new scale for all monetary measures. Combining data on incomes from before 2002 with data from 2002 and after requires us to harmonize the two variables (e.g., to measure the incomes from before 2002 in Euros). For various reasons, categorical variables are at high risk of nonidentical codings. This is because the coding of (most) categorical variables is somewhat arbitrary and can change between different waves.[5] Therefore, it is of outstanding importance to have a look at the coding (and scaling) of all variables before you combine two data sets.

Most statistical software packages try to control the process of appending and merging data by providing detailed information about each step and its outcome. Nevertheless, if an operation is logically possible, the computer will never be able to tell us which operation makes sense and which does not.

2.3 Three Case Studies on Poverty in Germany

In the last section we presented the basic operations we may use to build a working data set from raw panel data. In this section we will present some examples with real survey data and address a series of practical challenges we usually face when we generate and analyze panel data. As already stated in the introductory Chap. 1, panel data can be used for different types of analysis. We present three examples showing how panel data can be prepared for *cross-sectional analyses*, the *analysis of trends* (pooled cross-sections), and the *analysis of change* (panel analysis). All examples are based on raw data from the SOEP (see Example 2.2).

In the first case study (Sect. 2.3.1) we generate a data set allowing us to estimate the poverty rate of the German resident population in 2004. In the second case study (Sect. 2.3.2) we use data for the years 2004 to 2006 to estimate the trend in German poverty rates. Finally, in the third case study (Sect. 2.3.3) we will make use of the full potential of panel data and analyze the dynamics of poverty. To that end, we generate working data sets in wide and in long format. In Sect. 2.4, we will discuss issues of balanced and unbalanced panels as well as questions of cross-sectional and longitudinal weighting. To that end, we will reconsider the results of the following analyses and reestimate poverty rates with weighted data.

Example 2.2 (SOEP data) For the following examples we will use three waves from the German Socioeconomic Panel (2004–2006). We combine data from individual-level files with data from household-level files to investigate income poverty.

The SOEP survey started in 1984 and is based on a representative sample of the German resident population. The survey consists of the initial two samples (a sample of West Germany's resident population in 1984 and a sample of selected foreigner groups in West Germany in 1984) and several additional

---

[5]Data providers, of course, intend to use identical codings in all waves. However, there are always instances where single variables are coded differently.

samples that have been selected since 1984 (for instance, a sample of the East German resident population after reunification, a sample of immigrants in 1994/1995, a sample of high income households and some refreshment samples drawn from the overall population to compensate for panel attrition; see also the discussion in Sect. 1.1.2.1). The SOEP provides information about the sampled households and about all individuals living in these households. Presently, about 20,000 individuals from more than 11,000 households are interviewed each year.

The SOEP is one of the few studies that provides detailed information about adults and children (defined as individuals 16 years of age and younger). These children do not answer a questionnaire on their own. Instead, information on children is collected through a children questionnaire answered by the head of the household.

Note also that some population groups are overrepresented by design (among them the foreigner groups from the initial sample in 1984) and the selection probabilities are not the same between all subsamples of the SOEP. Since the following case studies focus on incomes, it is also important to remember that one subsample focuses on high income households and hence, the overall distribution of incomes in the SOEP overrepresents higher incomes, if this subsample is not excluded.

SOEP data are provided in a form that reflects the process of data collection. Since the data is collected on a yearly basis, the general structure of the raw data is cross-sectional. Therefore, SOEP data files are, first, differentiated by waves: The first letter of a data set's name indicates to which wave the data set belongs (a = 1984, b = 1985, ...). Second, within each wave, the data are differentiated with regard to the unit of analysis. The letters following the first letter indicate to which unit of analysis a data set pertains (h = household, p = person). Within each wave we find a file ?pbrutto for individuals (adults and children) and for households (?hbrutto). These files contain the gross sample population, i.e., all sampled units irrespective of whether they have participated or not. The data from the actually answered questionnaires (net population) is provided in three main files: one data set for households (?h), one data set for the directly interviewed household members (?p) and, as children are not directly interviewed, a separate file for the data from the children questionnaire (?kind). Thus, within each wave we find three files that make up the net population of the survey: the household file (?h) and the two person files on adults (?p) and children (?kind).

Moreover, within each wave, the SOEP provides additional data sets including specialized information. For instance, within each wave we find the data files ?pgen (for persons) and ?hgen (for households), which include variables that have been generated by the SOEP group. Finally, we can find two files containing the weights for cross-sectional and longitudinal analyses (phrf for individuals and hhrf for households). These files include the weighting factors for all years and, thus, are not wave-specific. For the analysis of income poverty in the examples below, we will have to combine data from most of these SOEP files.

The data files used in the examples are not the original SOEP files. For reasons of data privacy protection we altered some variables and use only a part of the full SOEP sample. We generated random identification variables and added a random error to the sensitive information on income. In total, we use 25 % of the full SOEP sample. Nevertheless, all the important features of real survey data are still in the data. The results produced with the example data are quite close to the results one produces with the full data.

Generally, you should take a serious amount of time to think about the process of generating your working data from the raw panel data before you actually begin with your computer work. Usually, you will have to carefully read the data documentation before you can start to build your working data set. At the beginning of this chapter, we introduced panel data as having three dimensions: time (t), units (i), and variables (v). When we build data sets using the raw data from a panel survey, we have to become clear about each of these three dimensions. What is the population we want to analyze, i.e., which *units* have to be included in the data set? Which period of *time* should be analyzed? Which *variables* do we need for the analysis? If we have decided which information (time points, units, and variables) we need, we can start to search for the right data.

Often, it is useful to first generate a data set including the population of interest. As in the examples below, this data set can be used as a *master data* set for all subsequent data management. The term *master data set* refers to a data set which is used to determine the basic structure (usually the unit and sometimes the time dimension) of the final working data set. It includes at least the key variables that we need in order to include the variables of interest into the master data set. The term *master data set* has already been used before, when we introduced Stata's append and merge commands. In Stata, the data set which is loaded into memory is also called the master data set. When data is appended or merged it is included into the master data set. The term *master data set* that we introduced now is related to this idea but has a theoretical component as well: the master data set includes one row for each relevant observation, i.e., it represents the target population. When including data from other files into our master data, we have to make sure that the basic structure (the target population) remains unaffected.

2.3.1 Case Study 1: How Many German Citizens Were Poor in 2004? A Cross-Sectional Analysis

In this case study we compute the poverty rate of the German resident population in 2004 using data from the SOEP (see Example 2.2). We estimate the at-risk-of-poverty threshold (60 % of median income) and create an indicator for the poverty status of each respondent. We define individual incomes on the basis of equivalized disposable household incomes. The complete working data set includes information from five raw data sets of the SOEP: `up`, `ukind`, `uhgen`, `uhbrutto` and `phrf`, all of them from 2004 (as indicated by the first letter u in the data set names).

At first sight, our example seems to be quite simple. We want to estimate the poverty rate of the German resident population in 2004. In principle, our final working data set consists of only one categorical variable indicating the individual poverty status. However, as you will see, we will have to do considerable data management before we can actually compute the poverty rate.

Let us start with the three dimensions mentioned above. The population we want to analyze is the *German resident population* in *2004*. The individual-level file from 2004 (`up`) includes only adults. Children are a part of the target population and are included in a separate data file (`ukind`). We have to combine both data sets (`up+ukind`) to obtain a sample of the German resident population in 2004. This combined data set will serve as the *master data set* for further data management, i.e., we will include the variables we need for our analysis into the master data set and keep its basic structure.

So, which are the variables we need? Ultimately, we want to measure the *individuals' poverty status*. Individuals are defined as poor if their income lies below a certain poverty threshold. Following a standard definition in poverty research, we will define incomes on the basis of *equivalized disposable household incomes* and the poverty threshold as 60 % of median equivalized disposable income. Hence, we need to calculate (i) each respondent's equivalized disposable income, (ii) estimate the median income, and (iii) compare each respondent's income to the respective poverty threshold (60 % of the median income).

What is an equivalized disposable household income? In many cases, a person's economic well-being not only results from the individual's income but also from the household in which the person lives. In fact, only a part of the population actually earns an income from employment. Therefore, individual incomes from employment, as well as other incomes like social transfers, are reallocated within households. Thus, we have to refer to the household context if we want to measure an individual's actual standard of living.

As an example, think of a family with two children and one employed parent who earns a net income of 4,000 €. If you want to describe the economic situation of these four people, you will certainly not assume that the working person has an income of 4,000 € while the other three persons have no income. Perhaps you will assume that each person has an income of 1,000 €. In other words, in order to describe the economic situation of the individuals, you will use a *per capita* measure of total household net income. The idea behind the concept of *disposable equivalized* incomes is to adjust per capita incomes within a household for scale effects due to its size and for different needs of adults and children. In doing so, the actual standard of living is comparable across households with different size and composition (in terms of adults and children).

For our examples we use the modified equivalence scale of the OECD. This scale gives a weight of 1 to the first household member, a weight of 0.5 to each additional

2.3 Three Case Studies on Poverty in Germany

household member 14 years of age or older and a weight of 0.3 to each household member younger than 14 years of age. Let us think about some household constellations to exemplify this. A woman living in a single-person household with a net income of 3,000 € has an equivalized disposable income of 3,000 € (=3,000/1). Suppose a partner moves into the household. This partner has no own income. Thus, the total household net income remains at 3,000 €. A simple per capita measure of the household income would be 1,500 € and suggests that the woman's standard of living is half as good as it was before. Obviously, this assumption is unrealistic. Many items within the household (and also the living space itself) can be used by several persons without additional monetary expenses. The equivalized disposable income would be 2,000 € (=3,000/(1 · 1 + 1 · 0.5)) and accounts for these scale effects: the standard of living of the two persons in one household should equal that of a single person with an income of 2,000 €.

Suppose nine months after the partner moved into the household a child is born. Luckily, the woman's partner found work. Let us say the total household net income, now, is 4,000 €. The equivalized income, then, would be about 2,222 € (=4,000/(1 · 1 + 1 · 0.5 + 1 · 0.3)), while a simple per capita measure would be about 1,333 €. The equivalized disposable income suggests that the three persons living in one household have a standard of living that is equal to that of a single person with a net income of 2,222 €.

To summarize, the disposable equivalized household income is a measure of the total net household income which has been divided by the weighted number of individuals living in the household. The weights, first, account for the different levels of need between children and adults; the modified OECD scale gives a weight of 0.3 to children under 14 years of age. Second, the weights account for scale effects; in the modified OECD scale only the first person in a household is weighted by 1. All other persons have a weight smaller than one.

To calculate the disposable equivalized income with the SOEP data we need (i) the total disposable income of each household, (ii) the number of household members younger than 14 years of age, and (iii) the number of household members 14 years of age or older. In the SOEP, disposable (net) household incomes (ahinc04) can be found in the file uhgen. The total household size (uhhgr) and the number of household members younger than 14 years of age (unhmu14[6]) are included in the data set uhbrutto. The number of household members 14 years of age and older can be calculated by subtracting the number of household members younger than 14 years of age from the total number of household members. Finally, we will also include the cross-sectional weights for 2004 (uphrf) from the data file phrf.

Now, we have identified all data sets we need to combine: up, ukind, uhbrutto, uhgen and phrf. The files up and ukind have to be appended to each other. Together they include the (realized) sample of the population of interest and

---

[6]The variable unhmu14 (number of household members under 14 years of age) is not included in the original SOEP files, but can be generated from the information included in the data. For reasons of simplicity, we generated this variable in advance and stored it in the file uhbrutto.

make up the master file. The other three files include variables we need. These data sets have to be merged with the master file.

Before we can start to combine the data we need to think about the key variables that are necessary to merge the data. The final working data set will contain individuals as units. However, in the data sets uhbrutto and uhgen units are households. The units in a household data set are uniquely identified by their household identification number (uhhnr = household number for 2004). Hence, our master data set (up + ukind) needs to include two key variables: a household identifier (uhhnr) to merge data from household-level files and an individual identifier (persnr) to merge data from individual-level files. Now, we have planned the complete process of data management and can start work.

Our first step is to generate a master data set by appending up and ukind. In theory, we can drop all variables except the required key variables (uhhnr and persnr). However, as an explanatory variable for additional analyses we keep the variable gender. The Stata syntax is quite simple and should be known to you from the previous sections:

```
. use ukind
. keep uhhnr persnr uksex
. rename uksex sex
. save children04
. use up
. keep uhhnr persnr up13901
. rename up13901 sex
. append using children04
. * some lines omitted *
. list in 1/5
```

	uhhnr	persnr	sex
1.	2	21	woman
2.	4	41	man
3.	11	111	man
4.	12	121	woman
5.	29	291	woman

First, we load the data set including children (use ukind). Then, we reduce the number of variables to those we actually need (keep uhhnr persnr uksex). Before appending the two data sets we have to make sure that the variables in both data sets have the same names. Otherwise, the variables cannot be matched correctly. For the key variables (persnr and uhhnr) this is the case. The variable gender, in contrast, has a different name in the file up than it has in the file ukind. Hence, we have to rename it (rename uksex sex).

After saving the children file in its current form we can start to prepare the adult data set. Again, we drop all variables except the two key variables and gender (here named up13901). Before we finally combine the data, we rename the variable up13901 and give it the same name as in the children file (sex). Now, both data

2.3 Three Case Studies on Poverty in Germany

sets include exactly the same variables (uhhnr, persnr, and sex) and can be combined using the append command. After labeling, ordering and sorting the data (we do not present this syntax here), we look at the first five rows of the master data (list in 1/5).

Now that we have the master data, we can start to include the variables we need to compute equivalized disposable incomes. The total household size (uhhgr) and the number of children younger than 14 years of age (unhmu14) can be found in the household-level file uhbrutto. First, we merge the two variables uhhgr and unhmu14 into our master data. In a second step, we merge the total net household income (ahinc04) into the master data. It can be found in the household-level file uhgen. As in Sect. 2.2.3, we have to perform a one-to-many merge and use the required option uniqusing because we merge individual-level with household-level data. We also use some other options which reduce the necessary steps for merging the data sets:

```
. merge uhhnr using uhbrutto, sort uniqusing keep(uhhgr unhmu14)
variable uhhnr does not uniquely identify observations
in the master data
. tab _merge
    _merge |      Freq.     Percent        Cum.
-----------+-----------------------------------
         2 |        297        4.33        4.33
         3 |      6,568       95.67      100.00
-----------+-----------------------------------
     Total |      6,865      100.00
. drop if _merge == 2
(297 observations deleted)
. drop _merge

. merge uhhnr using uhgen, sort uniqusing nokeep keep(ahinc04)
variable uhhnr does not uniquely identify obs. in the master data
. drop _merge
```

In Sect. 2.2.2 we said that two data sets which should be merged have to be sorted by their key variable(s). In the output above we use the sort option of the merge command. This option automatically sorts both data sets by the key variable(s). Hence, we do not need to sort both data sets in advance of the data merging. The option keep(uhhgr nhmu14) specifies which variables from the using data set ought to be included in the new data set. Without this option, Stata would include all variables from the data set uhgen into the master data set.

After each merge, Stata generates a variable _merge. This variable provides information about the fit of the two data sets. The variable can take three different values indicating whether an observation appeared only in the master data set (_merge = 1), whether an observation appeared only in the using data set (_merge = 2), or whether an observation appeared in both data sets (_merge = 3). Fortunately, a frequency table of the variable (tab _merge) indicates that 95.67 % of the observations in the combined data set were included in both data sets (_merge = 3), i.e., the household income could be merged successfully with the individual-level data.

There are also 297 cases in which the household-level data set includes an observation that is not included in the master data set (_merge = 2). The number of observations which appeared in the household-level file but not in the individual-level file is that high because the file uhbrutto includes a record for each sampled household even if the household did not participate in the respective year. However, using the command drop if _merge == 2, we delete all observations which have not been included in the master data set. We do so because the master data includes, by definition, the sample of our target population. We also delete the variable _merge because Stata would not execute the following merge commands if a variable _merge already exists.

In the second part of the syntax, we include the total net household income (ahinc04) into the master data set. We use the same options as before but add the option nokeep. This option forces Stata to keep only observations which are already included in the master data set. Using this option is equivalent to the separate command drop if _merge == 2 that we used in the first merge procedure. Again, we do so because we want the master data set to remain unaffected with regard to the included units (its rows). The resulting data set includes the variables uhhnr, persnr, sex, ahinc04, uhhgr, and unhmu14.

For later use in the section on weighting (Sect. 2.4), we finally include the cross-sectional weight for 2004 (uphrf from the file phrf):

```
. merge persnr using phrf, sort nokeep keep(uphrf)
. drop _merge
```

The using data set is an individual-level data set and therefore the observations in both data sets are uniquely identified (*one-to-one merge*). Again, we include only the required variable by using the option keep. After generating a variable that measures the number of household members 14 years of age or older (nahm), our data set includes all the variables we need to compute equivalized disposable incomes and we can start to prepare our data set for the final analysis. The complete data set, now, looks like this:

```
. list in 381/390
```

	uhhnr	persnr	sex	uhhgr	unhmu14	nahm	ahinc04	uphrf
381.	895	8951	man	4	1	3	2414	4140.28
382.	895	8952	woman	4	1	3	2414	3931.68
383.	895	8953	man	4	1	3	2414	3798.59
384.	895	8954	woman	4	1	3	2414	3874.96
385.	897	8971	woman	2	0	2	4780	1669.89
386.	897	8972	man	2	0	2	4780	1664.87
387.	905	9051	man	1	0	1	2942	4794.66
388.	907	9071	woman	4	0	4	5562	2068.16
389.	907	9072	man	4	0	4	5562	2074.76
390.	907	9074	man	4	0	4	5562	2087.41

2.3 Three Case Studies on Poverty in Germany

A simple per capita measure of household income would be obtained by dividing the total disposable household income (ahinc04) by the total number of household members (uhhgr). For example, for the members of household uhhnr = 895 we would get a per capita income of $2{,}414/4 = 603.5$ €. However, for the reasons explained above, we calculate an equivalized disposable income to get a better measure of an individual's standard of living. The equivalized disposable income of each member of household uhhnr = 895 amounts to $2{,}414/(1 \cdot 1 + 2 \cdot 0.5 + 1 \cdot 0.3) \approx 1{,}050$ €. We generate the equivalized disposable income as a new variable (edincome) using the command generate edincome = ahinc04/(1+.5*(nahm-1)+.3*unhmu14). This command divides the variable ahinc04 by the *weighted* number of household members.

The last task, now, is to generate a variable which measures individual poverty status. To that end, we need to estimate the poverty threshold. The median of the variable edincome is 1,422.22 €. The poverty threshold, then, equals $0.6 \cdot 1{,}422.22 = 853.33$ €. We can generate the poverty indicator using a simple syntax which sets the new variable to 1 if the variable edincome is lower than 853.33 and to 0 if it is larger or equal to 853.33:[7]

```
. generate poverty = 1 if edincome < 853.332
. replace poverty = 0 if edincome >= 853.332 & edincome < .
. list uhhnr persnr edincome poverty in 180/184

      +----------------------------------------+
      |  uhhnr    persnr    edincome   poverty |
      |----------------------------------------|
 180. |    517      5171    583.3333         1 |
 181. |    517      5172    583.3333         1 |
 182. |    521      5211    2378.333         0 |
 183. |    521      5212    2378.333         0 |
 184. |    521      5213    2378.333         0 |
      +----------------------------------------+

. tab poverty
   poverty |
  ind. [0,1] |     Freq.      Percent        Cum.
-------------+-----------------------------------
           0 |     5,649        89.04       89.04
           1 |       695        10.96      100.00
-------------+-----------------------------------
       Total |     6,344       100.00
```

The resulting poverty rate for 2004 is 10.96 % (using unweighted data). In Sect. 2.4 we will compare this poverty rate to the poverty rate that we get if we use cross-sectional weights. For now, we will assume that the German poverty rate in 2004 was about 11 %.

---

[7] In Stata, missing values are treated as infinitely large numbers. A logical operation which uses the operator ">" includes missing values. Therefore, we have to add "& edincome < ." in the replace command.

2.3.2 Case Study 2: Did Poverty Increase in Germany After 2004? An Analysis of Pooled Cross-Sections

In this second case study, we calculate the trend in German poverty rates from 2004 to 2006 using data from the SOEP (see Example 2.2). We use a poverty threshold of 60 % of median equivalized disposable incomes and create an indicator for the poverty status on the individual level. The analysis is based on 13 raw data sets from the SOEP: up, ukind, uhgen, uhbrutto, vp, vkind, vhgen, vhbrutto, wp, wkind, whgen, whbrutto and phrf, all of them from 2004–2006 (as indicated by the first letters u, v, and w in the data set names).

In the last section we estimated the poverty rate for 2004. Now, we extend the scope of the analysis to the time period 2004 to 2006. The aim of this analysis is to obtain an estimate of the trend in poverty rates, which from a methodological point of view resembles an analysis of *pooled cross-sections*. As in the previous example, we do not necessarily need panel data for this purpose. We do not need to pool the cross-sections to obtain an estimate of the time trend. Alternatively, we could use three single cross-sections and estimate the poverty rates with each of these data sets separately. However, once we have built a trend data file this task is much easier and therefore, we will actually pool the cross-sections. Nevertheless, it is important to notice that the resulting data will still be a panel data set and not a pooled file including independent cross-sections. This is because the data we combine include repeated observations of the *same* units over time no matter how we combine these data.

Again, we should take some time to think through how we will manage the complete data before we begin to work with the data. We start by defining the three dimensions of our data cube: The units (German resident population) and variables (individual poverty status) are, in general, identical to those used in the previous case study. However, the time dimension differs from the last example. We aim to estimate the poverty rates of the German resident population from 2004 to 2006. For this trend analysis we need a sample of the German resident population for each year.

We can think of two general approaches to generate the working data set. First, we can repeat the steps from the previous case study for the data sets from 2005 and 2006 and pool the three resulting cross-sections. Second, we could start by appending the adult and children files for all years (up, ukind, vp, vkind, wp, and wkind) to come up with a master file representing the target population (the German resident population 2004–2006). In the next step we would merge all required variables into this master data set. Both ways result in exactly the same working data set. The second approach would require us to use the time points t as an additional key variable, because the master file would include three observations per individual and at least three observations per household. In this case the observations are only uniquely identified by the combination of identification number and time point.

Both ways are reasonable. Since we already wrote the syntax for constructing the cross-section 2004, we can easily adjust it to the data from 2005 and 2006. This is done by changing the first letter of all data sets (remember that the first letter indicates the wave) and substituting the variable names from the 2004 data by the

2.3 Three Case Studies on Poverty in Germany

Table 2.2 Pooled cross-sections

Persnr	Wave	Edincome	Poverty	Sex
41	2004	480	1	man
...	...	...	...	...
18,723	2005	2,159.2	0	man
...	...	...	...	...
15,342	2006	801.33	1	woman

Source: Unweighted SOEP data (see Example 2.2)

corresponding names in the following waves. We do not present the syntax here but we come up with three data sets named upoverty, vpoverty and wpoverty (we stay with the SOEP naming system and use the first letter of a data set's name as an indicator for the time point). The data set upoverty results from the last example; the other two data sets have been generated with the adjusted syntax. All three data sets include the variables gender (sex), income (edincome), poverty (poverty), a cross-sectional weight (?phrf), and the key variables (persnr, ?hhnr).

However, the three cross-sections cannot be pooled together right away. Again, we have to make sure that all variables have exactly the same names. In the current form each data set includes two variables with wave-specific names (uhhnr and uphrf in the file upoverty, vhhnr and vphrf in the file vpoverty, and so on). Moreover, as we want to differentiate between years, we need an additional variable indicating to which wave an observation belongs. We have to create this variable before appending the data sets because we would not be able to identify the observations of one time point after pooling the data. In the Stata output below we present only the syntax for preparing the 2006 and 2004 data. We have to do exactly the same with the data sets from 2005 (vpoverty). Then, we can append the data sets to finally obtain our pooled cross-section data set:

```
. use wpoverty
. generate wave = 2006
. rename whhnr hhnr
. rename wphrf phrf
. save wpoverty, replace

* syntax for 2005 omitted

. use upoverty
. generate wave = 2004
. rename uhhnr hhnr
. rename uphrf phrf
. append using vpoverty wpoverty
. save pcspoverty
```

An excerpt of the resulting data set is presented in Table 2.2. Since the poverty indicator is coded as a dummy variable (0, 1), we can calculate the poverty trend by computing the mean of the poverty indicator for each year. The results are presented in Fig. 2.5. The estimated poverty rate increased between 2004 (10.96 %) and 2006 (11.66 %).

Fig. 2.5 Unweighted German poverty rates (2004–2006)

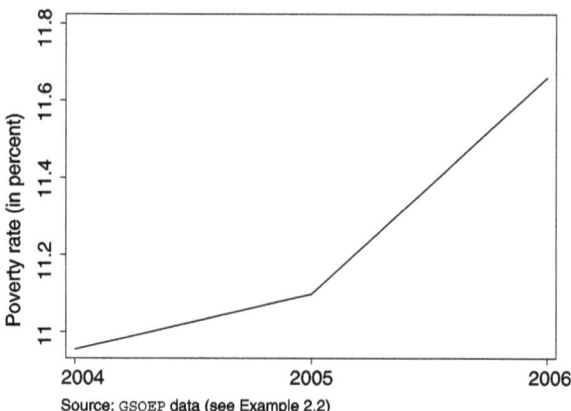

As already indicated above, the data set used for this analysis is a *panel data* set. Currently it is organized in the form of pooled cross-section data, with the first 6,568 observations being from 2004, the next 6,248 observations from 2005 and so forth, but sorting the data by the key variables persnr and wave would put it in the typical form of panel data in long format. So, what is the difference between panel data analysis and the analysis of pooled cross-sections? In pooled cross-sectional analyses like the present trend analysis, we ignore the fact that units are repeatedly observed over time. In a panel analysis we make use of the information provided by the key variables, i.e., we want to know which observations belong to the same units. This is the topic of the next section which analyzes individual change in poverty states. This analysis starts again with the pooled cross-section data that we have just generated.

2.3.3 Case Study 3: How Large Is the Risk of Becoming Poor in Germany? A Panel Analysis

In this case study we calculate the percentage of individuals moving between poverty and non-poverty in the period from 2004 to 2006. The target population is the German resident population. Again we use SOEP data and a poverty threshold of 60 % of median equivalized disposable incomes. The case study illustrates how to create panel data in long and in wide formats. The panel data set in long format will be generated using the data set that we constructed in the last example (pcspoverty). The working data set in wide format will be generated from 13 raw data sets from the SOEP: up, ukind, uhgen, uhbrutto, vp, vkind, vhgen, vhbrutto, wp, wkind, whgen, whbrutto, and phrf, all of them from 2004 to 2006 (as indicated by the first letters u, v, and w in the data set names).

In this case study we want to make use of the full potential of panel data and analyze the dynamics of poverty between 2004 and 2006. In Sect. 2.3.2 we almost generated a panel data set in long format. We used three single cross-sections and pooled them together. Sorting these data by the unit identifier (persnr) and the years (wave) transforms them into the usual long format of panel data:

2.3 Three Case Studies on Poverty in Germany

```
. use pcspoverty
. sort persnr wave
. list persnr wave edincome poverty sex in 1/7, sepby(persnr)
     +---------------------------------------------------+
     |  persnr    wave   edincome   poverty     sex      |
     |---------------------------------------------------|
  1. |      21    2004       1245         0   woman      |
  2. |      21    2005       1275         0   woman      |
  3. |      21    2006        814         1   woman      |
     |---------------------------------------------------|
  4. |      41    2004        480         1     man      |
  5. |      41    2005        502         1     man      |
  6. |      41    2006        524         1     man      |
     |---------------------------------------------------|
  7. |      51    2006       1246         0   woman      |
     +---------------------------------------------------+
```

Is this data set a panel data set? In principle yes, but the target population is not completely identical to the population we want to have in a genuine panel data set. In the last example, we pooled three cross-sections without considering the data to be panel data. The target population of the last analysis was the German resident population 2004 to 2006. As you can see in the data excerpt above, the data set includes units which have not been observed in 2004 (see individual persnr = 51 which enters the panel in 2006). In our panel analysis we want to follow an initial panel population over time. Individuals that entered the panel in 2005 or after are not part of the longitudinal population from 2004. A panel population, of course, can be also defined in a different way. Many studies use unbalanced designs and define late entries as members of the population. However, in this example we will use the longitudinal population from 2004 as our target population.

Table 2.3(a) shows the observed response patterns in the data.[8] In the current form, only 70.7 % of the units were observed in all three years. About 14.4 % of the units enter the panel in 2005 or later. We will discuss some general issues of balanced and unbalanced panels in Sect. 2.4.2. For now, you should simply notice that the population represented in the data is not the longitudinal population from 2004. We obtain a panel data set with the longitudinal population if we delete all units which have not been observed in 2004. The syntax needed for this task is Stata-specific and not presented here. However, Table 2.3(b) presents the response patterns in the data after deleting the units which are not part of the target population. The data set is still unbalanced because it includes units which have not been observed in all the years following 2004.

As mentioned at the beginning of this chapter, it is more difficult to analyze the relationships between measurements at different points in time if the data is in long format. This is because we need to combine values from different rows of the data

---

[8] Each row in these tables represents an observed response pattern and gives the number of units which show the respective pattern. For instance, the row with the pattern 111 shows the number of units which have been observed in all three waves; the row ..1 shows the number of units which have been observed only in the last wave (2006) but not in 2004 and 2005, etc.

Table 2.3 Response patterns in unbalanced panel data

(a) Cross-sectional population

Pattern	Freq.	Percent
111	5,419	70.67
..1	877	11.44
11.	606	7.90
1..	445	5.80
.11	188	2.45
1.1	98	1.28
.1.	35	0.46
xxxx	7,668	100.00

(b) Longitudinal population

Pattern	Freq.	Percent
111	5,419	82.51
11.	606	9.23
1..	445	6.78
1.1	98	1.49
xxxx	6,568	100.00

Source: SOEP data (see Example 2.2)

matrix to calculate statistics. If we want to compute the percentage of individuals that moved between poverty and non-poverty, we have to create the so-called *lag* of poverty. The lag of poverty is a variable that takes, at time point t, the value of poverty at $t-1$ ($Ly_t = y_{t-1}$). Therefore, the software needs to know which rows belong to the same units. In Stata we can create the lag of poverty like this:

```
. tsset persnr wave
        panel variable:  persnr (unbalanced)
         time variable:  wave, 2004 to 2006, but with gaps
                 delta:  1 unit
. generate lpoverty = L.poverty
. list persnr wave poverty lpoverty in 192/197, sepby(persnr)
     +------------------------------------+
     |  persnr   wave   poverty   lpoverty |
     |------------------------------------|
192. |   2351    2004       0          .  |
193. |   2351    2005       0          0  |
194. |   2351    2006       0          0  |
     |------------------------------------|
195. |   2382    2004       0          .  |
196. |   2382    2005       1          0  |
197. |   2382    2006       0          1  |
     +------------------------------------+
```

The command `tsset persnr wave` declares the data as panel data, with `persnr` being the unit identifier and `wave` being the time identifier. Stata responds with a summary of the characteristics of the panel (unbalanced, including observations from 2004 to 2006). The lag operator L. then is used to generate the lag of poverty. This lag operator does work only after the data set has been declared as a panel data set. The value of the lag variable (lpoverty) is, of course, missing for all observations from 2004. This is because in the data set there is no data available for 2003 that could have been used to fill these cells of the data matrix. In all other cells, the lag of poverty at time point t equals the poverty status at $t - 1$. Now, the data is prepared for an analysis of individual poverty dynamics. We only present the syntax for analyzing change in poverty status between 2004 and 2005:

```
. tab lpoverty poverty if wave == 2005, nofreq row

          |  poverty ind. [0,1]
 lpoverty |      0           1  |    Total
----------+----------------------+----------
        0 |  95.95        4.05  |   100.00
        1 |  33.71       66.29  |   100.00
----------+----------------------+----------
    Total |  89.30       10.70  |   100.00
```

The table is a simple cross-tabulation that shows individual poverty status 2004 in the rows and poverty status 2005 in the columns.[9] Of those individuals that were not poor in 2004, 95.95 % are also not poor in 2005, while 4.05 % became poor in 2005. Of those individuals that were poor in 2004, 33.71 % are not poor in 2005, while 66.29 % stayed in poverty. However, the lesson to learn here is that we need special time-series operators if we use data in long format. Statistical software packages that can deal with time-series data provide a number of different time-series operators. All of these operators have in common that they interrelate different rows of the data matrix.

The data set from the analysis above was in long format because it was built from data which has been prepared for an analysis of pooled cross-sections (case study 2 in Sect. 2.3.2). However, to come up with a panel data set we had to adjust the sample population to the target population of our panel analysis. Therefore we had to delete the units which are not part of the longitudinal population from 2004. This is an intricate way of generating a panel data set. How would you construct a panel data set that directly includes only those units that actually belong to this target population?

Well, we would use the cross-section of 2004 (up + ukind) as the master data set and merge each year's variables into this master data. The resulting data set would be in wide format. In the following paragraphs we will demonstrate how to construct such a data set in wide format. We will also include longitudinal weights. These weights can be found in the same file as the cross-sectional weights (phrf). We will use these weights in Sect. 2.4. At this stage we simply have to notice that we

---

[9]The options nofreq and row are used to suppress absolute frequencies and show row percentages only.

need three variables to create longitudinal weights, i.e., the cross-sectional weight for 2004 (uphrf) and the inverse of the dropout probability for 2005 (vpbleib) and 2006 (wpbleib).

Now, there is one point left to be mentioned. When we generate panel data sets in wide format, it makes sense to distinguish between time-constant and time-varying variables. In our example, gender is a time-constant variable while all other variables are theoretically time-varying. In this case, our final data set does not need to include three variables measuring gender (sex2004, sex2005, sex2006). In wide format, only time-varying characteristics have to be stored as separate variables for each time point t. Creating three variables that measure gender would not be harmful but highly inefficient. By including only one variable for each time-constant characteristic we can reduce the number of variables. Now, we have a plan of what we need and can start to work.

We generate the master data set by appending up and ukind. We do not present the syntax here because the resulting data set is completely identical to the master data set of the first case study. The master data include a key variable identifying the household a person lived in 2004 (uhhnr), a key variable identifying individuals (persnr) and the variable gender (sex). An excerpt of the data is presented after the list command in the following output. Then, we merge household identification numbers for 2005 and 2006 into the master data:

```
list in 1/5
     +------------------------+
     | uhhnr   persnr    sex  |
     |------------------------|
  1. |     2       21  woman  |
  2. |     4       41    man  |
  3. |    11      111    man  |
  4. |    12      121  woman  |
  5. |    29      291  woman  |
     +------------------------+
. merge persnr using vp, sort nokeep keep(vhhnr)
. drop _merge
. merge persnr using vkind, sort nokeep keep(vhhnr) update
. drop _merge
. merge persnr using wp, sort nokeep keep(whhnr)
. drop _merge
. merge persnr using wkind, sort nokeep keep(whhnr) update
. drop _merge
. order persnr ?hhnr
. sort persnr
. list in 140/144
     +--------------------------------------------------+
     | persnr   uhhnr    vhhnr    whhnr         sex     |
     |--------------------------------------------------|
140. |  4171     417      417      417          man     |
141. |  4172     417    10740    10740        woman     |
142. |  4201     420      420      420          man     |
143. |  4202     420      420      420        woman     |
144. |  4221     422      422      422          man     |
     +--------------------------------------------------+
```

2.3 Three Case Studies on Poverty in Germany   47

The resulting data set includes the household identification numbers for each wave. We need these variables because the household identification number is not a time-constant characteristic (see individual persnr = 4172). A respondent may move into another household (e.g., due to a marriage) or may found a new household (e.g., after finishing school and getting a job). Thus, we need the household identification number for each year in order to merge the corresponding household-level data into the master data. The second and the fourth merge command in the output above differ slightly from those used in the previous examples. After including the household numbers from the adult file vp (first merge command), we have to specify the option update in order to include the household numbers from the children file (vkind). This is necessary because Stata, by default, does not allow us to overwrite the values of a variable that already exists in the master data (in this case vhhnr). An alternative to this approach would be to append the adult and children files in advance. Using these pooled data we could include household numbers with a single merge command. In the next step we can now merge the required variables into the master data set. We present only the syntax for merging the variables from 2004 (ahinc04, uhhgr, and unhmu14) and from 2006 (ahinc06, whhgr, and wnhmu14). We also include the variables needed to construct longitudinal weights:

```
. merge uhhnr using uhgen, uniqusing nokeep keep(ahinc04) sort
. drop _merge
. merge uhhnr using uhbrutto, uniqusing nokeep keep(uhhgr unhmu14)
sort
. drop _merge

* syntax for 2005 omitted

. merge whhnr using whgen, uniqusing nokeep keep(ahinc06) sort
. drop _merge
. merge whhnr using whbrutto, uniqusing nokeep keep(whhgr wnhmu14)
sort
. drop _merge

. merge persnr using phrf, sort nokeep
keep(tphrf upbleib vpbleib wpbleib)
```

The first block is known to you from the first case study. It includes the three variables required to compute equivalized disposable incomes for 2004. After adjusting all wave-specific variable names (including the household identification number), we can use exactly the same syntax to merge the variables from 2005 and 2006 to our master data. The last merge procedure includes the weights we need. We do not present the syntax for computing equivalized disposable household incomes, but you should notice that we need to calculate three equivalized household incomes. After generating the income measure, deleting some variables, and reorganizing the data we obtain a file with the typical structure of panel data in wide format:

```
. list persnr sex income* in 1/5
  +------------------------------------------------------------+
  |   persnr      sex    income2004    income2005    income2006 |
  |------------------------------------------------------------|
1.|       21    woman          1245          1275           814 |
2.|       41      man           480           502           524 |
3.|      111      man          2734          2434             . |
4.|      121    woman          1195          1795          1595 |
5.|      291    woman          1047             .             . |
  +------------------------------------------------------------+
```

Although not shown in the former listing, the file also includes the cross-sectional weight for 2004 (now named `phrf2004`) and the estimated inverse drop out rates for 2005 and 2006 (now named `pbleib2005` and `pbleib2006`). We will come back to these weights in the section on weighting (Sect. 2.4.2). As in the previous example we create an indicator which is 1 for individuals with an equivalized disposable income below the poverty threshold and 0 otherwise. We did not calculate the poverty threshold from the data at hand. Instead we used the poverty thresholds that we have estimated with the pooled cross-section data. We do so because we want to relate the income of our longitudinal population (the German resident population of 2004) to the incomes of the population in each of the years 2004–2006. Since the data set is in wide format, we come up with three poverty indicators (`poverty2004`, `poverty2005`, and `poverty2006`), each based on the respective income variable. In the wide format we can easily analyze the statistical associations between the measurements at different points in time because each measurement is a variable on its own. We do not need lagged variables. A simple cross-tabulation, for instance between `poverty2004` and `poverty2005`, shows the percentage of people that moved into or out of poverty.

Table 2.4 summarizes the results of our analysis. In the column "$0 \Rightarrow 1$" the table shows the percentage of non-poor people (in t) who moved into poverty in $t + 1$. The column "$1 \Rightarrow 0$" shows the percentage of poor people (in t) who moved out of poverty in $t + 1$. In both waves about one third of the people living in poverty at t moved out of poverty in $t + 1$. Thus, the chance to move out of poverty was quite high. The risk of becoming poor was, in contrast, relatively low (about 4–5 %). Similar to the trend analysis (Sect. 2.3.2), the problem of poverty seems to get worse in the observation period. The trend analysis suggested an increasing poverty rate. The panel analysis, now, shows that the percentage of people moving into poverty rose between 2004 and 2006.

Table 2.4 Poverty dynamics in Germany (2004–2006, unweighted data)

Source: Unweighted SOEP data (see Example 2.2)

t	$t+1$	
	$0 \Rightarrow 1$	$1 \Rightarrow 0$
2004	4.05	33.71
2005	4.80	34.51

2.4 How to Represent a Population with Panel Data?

In the previous three examples we used unweighted data. Are the results good estimates of the true population parameters? In theory, one can infer population parameters from a *simple random sample* (SRS) of the population. A SRS has to meet the following assumptions:
- random selection of units,
- identical selection probability for all units,
- selection of one unit does not affect the selection probabilities of other units.

For different reasons social surveys seldom meet these assumptions. For example, some of the sampled units may refuse to participate in the survey or cannot be contacted (*unit non-response*). Some of the respondents may also refuse to answer certain questions (*item non-response*). Furthermore, social science surveys often use complex sampling designs. As a result, *selection probabilities* differ between units.

Many social science surveys, for example, apply some kind of multi-stage sampling selecting regional units in the first stage, then households within these regions in the second stage, and finally individuals within the selected households in the third stage. If one individual is selected randomly from each randomly selected household, selection probabilities of individuals living in single-person households equal 1, while selection probabilities of individuals living in multi-person households equal the inverse of the household size. Hence, selection probabilities differ at the last stage of this sampling design. Some surveys also draw samples, which intentionally *overrepresent* certain groups in order to provide enough cases to study these groups. For example, the SOEP includes a sample of immigrants and a sample of high income households. Both groups are overrepresented in the total SOEP sample and hence, have higher selection probabilities than the other sample members (compare Example 2.2).

Statistical weights can be used to compensate for different selection probabilities. Weights modify the relative importance of single units in the sample. In an unweighted analysis all units have the same influence, i.e., all units have an equal weight of 1. In weighted analyses, underrepresented units (i.e., units with below average selection probabilities) get a weight larger than 1 and, thus, count more than they do in an unweighted analysis, while overrepresented units (i.e., units with above average selection probabilities) get a weight smaller than 1 and, thus, count less than they do in an unweighted analysis. *Design weights* account for different selection probabilities due to the sampling design. Design weights are obtained as the inverse of the (known) sample selection probability. For example, if only one individual is sampled within each selected household, an individual coming from a two-person household would get a weight of 2 ($1/0.5 = 2$), while an individual coming from a one-person household would get a weight of 1 ($1/1 = 1$). These weights would account for the selection probabilities at the last stage of the sampling design and would be multiplied with the selection probabilities from the other stages. In addition to design weights, *population weights* (also called *redressment weights*) can be used to adjust the profile of the sample to the marginal distribution from official data and in doing so, account for unit non-response bias and sampling errors. In that case, the total weighting factor is obtained as the product of design and population weights.

When should we use weights? This depends on what we want to do with the sampled data. Do we want to estimate statistical models that test our theoretical propositions (e.g., whether poverty is related to family background as assumed by some poverty researchers; e.g., Bowles et al., 2008) or do we want to make representative statements about the population from which the sample has been drawn (e.g., about the percentage of citizens at risk of poverty in a given year)? In the latter case of descriptive inference, one should use weights if the sample does not have the same characteristics as a SRS. In the former case of analytical inference, one should specify the model in such a way that it controls for different selection and response probabilities. We will address this particular issue at the end of this section (Sect. 2.4.3).

However, before we give some general recommendations, we will show how to use cross-sectional and longitudinal weights in balanced and unbalanced panels. In Sect. 2.4.1 we will demonstrate how to use cross-sectional weights. We will reconsider the results of our trend analysis of German poverty rates in the second case study and compare weighted and unweighted results. In Sect. 2.4.2 we will address particular issues that arise if we use weights for longitudinal analyses. We will reconsider the analysis of individual change in poverty status from the third case study and again, compare weighted and unweighted results. We will also address the issue of balanced and unbalanced panel designs because both designs require different weighting procedures. In Sect. 2.4.3 we will give some general recommendations on weighting, in particular, on weighting in regression-type analyses.

2.4.1 Weighted and Unweighted Analysis of Cross-Sections

In the first and second case study we aimed at estimating the poverty rates of the German resident population from 2004 to 2006. The analysis was based on the available sample for each of the years from 2004 to 2006. The crucial question is whether the available sample of, say 2005, can be treated as a simple random sample of the German resident population from 2005. In this case we would get unbiased estimates from the unweighted analysis. But is that true?

In general, one can assume that the deviation between sample and (cross-sectional) target population increases from wave to wave. This is, first, because the composition of the sample might change due to panel attrition and, second, because the target population might change as well (see the discussion in Sect. 1.1.2.1). Refreshment samples counterbalance this process, but as they are not drawn yearly they are not able to completely compensate for the lack of coverage between sample and population. Moreover, each new refreshment sample, of course, suffers from the same problems as the initial sample, i.e., not all sampled units will participate.

The cross-sectional weights provided by the SOEP group account for different selection probabilities due to unit non-response and sampling design (including the fact that some groups—foreigners, high income households—are intentionally overrepresented in the SOEP). They adjust the profile of each years' total sample to

2.4 How to Represent a Population with Panel Data?

the respective population marginals using data from official statistics (i.e., the Microcensus, a 1 % sample of the German resident population where participation is mandatory by law).

When we built the data set for the trend analysis of poverty rates, we already included cross-sectional weights. Thus, we can easily repeat the analysis with weighted data. However, it is not sufficient to estimate the percentage of poor individuals by computing a weighted mean of our poverty indicator. We also have to reestimate the poverty threshold and create a new poverty indicator based on this threshold.

Using unweighted data from 2004, we computed a median equivalized disposable income of about 1,422 € and a respective poverty threshold of about 853 €. Using statistical weights for 2004 (uphrf), the estimated median income equals 1,348.33 € and the respective poverty threshold is $0.6 \cdot 1,348.33 \approx 809$ €. Thus, the estimated poverty threshold is lower if we use statistical weights. While an individual with an income of 830 € was considered poor using the unweighted poverty threshold, we would consider this individual non-poor if we use the weighted poverty threshold.

Why is the weighted poverty threshold smaller? Well, it is smaller because the estimated median income is smaller. One important reason for this outcome is that all respondents which are members of the high income sample have a weight of zero.[10] In other words, in the unweighted analysis we included the oversampled group from high income households, while we now exclude this group in the weighted analysis. Having computed the new poverty threshold for each year from 2004 to 2006, we can generate a new poverty indicator and use this variable to calculate the percentage of individuals that were poor. Figure 2.6 presents the results of the weighted and the unweighted analysis of the trend in poverty rates.

The first observation is that, in all years, the weighted poverty rate was higher than the unweighted poverty rate. This is surprising because the unweighted poverty threshold is larger than the weighted poverty threshold. Thus, we define less cases as poor if we use the weighted poverty threshold. So, why is the weighted poverty rate higher than the unweighted? This is because the group of poor people, on average, has a higher weight than the group of non-poor people. In our example the average weight of the group of poor people is 1.31, while the average weight of the group of non-poor people is 0.97. Hence, those individuals who are defined as poor gain importance through weighting while the non-poor lose importance through weighting. In other words, people near the bottom of the income distribution seem to be underrepresented in the sample, a result well known from survey research.

---

[10] Since the size of population of high income households (defined as having a monthly income of at least 3,835 €) was unknown, it is not possible to quantify the selection probability for the high income sample. Hence, it is impossible to correct the oversampling with design weights and the safest way to get unbiased income statistics is to exclude the high income sample with a "weight" of zero. Income statistics are then estimated with the remaining SOEP respondents whose selection probabilities are known.

Fig. 2.6 Weighted and unweighted German poverty rates (2004–2006)

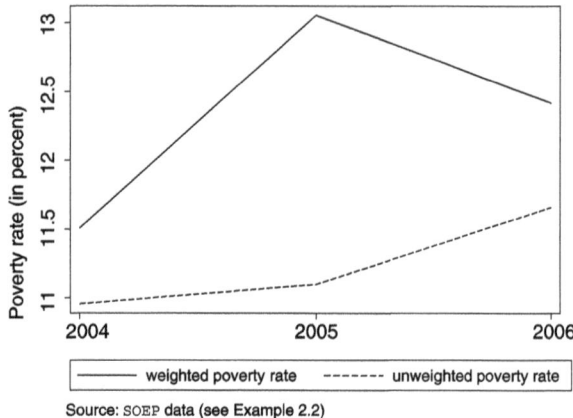

The second observation is that weighted and unweighted estimates lead to substantially different conclusions. Between 2004 and 2005, weighted and unweighted estimates showed an increasing poverty rate. However, between 2005 and 2006 weighted poverty rates decreased while unweighted poverty rates increased. Obviously, weighting can have a substantial effect on the results and we should put some thought in the question of whether we should weight our data or whether we should use unweighted data. In this example, the weighted estimates are much more trustworthy, because the unweighted estimates are biased due to the overrepresentation of high incomes. From a technical point of view, weighting cross-sections is easy if the weighting factors are provided with the data. We did not present the syntax here because weighting commands are somewhat software-specific, but you will be able to weight your data with cross-sectional weights.

2.4.2 Weighting in Balanced and Unbalanced Panels

Statistical weighting in panel analyses is much more complex than it is in cross-sectional analyses. In the third case study we performed a panel analysis of individual poverty dynamics. Can we use weights to improve the estimates from this analysis? What will these weights look like?—We have to use *longitudinal weights*. What is a longitudinal weight and how does it differ from cross-sectional weights? As explained in the last section, cross-sectional weights adjust the profile of the sample to the profile of the target population. Longitudinal weights do exactly the same. Remember that our target population in the third case study was the German resident population of 2004.

Consequently, we should use weights which account for differences between the sample and the German resident population of 2004. For observations from 2004 this weight is the cross-sectional weight for 2004 (phrf2004). What about using the 2004 cross-section weight for all waves? In principle, this is a good idea. As

2.4 How to Represent a Population with Panel Data?

long as all individuals who participated in 2004 stay in the panel we could use the cross-sectional weights from 2004 for each wave.

However, we face the problem of panel attrition. Some of the units that participated in 2004 drop out of the sample. Table 2.3(b) shows the observed response patterns in the longitudinal sample. Only 82.5 % of the sample members are observed over the complete period under investigation. Nearly one-fifth of the sample drops out between 2005 and 2006. If these dropouts are not distributed randomly, we get biased estimates from unweighted analyses. Longitudinal weights compensate for these dropouts. They are obtained by multiplying the cross-sectional weights with the inverse of the dropout probability.

As an example, assume that the members of a particular group, say men between 20 and 30 years of age, tend to drop out of the sample while all other groups stay in the panel. Thus, in each wave the percentage of men aged 20 to 30 years decreases while the percentage of all other individuals increases. Longitudinal weights compensate for these dropouts by giving higher weights to the remaining men between 20 and 30 years of age and lower weights to all the other cases.

To obtain the longitudinal weighting factor for time point $t+x$, we have to multiply the cross-sectional weight for time point t with all inverse dropout probabilities from $t+1$ to $t+x$. Drop out probabilities can be estimated using logistic regression models. The SOEP group provides estimates of the inverse dropout probability for each unit and each time point. Researchers who want to weight their panel data have to construct the respective longitudinal weights manually. As we included these variables in our data set, we can create the longitudinal weights using syntax like this:

```
gen lweight2004 = phrf2004
gen lweight2005 = lweight2004*pbleib2005
gen lweight2006 = lweight2005*pbleib2006
```

The three resulting longitudinal weights now can be used to analyze the transitions between poverty and non-poverty. However, before we repeat the analysis from the third case study with weighted cases, we have to consider the structure of our data set.

Table 2.3(b) shows the response patterns in the data set. Obviously, the panel is unbalanced. A balanced panel would consist of the 5,419 cases with the pattern 111. Longitudinal weights are easily applied to these cases.

The same is true for units that drop permanently out of the panel (patterns 11. and 1..). *Temporary* dropouts (pattern 1.1), however, create a problem because longitudinal weights equal the product of the cross-sectional weight for the starting wave t and all inverse dropout probabilities between the start of the panel t and the current time point $t+x$. By definition, a unit's inverse dropout probability is zero for the year the unit drops out temporarily. Consequently, the weight remains zero after the unit reenters the panel. The same is true for *late entries*, but in our current example these are excluded anyway. Thus, we cannot make use of the reentries (and late entries) if we use the weights from above. The data excerpt below exemplifies this:

```
+---------------------------------------------------------------------------+
| persnr   po~w2004   lwe~2004   po~w2005   lwe~2005   po~w2006   lwe~2006  |
|---------------------------------------------------------------------------|
|   9991          0    8246.89          0   8659.233          0    8919.01  |
|   9992          0    8023.71          0   8424.896          0   8677.643  |
|  10032          0    3605.76          .          0          .          0  |
|  57555          0    1772.03          .          0          1          0  |
+---------------------------------------------------------------------------+
```

The table shows the individual identification number (persnr), the poverty indicator for 2004 (here abbreviated as po~w2004) followed by the respective weight for 2004 (here abbreviated as lwe~2004), followed by the poverty indicator for 2005, and so on. The individuals persnr = 9991 and persnr = 9992 participated in all three waves. Individual persnr = 10032 dropped permanently out of the panel in 2005. The problematic case is individual persnr = 57555. This unit dropped out in 2005 but reentered the panel in 2006. As the inverse dropout probability for the unit is 0 in 2005, we get a longitudinal weight of zero for all waves after the temporary dropout.

There are different solutions to this problem. First, we can run an unweighted analysis as we did in Sect. 2.3.3. This would, however, lead to biased estimates if the sample does not have the characteristics of a SRS. Second, we can estimate longitudinal weights for each temporary non-response pattern. This is a complex and time-consuming task. Third, we can ignore the problem and disregard the observations of the temporary dropouts after they reentered the panel. In other words, we can treat the temporary unit non-responders as if they were permanent dropouts. In this case we assume that temporary non-responders do not differ from the other units in the sample.

In our example we will disregard all observations after and including a temporary dropout, because the number of units with such a response pattern is rather small ($n = 98$; see Table 2.3(b)). To exclude these observations, we simply perform the analysis with the weights that we have generated. As the weights are zero for all observations after and including the time point when a unit dropped out of the panel, all observations after and including a temporary dropout are excluded from the analysis.

The poverty thresholds we use for this analysis are based on weighted cross-sectional estimates of median equivalized disposable incomes. This is necessary because the poverty threshold should be related to the current income distribution. In other words, we used the thresholds from the weighted analysis of pooled cross-sections (Sect. 2.4.1). Table 2.5 presents the results of the weighted analysis of individual poverty dynamics.

As a final point in this section, we briefly consider some general issues of balanced and unbalanced panels which are independent of the weighting problem. In the ideal case, raw panel data are already balanced. In this case one does not have to decide how to treat attritors, late entries, and temporary dropouts. Nevertheless, all real panel data from social and economic panel surveys are unbalanced. Usually, we cannot afford to delete all cases which have not been observed over the complete

2.4 How to Represent a Population with Panel Data?

Table 2.5 Poverty dynamics in Germany (2004–2006, weighted data)

Source: Weighted SOEP data (see Example 2.2)

t	$t+1$	
	$0 \Rightarrow 1$	$1 \Rightarrow 0$
2004	4.91	26.17
2005	4.56	32.44

period of investigation. In the third case study we would have to delete 1,149 cases (compare Table 2.3(b)). This is nearly one-fifth of the complete three-year sample. One can easily imagine how many cases one loses if the panel has 10 or 20 waves. Thus, in most cases we will have to use unbalanced panel data.

When doing regression-type analyses, we have to use statistical software which is capable of dealing with unbalanced panel data. Fortunately, regression models for unbalanced panel data are a simple extension of regression models for balanced panel data and all statistical software packages that can estimate panel models can also deal with unbalanced data. Generally, we recommend using unbalanced panel designs. In most instances, the loss of cases is simply too large if we delete all units which have not been observed in all waves. As a consequence, the *efficiency* of our estimates will be significantly lower with balanced panel data, i.e., the standard errors of the estimates will increase if we reduce the sample to those cases which make up a balanced panel.

A more serious problem arises if those units which are not observed over the complete panel study (attritors, late entries, and temporary dropouts) differ systematically from the rest of the sample. In this case balanced and unbalanced panels will lead to biased estimates, but the strength of this bias would be much larger if we use balanced instead of unbalanced data. As we discussed in this section, weighting offers a solution to the problem of biased estimates due to unit non-response.

2.4.3 When to Use Weights?

In general, we recommend using statistical weights to counterbalance different selection probabilities due to non-response and sampling design. This is particularly true for representative statements about the population like those in our three case studies. However, when using weights in statistical models (e.g., panel regression models) one has to be aware of one drawback. The formulas for standard errors used by most statistical software packages assume a simple random sample. If we use statistical weights, the standard errors of the estimated coefficients will be biased and the statistical inference drawn from the sample might be invalid. However, before we discuss the particular issue of weighting in statistical modeling (analytical inference), we will briefly consider the use of weighted data to make representative statements about the population (descriptive inference).

Assume we want to do a simple descriptive analysis of one variable: income. Different selection probabilities due to sampling design are one reason for biased estimates. As an example, think about the overrepresentation of high income households in the SOEP. In the example above, the median income estimated with unweighted data was higher than the median income estimated with weighted data (see Sect. 2.4.1). If selection probabilities are related to the variable of interest, design weights *must* be used to account for different selection probabilities. Hence, if we are interested in the variable income and certain income groups are overrepresented by design, we must use design weights to get unbiased estimates of median incomes (and the respective poverty thresholds).

With regard to non-response, things are a little bit more complicated. In the best case, a smaller sample size due to non-response leads to less efficient but unbiased estimates, i.e., we will get the correct mean income but with a larger standard error. However, this ideal case is only true if non-response is distributed randomly (*missing completely at random*, MCAR). If non-response is selective, we will get not only less efficient but also biased estimates using unweighted data. In this case, missing data can be either *missing at random* (MAR) or *not missing at random* (NMAR).

Missing data is MAR if the probability of non-response depends on variables that are related to the variable of interest but not on the variable of interest itself. If the data is MAR and the distribution of the "related" variables in the population is known, we can use weights to obtain unbiased estimates. As an example, imagine that women are less likely to participate in a survey because they are not as willing as men to let strangers (interviewers) into their homes. Furthermore, assume that this reason is the only determinant of unit non-response. As a consequence, the gender ratio in the sample will deviate from the gender ratio in the population. Now assume that we want to estimate average body height in the population. As men are taller than women we would overestimate average body height in an unweighted analysis. The population weight obtained from a comparison of the gender ratio in our sample with the true gender ratio in the population would allow us to estimate the true population parameter by giving higher weights to women and lower weights to men in our sample.

However, weights *cannot* correct for unit non-response bias if the response probability depends on the variable we are interested in (NMAR). For example, if the non-response probability depends on income (poor people may be less inclined to participate in a survey) and not on other variables like age or gender, we cannot get an unbiased estimate of median income using population weights based on gender and age. Note that this example is different from the former example on high income households overrepresented in the SOEP.

Overrepresenting by survey design also results in different selection probabilities, but these selection probabilities are known ("designed") a priori. Hence, they can be controlled for by design weights.[11] Yet, if ("non-designed") non-response is related to the variable of interest (say, income), one cannot compute a selection probability because, by definition, information about the distribution of this variable in the population is not available (if that distribution would be known, there is no

---

[11] The fact that such selection probabilities could not be computed for the SOEP sample of high income households is no contradiction. The fact that they are overrepresented is known a priori and hence, they can at least be excluded with a design weight of zero so as not to bias the estimates.

need to do a statistical analysis). In this case other techniques are called for that are not considered in this textbook.

Let us now consider the case of statistical modeling, e.g., regression analysis. In an ideal case, we would prefer to get unbiased coefficients and unbiased standard errors. As indicated above, standard errors in regression analysis are biased if weights are used. On the other hand, coefficients might be biased if the data is not weighted. So, what do to? If the missing data process is MAR (respectively, if the selection probability is not directly determined by the variable of interest), a common strategy is to perform unweighted analysis but to control for the background variables that are related to the response probabilities. In other words, unweighted regression coefficients are unbiased if responders and non-responders do not differ systematically after controlling for the variables that determine the selectivity of participation.

As an example, think of the foreigner sample in the SOEP and assume that the rest of the SOEP sample satisfies the conditions of a SRS. A model including a dummy variable indicating whether the respondent is a member of the foreigner sample or not would control for the higher selection probability (overrepresentation) of foreigners and yield unbiased results. Of course, this statement is only true under the assumption that the rest of the model is correctly specified.

As background variables, like gender, age, education or immigrant status, should be controlled in social science applications in any case, one can argue that unweighted regression analyses should be preferred because the estimates of unweighted analyses are more efficient. Again, this is only true if the missing data process is determined solely by the background variables (MAR) and is not related to the variable of interest, i.e., if weights are solely a function of the X variables and not of the dependent variable Y.

Therefore, we recommend that you, first, try to specify models that control for the selectivity of participation. Once you find a model satisfying this condition, unweighted analysis should be preferred because it yields more efficient estimates. A formal procedure to test whether the X variables in your model can control for the selectivity of participation was proposed by DuMouchel and Duncan (1983). This recommendation is particularly important for panel data analysis, because weighting in panel data is quite complex and time-consuming if the panel includes late entries and temporary dropouts. It is also a good strategy to compare the results of weighted and unweighted regressions. If the unweighted regression coefficients are (almost) identical to the weighted coefficients you can assume that coefficients will not be biased and apply an unweighted analysis. Finally, you can also try to control for sample selection bias with a Heckman selection model (Heckman, 1979).

Nevertheless, if you have to use statistical weights in regression analyses, you should use the *heteroscedasticity consistent estimator* for the standard errors (White, 1980). Heteroscedasticity consistent standard errors (called *robust* standard errors) are not at risk of providing downwardly biased standard errors. Using robust standard errors, you can avoid drawing invalid inferences from weighted regression coefficients and their standard errors. The drawback of this strategy is that your estimates of the regression coefficients will be less efficient.

2.5 Conclusion and Further Reading

This chapter provided an overview of the work that has to be done *before* one can start analyzing panel data. The main purpose of this chapter was to show how working data sets are constructed from raw panel data. The emphasis was put on technical matters, in particular, the combination of single data files into a working data set and on different forms of organizing panel data (long and wide formats). We showed how to transform data from long to wide and from wide to long format. For the purpose of this demonstration, we used Stata, but the general principles of data management discussed in this chapter work the same way in all statistical program packages. You will find descriptions of the necessary data management tools in the documentation of the statistical software that you are using. For Stata we can recommend a book by Mitchell (2010) that gives a broad overview of all data management tools available in Stata. Kohler and Kreuter (2005) give a general beginner-level introduction to Stata.

We also addressed some other issues that one should think of when analyzing (panel) data. We discussed how to define populations for cross-sectional and longitudinal analyses. We showed how to include cross-sectional weights in a working data set. We also demonstrated how longitudinal weights are constructed from the information in the raw data.

However, we discussed the issue of weighting from a technical rather than from a theoretical statistical perspective; except in Sect. 2.4.3, where we gave some general recommendations on weighting based on statistical considerations. We recommended using weights when making representative statements about the population, but you should avoid, if possible, using weights in statistical modeling. Before using weights in regression analyses, you should try to control the possible selectivity of your sample by the model specification.

Readers who are unsure whether they should weight their data or not should start with an article by Winship and Radbill (1994). In this article you will find an easy introduction into the issue of weighting in regression analysis. You will also find a description of a formal test that you can use to investigate whether you need to weight your data or not. This test was originally proposed by DuMouchel and Duncan (1983). A more advanced article dealing with the issue of weighting in regression analysis is Pfeffermann (1993). If you are using weights, you should use heteroscedasticity consistent standard errors (White, 1980) in order to avoid drawing invalid inferences. Heckman selection models are an alternative method to correct for sample selection bias (Heckman, 1979). An easy introduction into sample selection bias can be found in Berk (1983), while more advanced readers might want to read Puhani (2000).

Readers who are interested in missing data (non-response) might want to have a look at the textbooks by Allison (2002) or Groves et al. (2002). Advanced readers who are interested in missing data might also want to read Little and Rubin (1987) and Rubin (1976). An overview of the particular issue of panel attrition, its consequences and some strategies to counterbalance the process of panel attrition can be found in Hirano et al. (2001).

2.5 Conclusion and Further Reading

A general introduction into the statistical intricacies that arise with complex survey designs can be found in Kish's classical textbook on survey sampling (Kish, 1965) or in Kalton (1987). Some more recent and general textbooks which do not go into details but give a broad overview of survey methodology (sampling, non-response, weighting) are Groves et al. (2004) and Bethlehem (2009).

Questions of item non-response and imputation have not been discussed at all. However, particularly with panel data, imputation is a common strategy to deal with missing values. Interested readers might want to start with Kalton (1986) or Pfeffermann and Nathan (2002) who describe how to deal with unit and item non-response in panel data. Readers interested in a general introduction into imputation should read De Leeuw (2001). Schafer and Olsen (1998) give an overview of multiple imputation techniques from the perspective of applied researchers. Originally, multiple imputation was proposed by Rubin (1987).

In the following chapters of this book we ignore the possible complex survey designs underlying the example data sets. Furthermore, we assume that missing information is missing at random and is controlled for by the variables in the statistical models. Hence, all following panel regression models will use unweighted data.

Describing and Modeling Panel Data 3

Statistical models for panel data are a rapidly growing field of methodological inquiry. Given the myriad techniques now available in statistical programs, it is difficult for novice users of panel data to make an informed choice about the methods that best suit their research questions. This chapter is intended to offer a basic orientation, before we introduce more specific methods in the following chapters. Another point of confusion concerns the different names under which these methods are discussed in the literature. In this chapter, we introduce the terminology for the statistical models presented in the next two chapters, and discuss the virtues and pitfalls of panel analysis in an informal way. Depending on the discipline and the application, various labels have been used for the specific statistical techniques involved in panel data analysis. Despite this terminological heterogeneity, most of them are special cases of a very general statistical model, and this overview clarifies their interconnections.

As a general guideline for this and the following chapters, we distinguish among the statistical models for panel data according to the type of dependent variable on which they focus. More specifically, we differentiate between models for *categorical* and models for *continuous* dependent variables, i.e., between variables having few discrete values and variables having values that—at least in principle—could be measured on a continuous scale. Employment status (employed, unemployed, out of the labor force), political attitudes (measured, e.g., on a 7-point scale of political liberalism), or number of children (0, 1, 2, ..., 5 and more) are examples of categorical variables. Income (in Dollars), firm size (e.g., number of employees), gross domestic product (in Dollars), or the amount of social expenditure (in Dollars) are examples of continuous variables. Variables having many discrete values cannot be efficiently treated as categorical variables. They must either be simplified (i.e., recoded into fewer categories) or be treated as continuous variables, if the underlying concept is continuous. Admittedly, this distinction between categorical and continuous variables is a simplification, but as a first orientation for choosing different statistical models, it is very helpful. Later on, Chap. 5 will introduce a more differentiated classification of categorical data.

Statistical methods for categorical variables typically focus on the probability of observing a certain value (category) of the variable of interest. "Does the probability of being unemployed change with labor force experience?" is a typical research question. This is not a feasible strategy for continuous variables, because of the many different values these variables can take on. Therefore, statistical methods for continuous variables focus on certain distributional characteristics of these variables, e.g., their expected values. "Does income—on average—increase with educational attainment?" would be a typical research question for such variables.

Since this chapter includes quite a lot of material, we have organized it alongside the typical questions a researcher will ask when beginning a statistical analysis of panel data.

1. What is the basic terminology that is used when analyzing panel data (Sect. 3.1)?
2. What is so specific about panel data that traditional statistical models for cross-section data are not suitable for their analysis (Sect. 3.2)?
3. What simple statistical tools are available to provide an initial and descriptive overview of the dependent variable (Sect. 3.3)?
4. Which typical explanatory variables can be used to explain the time path of the dependent variable (Sect. 3.4)?
5. What does a formal model that tests their effects actually look like (Sect. 3.5)?
6. How can the parameters of this model be estimated with panel data (Sect. 3.6)?

At the end of this chapter, you will have an idea of the specifics of panel data and how statistical models can be adapted in this regard. Section 3.7 concludes with an overview of the following chapters, in which these models will be explained in greater detail.

3.1 Some Basic Terminology

As described in the previous chapter (Chap. 2), panel data can be organized in a *data cube* having three dimensions: units $i = 1, \ldots, n$, measurements (panel waves) $t = 1, \ldots, T$, and variables $v = 1, \ldots, V$ (some time-constant, some time-varying). Units could be individuals, firms, nations, or other objects of analysis. In this chapter, we use the generic term *unit*, although in some instances, it may sound a little technical. If each unit is observed T times, the data are called a *balanced panel*. If there are missing data, the number of measurements, T_i, varies between individuals, and we are analyzing an *unbalanced panel*. Unbalanced panel data occur when some respondents do not participate at all points in time (*temporary unit non-response*), when some respondents drop out of the panel at some point in time (*panel attrition*), or when some respondents enter the panel at a later point in time (*late entry*). For simplicity, in this and many other chapters, we assume that we are dealing with *balanced* data. However, all methods presented in this book can easily be extended to unbalanced data. For example, focusing on balanced panel data simplifies many formulas, because the number of measurements, T, is the same for all units and we do not have to differentiate between the various T_i.

3.1 Some Basic Terminology

Throughout this text, we make the distinction between two types of panel data. The first type comprises panel data, where the number of units is much larger than the number of measurements ($n \gg T$), as is usually the case with large panel surveys of the resident population. A typical example would be the National Longitudinal Surveys (NLS), sponsored by the U.S. Bureau of Labor Statistics (BLS), which encompass a set of surveys designed to gather information at multiple points in time on the labor market experiences of groups of men and women. Vella and Verbeek (1998) use a subsample of these surveys to analyze the effect of union membership on wages. We will reanalyze these data in Example 3.1. The second type of panel data is characterized by data where the number of units is much smaller, sometimes even smaller than the number of measurements ($n < T$). However, compared to type I panels, the number of measurements over time is quite sizeable. A typical example is a time-series data set collected for a sample of countries (a panel of countries), such as in the study by Garrett and Mitchell (2001), in which the amount of public spending in 18 OECD countries over a period of 33 years (1961–1993) is analyzed. We will use these data in Example 4.1.

Obviously, type II panels include much more information in order to analyze the time dimension of the data and to test sophisticated models of the underlying process. This is often unfeasible with type I panels, because of their limited number of measurements over time. These kinds of panels have their strengths when it comes to the analysis of unit heterogeneity. From a statistical point of view, type I panels are data where T is usually assumed to be fixed so that distributional assumptions are derived as $n \to \infty$. Type II panels, on the other hand, are data where n is usually assumed to be fixed and the asymptotic theory depends on $T \to \infty$. Since the terms "type I" and "type II" are not self-explanatory, we tried to find some more evident terms, and came up with the distinction between micro (type I) and macro (type II) panels. The terms "micro" and "macro" refer to the fact that the former are mostly based on micro units (e.g., individuals), while the latter mostly include macro units (e.g., countries). The following chapters will focus primarily on statistical models for micro panels, because for macro panels, much more knowledge of time-series analysis is needed; also, in the case of countries, one may also need to control for the dependencies among those units (due to international trade and geographical proximity). Nevertheless, the methods presented in this textbook are a necessary starting point for these more refined methods. Note also that the distinction between micro and macro panels is not exactly the same as the distinction between type I and II panels, because there may be panels of macro units where the number of units is larger than the number of measurements ($n > T$).

Before we start our statistical analysis, we have to rearrange the three-dimensional data into a two-dimensional *data matrix* in order to put them into a computer program. Chapter 2 showed that this can be done in *wide format*, with one record for each unit that includes all measurements for all variables over time. This matrix has n rows and $T \cdot V$ columns. The number of records, N, in the corresponding data file equals the number of units ($N = n$). Alternatively, the data can be organized in *long format*, with one record for each measurement per unit that includes the values for all variables at that particular point in time. This particular matrix has

$n \cdot T$ rows and V columns. The number of records in the corresponding data file equals $N = n \cdot T$. This is exemplified in Table 2.1 in the previous chapter. The wide format has been the traditional way of organizing panel data, especially when the number of measurements has been small (e.g., $T < 4$). In the wide format, it is easy to correlate variables that have been measured at different points in time. The long format has now become the "modern" way of organizing panel data, although one needs additional data management tools, e.g., to correlate measurements over time (see the case study in Sect. 2.3.3). It is also easy to show that the long format often stores data more efficiently and facilitates the specification of statistical models. However, for data input, you can choose the format that is most convenient for you. As Chap. 2 showed, the specific format is not important, because all statistical program packages provide commands to convert data from wide to long format, and vice versa.

Because records (measurements) are nested within units, the long format is also a typical example of a *hierarchical* data set. Obviously, measuring each variable over time results in a small time series for each unit. Because all values from each time series belong to the same unit, they will probably have more in common with each other than with values from time series that belong to other units. In other words, the hierarchical nature of panel data organized in long format implies a grouping structure, with each group consisting of data from one unit, and possibly higher statistical associations within each group than between groups.

Panel data share this hierarchical feature with other data, such as cross-sectional surveys of pupils within classes, children within families, or respondents within countries. The only difference is the ordering of units within "groups". While panel data have a natural ordering with respect to time, pupils can be ordered in different ways within the class. As a consequence, adjacent panel measurements will have more in common than panel measurements several years apart, while measurements for one pupil can be correlated with measurements for any other pupil within the class. However, for any kind of hierarchical data, be it the aforementioned "grouped" cross-section data or the panel data in which we are interested in this textbook, the assumption of independent observations that is so often made for randomly selected cross-section data does not hold. This is one of the most important statistical problems that has to be dealt with in the following chapters (see also Sect. 3.2).

Data organized in long format are also called *pooled* data. The motivation for this term comes from the observation that panel data can be conceived of as consisting of many individual (unit-specific) time series that are put (pooled) together in one big data file. Alternatively, one could think of each panel wave as one cross-section, and of these different cross-sections as being put (pooled) together in one big file. Viewing panel data as a collection of n unit-specific time series or, alternatively, as a collection of T cross-sections, explains why economists and political scientists also call panel data *pooled time-series cross-section (TSCS) data*.

Some researchers argue that pooling increases the number of cases and hence the statistical power of the statistical analyses. Instead of n units, they say, the data now include $N = n \cdot T$ "cases". As mentioned in Sect. 1.1.1.5, this argument is often advanced in macro-economic, macro-sociological, or political science research that is interested in the characteristics of macro units, such as countries. Typically, in this context, the number of cases is limited. For example, much statistical information on countries is available from the OECD and the number of member states currently amounts to 34 countries. A statistical analysis of only 34 cases does not have much statistical power. Hence, methodologists recommend increasing the number of cases by using information from different points in time for each country. However, these over-time data are not independent observations, as they do not represent additional countries. This is obvious in the wide format. It explicitly shows that measurements over time belong to one and the same unit by putting them all into one record for each unit (in this case $N = n$).

TSCS data (panel data) should not be confused with data sets that pool *independent* cross-sections. For example, the European Social Survey (ESS, a harmonized survey in different European countries), the International Social Survey Programme (ISSP, a similar survey that includes countries all over the world), and the General Social Survey (GSS, a harmonized biennial survey in the USA) include pooled cross-sections, either from different countries (ESS, ISSP) or from different years (GSS). The cross-sections are sampled independently from another and, thus, do not include identical individuals. In these cases, the argument for increasing the sample size by pooling is much easier to justify, because the units are sampled independently of one another. Nevertheless, even in those cases, one could argue that individuals from one country (year) are more alike than individuals from other countries (years) and indeed, in cross-national research, it is quite common that within a country, homogeneity is controlled for using specific statistical techniques (multi-level analysis) that are also applicable to panel data (see Sect. 3.5.3.3).

3.2 Measurements over Time Are Not Independent

In this section, we want to illustrate the statistical dependencies inherent to panel data with an example that includes both a continuous and a categorical variable.

Example 3.1 (wagepan data) The data are taken from the National Longitudinal Survey (NLS Youth Sample) and contain observations on 545 males who have been observed continuously from 1980 to 1987 (i.e., $n = 545$, $T = 8$). The sample is a balanced panel, since it includes only those individuals that have provided information for each of the eight panel waves. Vella and Verbeek (1998) use these data to estimate the wage premium of union membership. The assumption is that the bargaining power of unions increases wages for workers, especially so for low wage earners. There are also indications that specific workers unionize, while others do not. Among other things, the data include information on hourly wages (in Dollars) and union membership status (a dummy variable). In this chapter, we will use hourly wages as an example of a continuous dependent variable, and membership status as an example of a categorical dependent variable.

Table 3.1 Log hourly wages 1980–1987

Year	1980	1981	1982	1983	1984	1985	1986	1987
	Arithmetic mean							
Wage	4.59	5.12	5.41	5.64	6.10	6.47	6.82	7.20
ln(*wage*)	1.393	1.513	1.572	1.619	1.690	1.739	1.800	1.866
No member	1.331	1.455	1.519	1.573	1.652	1.689	1.772	1.844
Member	1.580	1.686	1.724	1.760	1.805	1.915	1.904	1.931
	Standard deviation							
ln(*wage*)	0.558	0.531	0.497	0.481	0.524	0.523	0.515	0.467
	Pearson correlation coefficient							
1980	1.000							
1981	0.454	1.000						
1982	0.432	0.611	1.000					
1983	0.408	0.582	0.690	1.000				
1984	0.316	0.506	0.626	0.675	1.000			
1985	0.356	0.469	0.588	0.625	0.664	1.000		
1986	0.297	0.407	0.523	0.549	0.565	0.632	1.000	
1987	0.310	0.480	0.498	0.563	0.588	0.672	0.693	1.000
Mean	$\bar{r}_y = 0.5277$							

Source: wagepan data (see Example 3.1)

In the following, y_{it} denotes a single measurement of the dependent variable Y for unit i at time point t. That measurements over time are not independent can easily be shown for both variables by correlating the yearly measurements. This is a simple command in the wide data format, where each (yearly) measurement is a variable of its own. For data in long format, the software must be able to correlate data from different records, and it must also know which records belong to the same unit. Thus, data organized in long format need software that is able to identify single measurements within the unit-specific time series and to retrieve data from adjacent (preceding and following) measurements of the same unit (technically: *lags* and *leads* of the corresponding variable; see the case study in Sect. 2.3.3).

Table 3.1 shows the results of our analysis. Overall, mean hourly wages increase in the observation period from \$4.59 in 1980 to \$7.20 in 1987. The distribution of wages is positively skewed and we analyze the natural logarithm of hourly wages to obtain a more symmetrically distributed dependent variable. This is also in line with human capital theory, which, for theoretical reasons, models the logarithm of wages. To keep our notation simple, we still use y_{it} (and not ln y_{it}) to denote our dependent variable log hourly wages. Not surprisingly, log hourly wages show a similar positive trend, roughly indicating that hourly wages increase by about 5–7 % each year (and 13 % between 1980 and 1981). For example, the 13 % increase is calculated as

$\exp(1.513 - 1.393) = 1.127$.[1] The variation of log hourly wages, on the other hand, remains more or less the same (see the standard deviation of $\ln(wage)$). Finally, if we distinguish respondents with respect to membership status, we see that union members receive slightly higher wages on average than non-members. Thus, there is descriptive evidence for a wage premium of union membership.

Table 3.1 also shows the Pearson correlation coefficients between all yearly income measurements. To analyze the statistical dependencies over time, we focus on two types of these correlation coefficients. The first one, $\text{Corr}(y_{it}, y_{i,t-1})$, correlates (log) hourly wages from each year t with (log) hourly wages from the previous year $t-1$. These are the correlation coefficients directly below the main diagonal of the correlation matrix. For example, the correlation between (log) hourly wages observed in 1981 and (log) hourly wages observed in 1980 amounts to 0.45, a fairly high figure that increases to values of between 0.61 and 0.69 in the subsequent years. This is a clear indication of the statistical dependencies between the panel measurements. The second type of Pearson correlation coefficient, $\text{Corr}(y_{it}, y_{i1})$, correlates (log) hourly wages from each year t with (log) hourly wages from the first year $t=1$ (see the correlation coefficients in the first column of the correlation matrix). According to this measure, statistical dependencies decrease—as expected— the greater the time interval between $t=1$ and T. Similar decreasing trends can be found in the other columns of the correlation matrix. Obviously, adjacent panel measurements correlate between 0.5 and 0.7 (with the exception of 1980–1981, where the correlation equals 0.45), while measurements seven or eight years apart show a much lower association of about 0.3.

In the following, we will call $\text{Corr}(y_{it}, y_{i,t-1})$ and $\text{Corr}(y_{it}, y_{i1})$ serial correlation coefficients (of order 1 or $k=t$). Some people refer to them as autocorrelations, but autocorrelations have a very specific meaning in the context of time-series (and panel) analysis.[2] The term "serial correlation", on the other hand, makes clear what is at stake here: $\text{Corr}(y_{it}, y_{i,t-1})$ and $\text{Corr}(y_{it}, y_{i1})$ are measures of the temporal (serial) dependencies in the data, which are apparently quite high in the wage

---

[1] Note that the arithmetic mean of log hourly wages equals the *geometric mean* of hourly wages and hence, $\exp(1.393) \neq 4.59$ and $\exp(1.513) \neq 5.12$. Therefore, the increase from (the geometric mean) 1.393 to 1.513 measured in log Dollars (+13 %) is different from the increase from (the arithmetic mean) 4.59 to 5.12 measured in Dollars (+12 %).

[2] An autocorrelation coefficient, as the name suggests, is a correlation of a series of data values with itself. More specifically, if you have a series of T observations on the same variable, $y_1, y_2, \ldots, y_T$, you compute the first-order autocorrelation coefficient by dropping the first observation and correlating the remaining observations with the original series shifted by one period, i.e., you correlate $(y_2, y_3, \ldots, y_T)$ and $(y_1, y_2, \ldots, y_{T-1})$. In the wage example, this would mean that you take the time series of (log) hourly wages for *one* individual and compute the autocorrelation by comparing it with the same time series lagged one period. This is also a measure of serial dependence, but obviously based on very few (7) values and computed for only one unit in the data. Of course, it could be extended to all units by pooling all unit-specific time series, but that (overall) first-order autocorrelation—let us call it $\text{Acorr}(y_{it}, y_{i,t-1})$—is still not identical to the correlation coefficient $\text{Corr}(y_{it}, y_{i,t-1})$ in Table 3.1. $\text{Corr}(y_{it}, y_{i,t-1})$ in 1981, for example, is based on all observations from 1981 ($y_{i,1981}; i = 1, \ldots, n$) correlated with all observations from 1980 ($y_{i,1980}; i = 1, \ldots, n$). In other words, $\text{Corr}(y_{it}, y_{i,t-1})$ focuses on one specific transition (e.g., the transition from 1980 to 1981), while $\text{Acorr}(y_{it}, y_{i,t-1})$ uses all transitions.

Table 3.2 Union membership 1980–1987 (frequencies and percentages)

Year t	Union membership	n	State probability	First-order transition matrix $(t, t+1)$		Higher-order transition matrix $(t = 1980, t)$	
				No member	Member	No member	Member
1980	No member	408	74.86	88.97	11.03		
	Member	137	25.14	33.58	66.42		
1981	No member	409	75.05	88.02	11.98	88.97	11.03
	Member	136	24.95	33.09	66.91	33.58	66.42
1982	No member	405	74.31	92.10	7.90	85.29	14.71
	Member	140	25.69	27.14	72.86	41.61	58.39
1983	No member	411	75.41	92.21	7.79	86.52	13.48
	Member	134	24.59	21.64	78.36	42.34	57.66
1984	No member	408	74.86	94.61	5.39	84.80	15.20
	Member	137	25.14	27.01	72.99	45.26	54.74
1985	No member	423	77.61	94.56	5.44	86.76	13.24
	Member	122	22.39	24.59	75.41	50.36	49.64
1986	No member	430	78.90	87.44	12.56	86.27	13.73
	Member	115	21.10	22.61	77.39	56.93	43.07
1987	No member	402	73.76			82.11	17.89
	Member	143	26.24			48.91	51.09
Total	No member	3,296	75.60	91.12	8.88		
	Member	1,064	24.40	27.25	72.75		
High wage	No member	1,793	82.40	91.64	8.36		
	Member	383	17.60	39.69	60.31		
Low wage	No member	1,503	68.82	90.49	9.51		
	Member	681	31.18	20.47	79.53		

Source: wagepan data (see Example 3.1)

data (though decreasing with temporal distance between measurements). Later on, we will discuss more concretely possible reasons for these high serial correlations (Sect. 3.4.2).

A similar exercise can be performed for our categorical variable "union membership" (see Table 3.2). As a measure of trend, we have computed the percentage of union members for each year, which, in the observation period, is almost constantly at a value of about 25–26 % (except for the years 1985 and 1986, where it is about 21–22 %; see the fourth column in Table 3.2). From the totals in the bottom lines of Table 3.2 we also see that low wage earners are more often (31 %) union members than high wage earners (18 %).[3] Similar to the former arithmetic mean

---

[3] For this dichotomy, we took the average (log) hourly wage of each individual across all years and dichotomized these averages at their median.

3.2 Measurements over Time Are Not Independent

of log hourly wages, which is an estimate of the expected value of the (continuous) dependent variable, $E(y)$, this percentage is an estimate of the probability of being a union member. More generally, the numbers in the fourth column of Table 3.2 represent the probabilities, $\Pr(y_{it} = q)$, of observing at time point t the corresponding category q of the (categorical) dependent variable Y for unit i. This is also called a *state probability*.

To get an idea of the statistical dependencies in the union membership data, we compute conditional percentages, measuring how many of the union members at a given point in time are still members the following year and how many leave the union. According to the statistics in the fifth column of Table 3.2, about a third (33.58 %) of the 1980 members quit their union in the subsequent year. Similar questions can be posed regarding non-members: How many enter a union and how many prefer not to join a union? By 1981, a little more than one tenth (11.03 %) became union members, while the rest (88.97 %) remained out of the union (see the sixth column of Table 3.2). For each year, the four percentages constitute a matrix of *transition probabilities*. Each entry in this matrix estimates a (conditional) probability, $\Pr(y_{i,t+1} = q | y_{it} = p)$, of being in the following year $t + 1$ in a certain category q of the dependent variable Y (the *destination* state), given that unit i has been in category p of Y in the current year t (the *origin* state). This matrix of transition probabilities is the equivalent of the former first-order serial correlation coefficient.[4] Its diagonal elements give an indication of the persistence of union membership and non-membership. According to the statistics in Table 3.2, the probability of remaining in the same status is quite high (about 66–78 % for members and 87–95 % for non-members).

We have also computed an equivalent for $\mathrm{Corr}(y_{it}, y_{i1})$. These are the higher-order transition probabilities, $\Pr(y_{it} = q | y_{i1} = p)$, in the seventh and eighth column of Table 3.2. They are conditional on membership status in the first year (y_{i1}). We observe a similar result as in the continuous case. The greater the time interval between both measurements y_{it} and y_{i1}, the smaller the percentages in the diagonal cells of the transition matrix and hence the probability of constant membership status. For example, while two-thirds (66.42 %) of the union members are still organized in 1981, five years later, in 1986, only two-fifths (43.07 %) remained in their unions. In 1987, this number rises slightly to 51.09 %, but it is still lower than 66.42 %.

Both examples (wages and union membership) tell the same story. Irrespective of the nature of our dependent variable, be it continuous or categorical, repeated observations over time will almost always correlate with one another.[5] In other words,

---

[4] Some readers may wonder why, in the continuous case, one correlates y_{it} (from the present year) with $y_{i,t-1}$ (from the previous year), while in the categorical case, y_{it} (from the present year) is tabulated with $y_{i,t+1}$ (from the next year). This is mainly for historical reasons, and is of no practical relevance. For example, we could just as well compute $\mathrm{Corr}(y_{it}, y_{i,t+1})$.

[5] In all practical situations, this correlation would be positive, because above (below) average units tend to remain above (below) the average in the continuous case or to remain in the same state in the categorical case. A negative correlation would imply that change is more frequent than stability.

panel data usually do not include independent information. We have certainly more information than in a cross-section of n units, but not as much as in a cross-section of $N = n \cdot T$ units. Ignoring these statistical dependencies is potentially dangerous when applying regression models, because traditional estimation procedures assume independent observations. Thus, they will estimate standard errors that tend to be too low. As a consequence, test statistics will be too high and correspondingly, p-values too low, such that significance tests will lead to erroneous conclusions. Furthermore, although parameter estimates may be unbiased, they could be estimated more efficiently, if the statistical dependencies were explicitly modeled. Section 3.4.2 will show why panel measurements are serially correlated. Section 3.6 will discuss how to cope with these serial dependencies.

3.3 Describing the Dependent Variable

Before proceeding with a deeper understanding of these statistical dependencies, it is always a good idea to have a descriptive overview of the data at hand. This is already a difficult task with cross-section data, but what to do with panel data that are much more complex? Simple techniques are the ones used in the previous section. Means and standard deviations describe the overall trend and the spread of continuous variables. Proportions do the same job for categorical data. Correlating a continuous variable (cross-tabulating a categorical variable) over time informs us about short- and long-term serial dependencies. However, the longer the time interval τ between y_{it} and $y_{i,t+\tau}$, the more difficult it is to trace the trajectory of how the unit got from y_{it} to $y_{i,t+\tau}$. Consider, for example, the transition from 1980 to 1983 in the union data: 13.48 % of the non-members from 1980 were members of a union in 1983 (see the higher-order transition matrix for 1983 in the last columns of Table 3.2). Not all of them joined their union in 1983. Some may have done so in the preceding years, and some may have even joined and then temporarily left their union. A similar argument applies to the continuous variable "(log) hourly wages". Even though we know that (log) hourly wages—on average—increase over time (see the means in Table 3.1), individual income trajectories may not be trending positively all the time. At the individual level, there may be periods of wage decrease followed by periods of increasing wages. Thus, besides measures of trend, spread, and serial dependence, we would want to have more information about the unit-specific trajectories of the corresponding variables.

For categorical data, it is easy to make tabular summaries, because the number of combinations is limited (at least for short panels). Consider again the union example: If union membership is coded with a dummy variable (member coded 1, 0 otherwise), a series of zeros and ones summarizes the membership career of each

---

A union member would have a high probability of being a non-member next year and again a member in the year to follow, etc. Individuals with high wages this year would have low wages next year and again high wages in the following year, and so on. For many applications, like this one, such alternating processes do not make sense.

3.3 Describing the Dependent Variable

Table 3.3 Sequences of union membership status 1980–1983

Starting state	Sequence	n	%	Total
No union member in 1980	0000	308	56.5	408
	0001	16	2.9	
	0010	23	4.2	
	0011	16	2.9	
	0100	17	3.1	
	0101	7	1.3	
	0110	5	0.9	
	0111	16	2.9	
Union member in 1980	1000	33	6.1	137
	1001	3	0.6	
	1010	3	0.6	
	1011	7	1.3	
	1100	15	2.8	
	1101	6	1.1	
	1110	7	1.3	
	1111	63	11.6	

Source: wagepan data (see Example 3.1)

individual. For example, the series 0001 indicates that an individual has been out of the union in years 1980 to 1982 and member of a union in year 1983. Such a series is also called a *sequence*. With four measurements over time and a dichotomous dependent variable, there are $2^4 = 16$ different sequences. Table 3.3 shows the frequency of different sequences of union membership as they appear in the wagepan data. Only two sequences—constant membership (1111) and nonmembership (0000)—are observed more often ($n = 63$ and $n = 308$). The rest of the sample is scattered across other sequences and it is difficult to detect additional patterns of interest. In other words: While the former measures of trend, spread, and serial dependence may be too coarse, this technique provides information that is too detailed. Furthermore, the amount of information will increase exponentially with the number of measurements. For example, if we analyze the sequences of union membership over the whole observation period from 1980 to 1987, the computer finds 95 different patterns, many of which are observed for only one individual.

Therefore, it is a much more promising strategy to decompose the heterogeneity of individual sequences into simpler processes of change. If you go back to the first-order transition matrices in Table 3.2, a starting hypothesis could be that the pattern of stability and change that is summarized in these figures is more or less constant for the whole observation period. This implies the assumption that the differences between single transition matrices are simple random noise, and that an average of all eight matrices from 1980 to 1987 describes the process equally well. Finding the correct transition matrix that describes the observed process is at the heart of *Markov modeling*. Another approach is to look at a single transition

Table 3.4 Survival probability of union members from 1980

Year	Members	Exits	$S_1(t)$	$h_1(t)$
1980	137	46	0.6642	0.3358
1981	91	21	0.5109	0.2308
1982	70	7	0.4599	0.1000
1983	63	7	0.4088	0.1111
1984	56	10	0.3358	0.1786
1985	46	6	0.2920	0.1304
1986	40	6	0.2482	0.1500
1987	34	0	0.2482	0.0000

Source: wagepan data (see Example 3.1)

(e.g., the exit from a union) to model the process of leaving the origin state (in this case: union membership) and then perform a similar analysis for all the other origin states (in this case, involving a dichotomous categorical variable, there is only one left: non-membership). Finally, the information about the different transitions can be put together into one overall model of the whole process. This technique has become known as *transition analysis*. Alternative labels are *survival* or *event history analysis*.

As an example, let us focus on those 137 respondents that start as union members from the very beginning in 1980 (a similar example could be made for the 408 non-members in 1980). Table 3.4 shows how long these individuals remain union members. Of the original 137, 46 leave their union in the first year, 21 in the second year, and so on. In terms of the former sequences, we are now focusing on sequences that start with 1, and we follow these sequences up to the point at which the dependent variable changes from 1 to 0. These changes are also called *events*, and the sequences of 1s before the event are called *spells*. From the number of events, we can estimate the probability of remaining a union member for each year. Technically, this probability is called the *probability of survival* $S_p(t)$ (p being the origin state), and the whole time series of survival probabilities is called the *survivor function*. The survivor function $S_1(t) = \Pr(y_{it} = \cdots = y_{i2} = 1 | y_{i1} = 1)$ is computed conditional on the union members in 1980 (i.e., $y_{i1} = 1$) and looks at those individuals that remain members until time point t. $S_0(t) = \Pr(y_{it} = \cdots = y_{i2} = 0 | y_{i1} = 0)$ would be the survivor function for the non-members. According to the statistics in Table 3.4, about one quarter (24.82 %) of the original members remain in their union until 1987. Note that the probability of survival $S_p(t)$ is different from the transition probabilities $\Pr(y_{it} = p | y_{i1} = p)$, because the former refers only to those units that *constantly* remain in category p, while the latter also include those units between $t = 1$ and t that have been temporarily in another category (compare the numbers in the last column of Table 3.2 with those in the fourth column of Table 3.4).

The survivor function is always a monotonically decreasing function, and therefore, it is often hard to see from inspecting $S_p(t)$ whether the process changes over time. However, we often have hypotheses assuming that the probability of change varies over time. For example, in our case, one could hypothesize that union dropout decreases over time, because the agreement with union policies increases

3.3 Describing the Dependent Variable

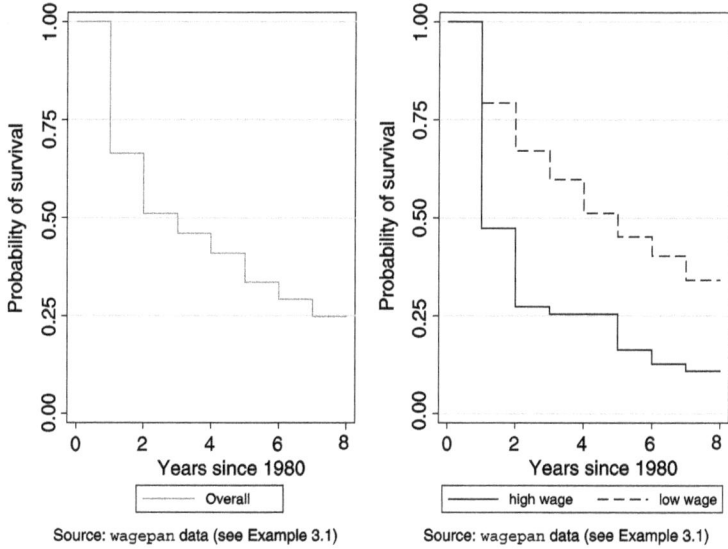

Fig. 3.1 Survival probability by income level

with membership duration. If this assumption is true, we expect the survivor function to decrease at a decelerating rate and this is indeed difficult to decipher from a plot of the survivor function. Hence, we need a statistic that captures the temporal variation of change. To this end, we estimate a *conditional transition probability*, $h_p(t) = \Pr(y_{it} \neq p | y_{i1} = \cdots = y_{i,t-1} = p)$, which measures the probability of experiencing a change to any of the other categories (states) $y_{it} \neq p$ of the dependent variable Y in year t given no change in the former years (i.e., y_{i1} up to $y_{i,t-1}$ have the value p). Since $\Pr(y_{it} \neq p | y_{i1} = \cdots = y_{i,t-1} = p)$ is a rather lengthy term, it is often abbreviated to $h_p(t)$, which is also termed a (discrete-time) *hazard rate*. Table 3.4 shows these conditional transition probabilities for the origin state of union membership ($p = 1$). For example, of the individuals that are still union members in 1982 ($n = 70$), 10 % leave the union up to 1983, and 90 % remain union members for another year. *Nota bene*, these conditional probabilities focus only on those individuals that have remained union members up to year t and hence, the denominator of the corresponding frequency ratio changes with each year.

As Fig. 3.1 shows, the survivor function is easily plotted on a two-dimensional graph, and by differentiating group-specific survivor functions, important information about possible explanatory factors can be obtained. For instance, if we differentiate between low and high wage earners, it appears that union membership is less stable among the higher income group than among the lower income group, because the probability of surviving declines more quickly for the high wage earners. The graph assumes a discrete-time process of change, in which events only happen at the end of the year and hence, the survivor function is a step function that decreases at the end of each year. Alternatively, we could assume a continuous-time process, in which change can happen at any point in time. However, if the concrete event dates are unknown, as is often the case with discrete panel measurements, we would have to make additional assumptions about the distribution of changes during the year.

Instead of plotting the survivor function, we could have plotted the hazard function (the time series of conditional transition probabilities resp. hazard rates). In most cases, however, it will show a much more irregular pattern, because its base—the number of units remaining in the origin state—is declining over time. Thus, at each time point, the conditional transition probability is based on an increasingly smaller number of cases, and, without controlling for sample size, it is hard to differentiate random from substantive changes over time. It can be shown, however, that the conditional transition probability should be the main variable of interest, if one models the change in categorical dependent variables. First of all, as we said, patterns of change may vary over time and this is best measured by a time-point specific statistic like $h_p(t)$. Furthermore, it can be shown that the hazard rate is the fundamental parameter, from which the survivor function can be computed as follows:

$$S_p(t) = \prod_{\tau=1}^{t} \left(1 - h_p(\tau)\right) \tag{3.1}$$

For example, the probability of "survival" in a union at the end of 1982 equals the following product:

$$S_1(1982) = \prod_{\tau=1}^{3} \left(1 - h_1(\tau)\right) = (1 - 0.3358) \cdot (1 - 0.2308) \cdot (1 - 0.1000) = 0.4599$$

For continuous data, simple presentations of unit-specific trajectories are more difficult. In most cases, tabular presentations are more or less data listings without any kind of data summary, because continuous variables assume many different values. Table 3.5 demonstrates that each income trajectory in the wagepan data is more or less unique unless one classifies wages. In principle, graphical displays are much more suitable for presenting this differentiated information, but for large sample sizes, graphical techniques usually fail because of heavy overprinting. Therefore, the left panel of Fig. 3.2 uses only a subsample of the original wage data to provide a graphical impression of individual differences in level and change of wages. We have selected the ten individuals with the highest and the nine individuals with the lowest average (log) hourly wages whose individual time series are also shown in wide format in Table 3.5.[6]

Figure 3.2 provides some interesting insights into panel data in general and the wagepan data in particular. Imagine, first, two different groups behind these ten respectively nine income trajectories. The two thick black lines in the left panel illustrate the linear wage growth observed in both groups. Obviously, the first group—on

---

[6]The unequal group sizes are due to the fact that one individual (nr = 813) with extremely low wages was dropped from the analysis.

3.3 Describing the Dependent Variable

Table 3.5 Log hourly wages for selected individuals 1980–1987

Id number	1980	1981	1982	1983	1984	1985	1986	1987	Mean
7,784	1.794	2.937	3.473	2.971	3.777	4.052	3.293	3.096	3.174
6,987	2.466	2.541	3.086	3.229	2.620	2.691	2.606	2.236	2.685
9,752	2.245	1.972	2.232	2.554	2.905	2.966	2.990	3.132	2.625
1,843	2.364	2.338	2.416	2.514	2.648	2.636	2.719	2.769	2.550
4,091	2.564	2.510	2.593	2.564	2.465	2.613	2.645	2.282	2.530
3,307	1.974	2.391	2.356	2.562	2.636	2.643	2.807	2.813	2.523
4,088	2.204	2.412	2.315	2.508	2.599	2.700	2.741	2.663	2.518
218	2.013	1.962	2.276	2.195	2.428	2.723	2.966	3.065	2.454
8,090	1.736	1.574	1.442	2.547	2.991	3.011	3.099	3.097	2.437
9,154	2.564	2.531	2.171	2.197	2.455	2.069	2.429	2.602	2.377
...	...	...	...	...	...	...	...	...	...
3,127	0.484	0.030	0.688	0.676	0.647	0.955	1.564	1.301	0.793
8,587	0.992	0.898	0.970	0.804	1.306	1.477	−0.981	0.791	0.782
6,020	0.676	0.541	0.195	0.383	1.104	1.066	1.202	1.046	0.777
6,025	0.774	0.289	0.906	0.339	1.213	1.502	1.277	−0.191	0.764
569	0.684	1.079	0.840	1.508	1.021	0.543	0.115	0.313	0.763
3,607	0.907	0.075	0.172	0.948	1.227	0.435	0.639	1.678	0.760
823	−0.373	0.606	0.865	0.512	0.684	1.000	1.435	1.177	0.738
10,570	1.056	−1.417	−0.670	0.703	0.713	0.820	1.069	1.039	0.414
2,147	−0.021	−0.149	0.181	−0.036	0.397	0.973	0.563	0.759	0.333

Source: wagepan data (see Example 3.1)

average—earns much higher wages than the second group, and it seems as if their wages grow a little more quickly during the observation period (the slope of the black line for the high wage group is slightly steeper). Thus, if the sample size is not too large and if the data are as clearly structured as the ones in Fig. 3.2, line plots provide important information regarding possible explanatory factors (similar to survivor plots in the categorical case). However, you can easily imagine what this line plot would look like, if we had used the data for all $n = 545$ individuals. Very likely, the entire plot area would have been hatched. In principle, such line plots could also be used for categorical data. But then we are faced with the opposite problem. The dependent variable includes only few distinct values, resulting in lots of lines plotted on top of one another.

Now imagine having estimated a linear growth model for each individual. This is shown in the right panel of Fig. 3.2. Similar to the two (thick black) group-specific lines, these unit-specific lines differ with respect to level and amount of growth. In principle, each unit has its own intercept and slope. At the moment, this plot simply illustrates the internal heterogeneity within each group. Later on, we will explore how to model variance of slopes and intercepts across units.

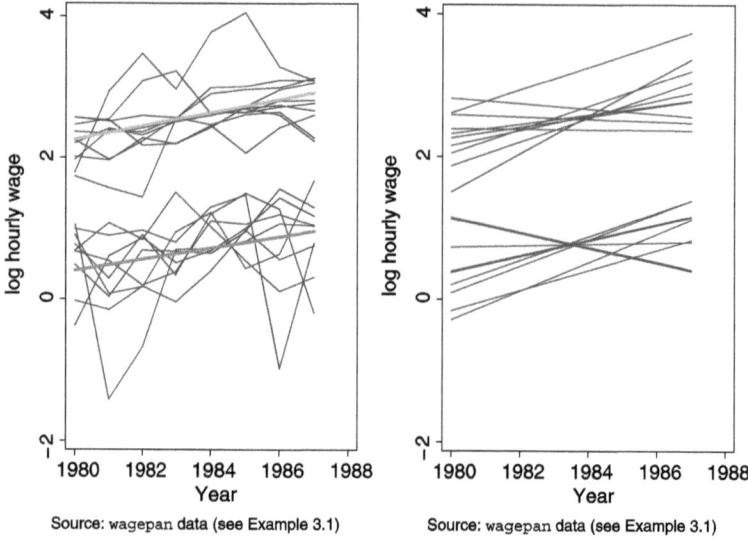

Fig. 3.2 Income trajectories of 19 individuals with very high and very low wages

Finally, Fig. 3.2 shows that the overall variation of wages derives from two sources: one is the heterogeneity of average wages between individuals (*between-unit variance*), the other is the heterogeneity of wages over time for each individual (*within-unit variance*). Let $\bar{\bar{y}}_{..} = \sum_{i=1}^{n}\sum_{t=1}^{T} y_{it}/(n \cdot T)$ be the overall arithmetic mean of all (log) hourly wages across individuals and time.[7] Using the $545 \cdot 8 = 4{,}360$ observations in the wagepan data, the overall average (log) hourly wage amounts to 1.649 (see the following Table 3.6 on p. 81). $\bar{y}_{i.} = \sum_{i=1}^{T} y_{it}/T$ is the corresponding mean for all wages measured over time for one specific individual i. For the 19 individuals shown in Fig. 3.2 mean (log) hourly wages can be found in the last column of Table 3.5. With this notation, estimates of both variance terms can be computed as follows:

$$\text{between-unit variance} \quad \hat{\sigma}_b^2 = \frac{\sum_{i=1}^{n}(\bar{y}_{i.} - \bar{\bar{y}}_{..})^2}{n-1} \quad (3.2)$$

$$\text{within-unit variance} \quad \hat{\sigma}_w^2 = \frac{\sum_{i=1}^{n}\sum_{t=1}^{T}(y_{it} - \bar{y}_{i.})^2}{n \cdot (T-1)} \quad (3.3)$$

The overall variance of the dependent variable is computed in the usual way:

$$\hat{\sigma}_o^2 = \frac{\sum_{i=1}^{n}\sum_{t=1}^{T}(y_{it} - \bar{\bar{y}}_{..})^2}{n \cdot T - 1} \quad (3.4)$$

---

[7] A period in the index indicates the dimension over which summation takes place. $y_{..}$, for instance, indicates that summation takes place for all individuals i and time points t.

3.3 Describing the Dependent Variable

For the total sample of all $n = 545$ individuals, we estimate $\hat{\sigma}_b^2 = 0.3907^2$ and $\hat{\sigma}_w^2 = 0.3872^2$ (see also the following Table 3.6 on p. 81), indicating that in this sample, the variance between units is nearly as large as the variance within units. In other words, wage heterogeneity across different individuals does not differ very much from wage heterogeneity across different years (taking into account the average level of wages for each individual).

Furthermore, both variance estimates can be used to compute an overall estimate of serial dependence, which is known as the *intra-class correlation (ICC) coefficient* ρ:

$$\hat{\rho} = \frac{T\hat{\sigma}_b^2 - \hat{\sigma}_w^2}{T\hat{\sigma}_b^2 + (T-1) \cdot \hat{\sigma}_w^2} \tag{3.5}$$

Given certain assumptions about the underlying process (see the following Textbox 3.1), it can be shown that it measures the correlation between *any* two measurements y_{it} and y_{is} ($t \neq s$) within the unit-specific (thus "intra-class") time series. Note that the time lag $(t - s)$ between both measurements does not matter. The ICC coefficient assumes that the correlation between all time points is the same. One could also say, it measures the "closeness" of measurements on the same unit relative to the closeness of measurements between different units. That is why we call it an *overall* measure of serial dependence. For the wage data, it is estimated as $\hat{\rho} = 0.4718$, indicating again a fairly high degree of overall serial dependence. It is slightly lower than the arithmetic mean ($\bar{r}_y = 0.5277$) of all the pairwise serial correlation coefficients in Table 3.1.

Textbox 3.1 (Intra-class correlation coefficient) The intra-class correlation (ICC) coefficient is based on the idea that each data point, y_{it}, is the result of three different effects: the overall level α_0 of Y, a unit-specific effect u_i, and finally for each unit, a measurement-specific effect e_{it}:

$$y_{it} = \alpha_0 + u_i + e_{it} \tag{3.6}$$

If the two effects u_i and e_{it} (i) are independent of each other, (ii) have variance σ_u^2 and σ_e^2, and if (iii) e_{it} is not serially correlated over time, it is easy to derive variances, covariances, and correlations of the measurements y_{it} (see Sect. 7.1):

1. If unit- and measurement-specific effects are independent of each other, the variance of the measurements y_{it} equals the sum of both variances: $\text{Var}(y_{it}) = \sigma_u^2 + \sigma_e^2$.
2. If both effects are independent of each other and not serially correlated, measurements from different time points, y_{it} and y_{is} ($t \neq s$), will covary, because u_i is constant over time and hence part of each y_{it}. The covariance equals the variance of u_i: $\text{Cov}(y_{it}, y_{is}) = \sigma_u^2$ ($t \neq s$).

3. Finally, a correlation coefficient is a "standardized" covariance, i.e., a covariance that is standardized by both variables' standard deviation:

$$\operatorname{Corr}(y_{it}, y_{is}) = \frac{\operatorname{Cov}(y_{it}, y_{is})}{\sqrt{\operatorname{Var}(y_{it})} \cdot \sqrt{\operatorname{Var}(y_{is})}} = \frac{\sigma_u^2}{\sigma_u^2 + \sigma_e^2}, \quad t \neq s \quad (3.7)$$

Equation (3.7) is a very general formula that describes the correlation between any two measurements y_{it} and y_{is} ($t \neq s$), if the assumptions (i) to (iii) are true. In that case, indeed, serial correlations will be identical irrespective of the time lag $(t - s)$ between measurements y_{it} and y_{is}. In other words, assumptions (i) to (iii) imply an *equal* correlation structure among the measurements.

In order to estimate (3.7), we need estimates of both variance terms σ_u^2 and σ_e^2. This is easy for the measurement-specific effects e_{it}, whose variance is estimated by (3.3):

$$\hat{\sigma}_e^2 = \hat{\sigma}_w^2 = \frac{\sum_{i=1}^{n} \sum_{t=1}^{T} (y_{it} - \bar{y}_{i.})^2}{n \cdot (T - 1)} \quad (3.8)$$

For the unit-specific effects u_i, the answer depends on how we consider the u_i. Do we assume that they are realizations of a random variable (i.e., random effects)? Or do we assume that they are specific to the units in the data set and, therefore, that they should be treated as fixed effects? In the latter case, the measurement-specific effects e_{it} are the only source of randomness, but within each unit they sum to zero by definition. Therefore, we can simply use the observed between-unit variance (3.2) as an estimate of σ_u^2:

$$u_i \text{ fixed effects} \quad \hat{\sigma}_u^2 = \hat{\sigma}_b^2 = \frac{\sum_{i=1}^{n} (\bar{y}_{i.} - \bar{\bar{y}}_{..})^2}{n - 1} \quad (3.9)$$

In case of random effects, however, we have two sources of randomness, and the measurement-specific effects e_{it} do not necessarily sum to zero within units. Therefore, an estimate of σ_u^2 is a function of both the observed between-unit variance (3.2) and the observed within-unit variance (3.3):

$$u_i \text{ random effects} \quad \hat{\sigma}_u^2 = \hat{\sigma}_b^2 - \frac{\hat{\sigma}_w^2}{T}$$

$$= \frac{\sum_{i=1}^{n} (\bar{y}_{i.} - \bar{\bar{y}}_{..})^2}{n - 1} - \frac{\sum_{i=1}^{n} \sum_{t=1}^{T} (y_{it} - \bar{y}_{i.})^2}{n \cdot (T - 1) \cdot T} \quad (3.10)$$

The ICC coefficient assumes random effects, and if you insert (3.10) and (3.8) into the formula (3.7) for the correlation coefficient, you arrive at the former formula (3.5) for the ICC coefficient. This and other types of ICC coefficients are discussed in Shrout and Fleiss (1979).

3.4 Explaining the Dependent Variable over Time: Typical Explanatory Variables

Now that we have an idea of what our dependent variable looks like, we can turn to the question of how to explain its distribution over time and between units. Why do some individuals earn higher wages or have higher probabilities of union membership? Why do wages increase more steeply for some individuals? Why is union membership rather stable for some individuals, but a more transient phenomenon for others? These questions relate to the level and change of the dependent variable. In the following, our primary interest is not in substantive arguments concerning the two examples (which would imply an excursus into human capital theory or rational choice models of political action). Instead, we will provide a synopsis of the typical explanatory variables that are used to answer these kinds of questions (Sect. 3.4.1). Once we have a better understanding of the factors that may explain why the dependent variable shows a certain pattern over time, we will then return to the question of serially correlated observations (Sect. 3.4.2).

3.4.1 Time-Constant and Time-Varying Variables

Figure 3.3 includes a random sample of wage trajectories from the wagepan data (similarly, we could plot the membership sequence of these individuals). The question is now: Why do the time trajectories of the dependent variable (wages, union membership or any other Y) behave the way they do? A very general answer to this question is that not all units are alike, and that units and context change over time. By context we mean the environment in which a unit is observed. In a micro panel of employees, this could be the economy of the country (or region) in which they live or the economic situation of the companies at which these individuals work. Note, in passing, that if we view panel data as hierarchical data, contexts introduce another (third) level to the data: it includes measurements within units within contexts.

More formally, this suggests three types of explanatory variable that are either located at the level of units or the level of contexts:

1. *Time-constant variables* Z characterizing the unit or the context: Typical time-constant variables are ethnicity or gender, if the unit is the individual, and geographical location or type of government, if the context is the country. It is easy to think of examples in which these variables function as explanatory factors in social and political research. For instance, it is well known that immigrants and women—because of segregation and discrimination—earn lower wages than native-born and male employees.
2. *Time-varying variables* X characterizing the unit or the context: Examples include variables like labor force experience and on-the-job training, if the unit is the individual, and economic growth and amount of public spending, if the context is the country. Again, it is quite obvious why these variables can be used as explanatory factors in different settings. For example, human capital theorists assume that wages increase with labor force experience and on-the-job training.

Fig. 3.3 Explaining panel data

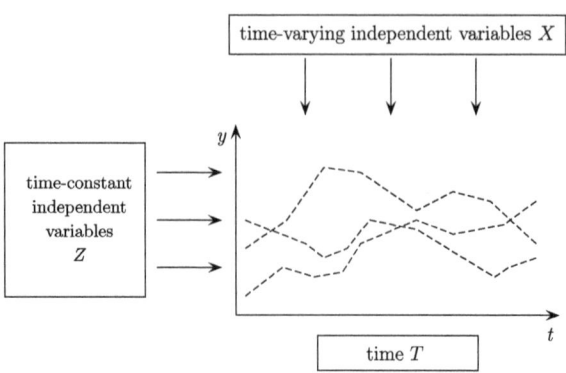

3. *Time T*: Finally, many researchers also mention "time" as an explanatory variable. However, it is questionable whether time itself is a real explanatory variable. In many cases, time is only an indicator for other characteristics that change over time (such as labor force experience or economic growth). Sometimes, however, it is not clear what these time-varying characteristics might be or the necessary data to control for them are absent. In these cases, it is a good idea to control for possible time trends in the data by including a variable "time" in the statistical model, although time is no causal factor on its own.

Depending on the research interest, it is also advisable to think about a good definition of time. In an analysis of job histories and earnings, there are several options: it could be chronological time (e.g., as an indicator of the business cycle), the time point of a certain event (e.g., first entry into the labor market as an indicator of the starting conditions of a career), or time elapsed since an event (e.g., time since entry into the labor force as an indicator of labor force experience). In the methodological literature of *cohort analysis*, these different definitions of time are referred to as *period*, *cohort*, and *age* effects. With panel data it is possible simultaneously to include different time variables into the same model (e.g., year of labor market entry, and labor force experience) as long as these time variables are not linearly dependent (see also the discussion in Sect. 1.1.1.2). This would be the case, for instance, if one also included chronological time into the model, because (period: chronological time) = (cohort: year of entry) + (age: time since entry).[8]

A necessary condition for all kinds of explanatory variables is that they show some variation. Put differently, constants are useless when it comes to explaining variation in the dependent variable. Technically, time-varying variables X vary across units and measurements, time-constant variables Z vary only across units, and chronological time T varies only across measurements.[9] There may be a problem with time-constant variables describing the context, because usually we have

---

[8] For an application of cohort analysis with panel data illustrating these identification problems, see Sect. 4.2.3.

[9] More specifically, chronological time (the period effect) does not vary across units. Other definitions of time, like age and cohort, pertain to the unit and, hence, show variation across units.

3.4 Explaining the Dependent Variable over Time: Typical Explanatory Variables 81

Table 3.6 Descriptive statistics of selected variables

Variable	Source	Mean	Std. Dev.	Min	Max	Observations
Log hourly wages (Log Dollars)	overall	1.6491	0.5326	−3.5791	4.0519	$N = 4{,}360$
	between		0.3907			$n = 545$
	within		0.3872			$T = 8$
Union membership (yes =1)	overall	0.2440	0.4296	0	1	$N = 4{,}360$
	between		0.3294			$n = 545$
	within		0.2950			$T = 8$
Afro-American (yes = 1)	overall	0.1156	0.3198	0	1	$N = 4{,}360$
	between		0.3200			$n = 545$
	within		0.0000			$T = 8$
Hispanic (yes = 1)	overall	0.1560	0.3629	0	1	$N = 4{,}360$
	between		0.3632			$n = 545$
	within		0.0000			$T = 8$
Education (years)	overall	11.7670	1.7462	3	16	$N = 4{,}360$
	between		1.7476			$n = 545$
	within		0.0000			$T = 8$
Experience (years)	overall	6.5147	2.8259	0	18	$N = 4{,}360$
	between		1.6549			$n = 545$
	within		2.4495			$T = 8$
From year 1981 (yes = 1)	overall	0.1250	0.3308	0	1	$N = 4{,}360$
	between		0.0000			$n = 545$
	within		0.3536			$T = 8$

Source: wagepan data (see Example 3.1)

fewer contexts than units (in the extreme case, there is only one context for all units, e.g., a sample of individuals from one country). Thus, if a context variable is time-constant, several contexts must be available in the data in order for us to observe some variation. Time-varying variables characterizing the context usually pose no problems. In the extreme case (only one context), there is at least some variation over time (but not over units).

To illustrate the distinction between different kinds of explanatory variables, we refer again to the wagepan data. The data set includes the typical variables from human capital theory: years of schooling and labor force experience. Additionally, it includes controls for ethnicity, health and family status, place of residence, and characteristics of the current job (occupation, industry). Most of these variables are time-varying, except ethnicity and years of schooling.

This can easily be verified with the descriptive methods discussed in the previous section. Table 3.6 shows the results for the variables (log) hourly wages (lwage), union membership (union), ethnicity (represented by two dummies: black and hisp), schooling (educ), and experience (exper). union, black and hisp are dummy variables, as can be seen from the overall minimum and maximum values (0 and 1). The time-constant variables black, hispanic and educ show no within variance ($\hat{\sigma}_w = 0$). That means that all eight observations for each individual are the same. However, all three variables vary between individuals, as can be seen from the between standard deviations (black: $\hat{\sigma}_b = 0.3200$, hispanic: $\hat{\sigma}_b = 0.3632$, educ: $\hat{\sigma}_b = 1.7476$).[10] The same is true for the time-varying variable exper ($\hat{\sigma}_b = 1.6549$), but additionally, this variable shows a fair amount of within variance ($\hat{\sigma}_w = 2.4495$) indicating that labor force experience varies (increases) for each individual during the observation period. Finally, instead of a continuous variable T, the dataset includes seven year dummies (d81,..., d87) to model a discontinuous-time trend. As an example, Table 3.6 shows the results for 1981 dummy d81, and we see that time-varying variables characterizing the context (like indicators of chronological time) show only within variance ($\hat{\sigma}_w = 0.3536$), but no between variance ($\hat{\sigma}_b = 0$).

Before we conclude this section, there are two final points worth mentioning in order to avoid any misunderstanding of Fig. 3.3. First of all, text and figure (especially the horizontal arrows) may suggest that time-constant Z only affect the level of the dependent variable Y ($Z \rightarrow Y$), while time-varying X only affect the change of Y ($\Delta X \rightarrow \Delta Y$).[11] Indeed, it is hard to conceive that the change (ΔY) of a dependent variable Y is causally related to something *constant* in time like Z ($Z \rightarrow \Delta Y$). However, possible relationships between X, Z, and Y are much more general and, therefore, this discussion is deferred to Sect. 3.5. At the moment, the horizontal arrows should simply indicate that Z remains constant over time, while the vertical arrows should indicate that X can change from one point in time to the next.

Second, few explanatory variables are time-constant by nature. Remember the aforementioned example "type of government", which was supposed to be time-constant. In the long run, it may very well change. Germany, changing from a totalitarian regime to a parliamentary democracy after World War II, is a good example. Seemingly, in the wagepan data, education is also a time-constant variable (see Table 3.6), although the qualifications of the respondents may change as a result of secondary and tertiary education. Obviously, this over-time information is missing in the data. Hence, some variables are *treated* as time-constant, either because changes are so rare that the corresponding variable is more or less a stable characteristic, or because we lack the necessary longitudinal information to measure the changes over time.

---

[10] Standard deviations (see Table 3.6, column 4) have been computed according to (3.2)–(3.4). Overall and between standard deviations are based on different observations (overall: $n \cdot T$, between: n) and hence, in case of time-constant variables (when $y_{it} = \bar{y}_{i.}$), have slightly different values. Note also that computer programs may use similar formulas, however with slightly different denominators.

[11] We use "($\cdots \rightarrow \cdots$)" as a shorthand for the statistical relationship that is of interest. The independent (explanatory) variable is indicated on the left side of the arrow and the dependent variable on the right side. At this point it is not necessary to specify the functional form of this relationship.

3.4.2 Serially Correlated Observations

Having discussed the possible explanatory factors for our dependent variable, we can now return to the question of why measurements over time show high degrees of serial correlation. The answer is now quite obvious:

1. First of all, having certain characteristics Z that do not change over time causes the dependent variable to have similar values during the next period. A migrant having a low income this year will not have a very different income next year (all other factors held constant). If migrants are more often members of a union in a given year, then they will probably still be members in the following year. The situation closely resembles the spurious causal correlation problem of which we are aware from the analysis of cross-sectional data. Two variables A and B may be correlated, because both of them are associated with a third variable C. With panel data, measurements $y_{i1}, y_{i2}, \ldots, y_{it}$ will be correlated, because all of them are correlated with a "third" variable Z that characterizes the unit.
2. A second source of serial dependence are the time-varying variables X. This may be a bit surprising, because we have just argued that the fact that these variables change over time explains why our dependent variable changes as well (and is *not* the same next year). However, this does not mean that the time-varying X change every year and/or to a large amount. Although they may not be identical next year, they may nevertheless correlate over time (similar to how the dependent variable does). If the X now influence Y, the serial correlation among the time-varying variables X leads to statistical dependencies among the measurements of the dependent variable Y.
3. Finally, having a certain value for the dependent variable this year may have a direct impact on the dependent variable's value next year. Consider, e.g., the case of union membership: Being member of a union implies having contact with other people of similar opinions and being exposed to political ideas of the organization, which may increase positive sentiments about the union movement and thus increase membership stability.

With these arguments in mind, it becomes obvious why measurements of the dependent variable over time must correlate. While the first two arguments are purely statistical (often termed *spurious state dependence*), the latter argument provides a substantive cause of serial correlation (often termed *true state dependence*). To the extent that we are able to control for all relevant time-constant and time-varying explanatory variables (Z, X and lagged values of Y), measurements of the dependent variable should be independent over time. Thus, if we regress the dependent variable Y on all necessary Z, X and lagged Y, the residuals of this regression will no longer be serially correlated.

This can be shown with the wagepan data. Having made the distinction between time-constant and time-varying explanatory variables (including time itself), we observe that the serial correlations of Y, computed earlier, have two components: one is due to the time-constant between-unit variation (on average, some individuals have, for various reasons, higher wages than other individuals), and the other is due to variation in the time dimension. Time-constant between-unit variation is

Table 3.7 Log hourly wages 1980–1987 (demeaned data)

Year	n	Arithmetic mean		Standard deviation	First-order serial correlation
		y	ln(y)		
1980	545	0.77	−0.256	0.452	
1981	545	0.87	−0.136	0.357	0.0251
1982	545	0.93	−0.077	0.292	0.0315
1983	545	0.97	−0.030	0.270	0.0620
1984	545	1.04	0.041	0.312	0.0275
1985	545	1.09	0.090	0.304	0.0335
1986	545	1.16	0.151	0.336	0.0346
1987	545	1.24	0.217	0.295	0.2514

Source: wagepan data (see Example 3.1)

easily controlled for by subtracting from each data value y_{it} the unit-specific mean $\bar{y}_{i.} = \sum_{i=1}^{T} y_{it}/T$. Let us call the transformation $\ddot{y}_{it} = y_{it} - \bar{y}_{i.}$ *demeaning*.[12] Table 3.7 shows the first-order serial correlations computed on the demeaned wage data, which are all much smaller than those computed on the original data (see Table 3.1).[13] Hence, most of the serial correlation in the wagepan data is due to different levels of wages *between* individuals, which in principle can be controlled by time-constant explanatory variables Z.

As a final point in this section, let us analyze why serial correlations get smaller with increasing time lags between measurements. The answer is quite easy. Let us begin with an extreme case: If only time-constant explanatory variables Z determine the process, and if they only affect the level of the dependent variable (effects being constant over time), the dependent variable will have the same values at each point in time, and nothing changes at all (resulting in serial correlations that are all equal to 1). If the world would look like this, we would not need panel data at all. A single cross-section would provide us with the same information as a panel data set. However, if the effects of the time-constant explanatory variables change over time, if time-varying explanatory variables X come into play, or if a random error exists, then the values of Y are not the same next year, and these "noise factors" might accumulate over time. All of this results in decreasing higher-order serial correlations.

Table 3.8 illustrates these ideas with three different prototypical correlation structures. The upper part of the table shows what the serial correlations would look like, if the measurements over time were independent of one another. Certainly, this assumption is not very realistic for panel data. The middle part of the table shows what the serial correlations would look like, if all the serial dependencies were due to time-constant characteristics of the units. This assumption is the basis of our over-

---

[12] Some scholars also use the term "centering". But you should note that this transformation is a specific form of centering. It subtracts the *unit-specific* means and not the overall mean of a variable, which is the transformation one usually thinks of when talking about centered variables.

[13] Only Corr($\ddot{y}_{i,1987}, \ddot{y}_{i,1986}$) = 0.251 is surprisingly large, but still much smaller than the serial correlation of the original data: Corr($y_{i,1987}, y_{i,1986}$) = 0.693 (see Table 3.1).

3.4 Explaining the Dependent Variable over Time: Typical Explanatory Variables

Table 3.8 Three different prototypical correlation structures

Year	1980	1981	1982	1983	1984	1985	1986	1987
a) Serially independent observations							
1980	1.0000							
1981	0.0000	1.0000						
1982	0.0000	0.0000	1.0000					
1983	0.0000	0.0000	0.0000	1.0000				
1984	0.0000	0.0000	0.0000	0.0000	1.0000			
1985	0.0000	0.0000	0.0000	0.0000	0.0000	1.0000		
1986	0.0000	0.0000	0.0000	0.0000	0.0000	0.0000	1.0000	
1987	0.0000	0.0000	0.0000	0.0000	0.0000	0.0000	0.0000	1.0000
b) All serial correlations due to time-constant variables							
1980	1.0000							
1981	0.4713	1.0000						
1982	0.4713	0.4713	1.0000					
1983	0.4713	0.4713	0.4713	1.0000				
1984	0.4713	0.4713	0.4713	0.4713	1.0000			
1985	0.4713	0.4713	0.4713	0.4713	0.4713	1.0000		
1986	0.4713	0.4713	0.4713	0.4713	0.4713	0.4713	1.0000	
1987	0.4713	0.4713	0.4713	0.4713	0.4713	0.4713	0.4713	1.0000
c) All serial correlations due to time-constant and time-varying variables							
1980	1.0000							
1981	0.6265	1.0000						
1982	0.3925	0.6265	1.0000					
1983	0.2459	0.3925	0.6265	1.0000				
1984	0.1541	0.2459	0.3925	0.6265	1.0000			
1985	0.0965	0.1541	0.2459	0.3925	0.6265	1.0000		
1986	0.0605	0.0965	0.1541	0.2459	0.3925	0.6265	1.0000	
1987	0.0379	0.0605	0.0965	0.1541	0.2459	0.3925	0.6265	1.0000

Source: wagepan data (see Example 3.1)

all measure of serial correlation, the ICC coefficient (3.5). Textbox 3.1 explains the statistical reasoning behind this equal correlation structure: according to (3.6), each observation y_{it} is conceived of a time-constant unit-specific part and a measurement-specific part that is pure random noise (i.e., is independent between observations). Finally, the lower part of the table shows what the serial correlations would look like, if the serial dependencies were due to time-constant and time-varying characteristics of the units. More specifically and contrary to the ICC coefficient, it is assumed that part of the measurement-specific influences carry over from one point in time to the next.

The correlation structure (a) implying totally independent measurements is typical for linear regression models applied to cross-sectional data. *Generalized estimating equations* (GEE) provide a very flexible approach, both to generalize this linear model to different kinds of (continuous and categorical) variables, and to model all kinds of covariance structures among clustered data such as panel data, where measurements over time are clustered within units (Hardin and Hilbe, 2012). The correlations in Table 3.8 have been estimated with a program for GEE assuming independent (upper part), exchangeable (middle part), and first-order autoregressive correlations (lower part). Because of the GEE methodology, the (exchangeable) correlations in the middle part differ slightly from our former estimate of the ICC coefficient ($\hat{\rho} = 0.4718$).

3.5 Modeling Panel Data

In this section, we discuss how the independent variables may influence the dependent variable and how we can formalize this. When we say "formalize", we mean a mathematical model (a regression function) that tells us numerically how our expectations about the values of the dependent variable are related to the values of the independent variables. As discussed in the Introduction (Chap. 1), panel data have been used for the analysis of trends (models in levels) and for the analysis of individual change (models of change). Therefore, we discuss models either for the level (Y) or the change (ΔY) of the dependent variable. In these models, we can use—besides time T—either the level (X, Z) or the change (ΔX; $\Delta Z = 0$ by definition) of our independent variables as explanatory factors. A few examples will illustrate this:

- $X, Z \rightarrow Y$: For example, human capital theory asserts that higher levels of labor force experience are associated with higher earned incomes, suggesting a (positive) association between the level of income (Y) and the level of a time-varying variable measuring employment duration (X). Rational choice models assume that the benefits of union membership are higher for low-status employees, suggesting a (negative) association between the level of school education (Z, assumed to be time-constant after labor market entry) and the level of union membership (Y). As you may have noticed, these are also the kinds of relationships we normally use in the analysis of cross-sectional data. We could survey a random sample of employees and collect information on hourly wages, union membership, labor force experience and schooling at the date of the survey. Using the between-unit variation of experience and schooling in this cross-section, we could test the hypotheses of human capital theory and rational choice models of political action. Obviously, these relationships can also be analyzed with panel data. However, one may question why we should use these more complicated, and possibly more costly, data to answer simple cross-sectional questions. As we shall see later on (Sect. 3.6), panel data provide additional information that allows us to avoid some of the specification errors that are typical of cross-sectional data.
- $\Delta X \rightarrow \Delta Y$: When we think about processes of change, this kind of relationship always comes to mind. The dependent variable changes, because something changes for the unit or the context. For example, as an employee gains more

3.5 Modeling Panel Data

and more labor force experience during his career, his or her wage should gradually increase. Or if, during a business cycle, private enterprises increase their profits, this should also increase employees' wages. Contrary to the model in levels $(X, Z \rightarrow Y)$, one now needs longitudinal data. One uses the over-time (within-unit) variation to test the two hypotheses. But you should note the connections with relationships of the first type $(X, Z \rightarrow Y)$. For example, if we interpret regression coefficients (β) from cross-sectional models, we say that a one unit change of the independent variable causes a change of β units of the dependent variable. Thus, if models of type $(X, Z \rightarrow Y)$ are true, models of type $(\Delta X \rightarrow \Delta Y)$ are true as well. Later on, we will see more formally that it is easy to transform models of type $(X, Z \rightarrow Y)$ into models of type $(\Delta X \rightarrow \Delta Y)$. In other words, they are conceptually equivalent. Nevertheless, it is important to make a distinction between them, because estimates based on data in levels generally will not be identical to estimates based on changes.

In the following, we discuss models for levels (Sect. 3.5.1) and change (Sect. 3.5.2) of the dependent variable. We conclude with a brief discussion of some more complex models, including lagged variables, interaction effects among the independent variables, and random coefficients (Sect. 3.5.3). Throughout the whole discussion, we will make a distinction between continuous and categorical variables. As already mentioned, models for continuous variables focus on the expected value, $E(y)$, of the dependent variable, while models for categorical variables focus on the probability, $\Pr(y_{it} = q)$, of observing a certain category, say q, of the dependent variable.

3.5.1 Modeling the Level of the Dependent Variable

3.5.1.1 Continuous Dependent Variables

For continuous dependent variables, a typical panel regression model in levels is a simple extension of the well-known linear regression model for cross-section data. The following linear model regresses the expected value of a continuous dependent variable Y on time T and a set of independent variables, which—according to the discussion in the previous section—may be either time-constant (Z) or time-varying (X). By using the appropriate subscripts, we take care of the time dimension in the data:

$$E(y_{it}) = \underbrace{\beta_0(t) + \beta_1 x_{1it} + \cdots + \beta_k x_{kit}}_{\text{time-dependent part}} + \underbrace{\gamma_1 z_{1i} + \cdots + \gamma_j z_{ji}}_{\text{time-constant part}} \qquad (3.11)$$

Subscript i refers to the $i = 1, \ldots, n$ units, which have been observed at $t = 1, \ldots, T$ equidistant points in time. y_{it} denotes the value of the dependent variable Y for individual i at time point t. Its expected value is modeled as a linear function of the values of j time-constant $(z_{1i}, \ldots, z_{ji})$ and k time-varying independent variables $(x_{1it}, \ldots, x_{kit})$. $\gamma_1, \ldots, \gamma_j$ and $\beta_1, \ldots, \beta_k$ denote the corresponding regression coefficients.

The term $\beta_0(t)$ determines the overall level of the dependent variable. Since its level may change over time, $\beta_0(t)$ can be any function of time to control for possible time trends. In other words: $\beta_0(t)$ specifies the *growth* of the dependent variable and as we will later see (Sect. 3.5.3.3), the growth can be a function of other explanatory variables. If there is no time trend, $\beta_0(t)$ reduces to the familiar regression constant $\beta_0(t) = \beta_0$. Possible functions of time include linear ($\beta_0(t) = \beta_0 + \beta_1 t$), quadratic ($\beta_0(t) = \beta_0 + \beta_1 t + \beta_2 t^2$), exponential (if we model $\ln y_{it}$) and discontinuous functions of time (e.g., $\beta_0(t) = \beta_0 + \beta_1 d_1 + \cdots + \beta_l d_l$ with l dummy variables, $d_1, \ldots, d_l$, measuring different time periods). Naturally, if the term $\beta_0(t)$ includes—besides β_0—l parameters (and time variables), the time-varying X and their effects β are indexed from $(l+1)$ to k. For example, a model including a quadratic time trend looks like this:

$$E(y_{it}) = \beta_0 + \beta_1 t + \beta_2 t^2 + \beta_3 x_{3it} + \cdots + \beta_k x_{kit} + \gamma_1 z_{1i} + \cdots + \gamma_j z_{ji}$$

Readers familiar with the linear regression model might miss an error term in the equation. They should remember, however, that at this point of our discussion, we are only looking at the expected value of the dependent variable and, thus, the systematic part of the regression model. Later, when we discuss the estimation of the parameters $\beta_0, \beta_1, \ldots, \beta_k$ and $\gamma_1, \ldots, \gamma_j$, it will be necessary to consider an error term (see Sect. 3.6).

3.5.1.2 Categorical Dependent Variables

The approach in Sect. 3.5.1.1 is easily transferred to categorical dependent variables. If q denotes the category of Y we are interested in (e.g., being member of a union), then the probability, $\Pr(y_{it} = q)$, of observing category q for unit i at time point t equals the following expression:

$$\Pr(y_{it} = q) = G\big(\beta_0(t) + \beta_1 x_{1it} + \cdots + \beta_k x_{kit} + \gamma_1 z_{1i} + \cdots + \gamma_j z_{ji}\big) \qquad (3.12)$$

This is also known as a *discrete response model*.

$G(\cdot)$ is a suitable distribution function (e.g., the normal or the logistic). By using a distribution function (instead of the linear-additive function (3.11)), we make sure that the right-hand side of the equation provides values that are within the proper limits of probabilities (i.e., $0 \leq \Pr(y_{it} = q) \leq 1$). Depending on the choice of the distribution function $G(\cdot)$, we arrive at either the *logistic* (3.13) or the *probit regression model* (3.14):

$$\Pr(y_{it} = q) = \frac{\exp(\beta_0(t) + \beta_1 x_{1it} + \cdots + \beta_k x_{kit} + \gamma_1 z_{1i} + \cdots + \gamma_j z_{ji})}{1 + \exp(\beta_0(t) + \beta_1 x_{1it} + \cdots + \beta_k x_{kit} + \gamma_1 z_{1i} + \cdots + \gamma_j z_{ji})} \qquad (3.13)$$

$$\Pr(y_{it} = q) = \Phi\big(\beta_0(t) + \beta_1 x_{1it} + \cdots + \beta_k x_{kit} + \gamma_1 z_{1i} + \cdots + \gamma_j z_{ji}\big) \qquad (3.14)$$

In (3.14), $\Phi(\cdot)$ represents the standard normal distribution function.

3.5.2 Modeling Change of the Dependent Variable

3.5.2.1 Continuous Dependent Variables

Both regression functions (3.11) and (3.12) model the level of the dependent variable over time. In this section, we focus on the change of the dependent variable and how it can be explained by various characteristics of the unit and the context. Change in a continuous dependent variable can be operationalized in different ways: as absolute change $(y_{it} - y_{i,t-1})$ or as relative change $(y_{it}/y_{i,t-1})$. In the latter case, one often models the logarithm of relative change, which is also a simple difference: $\ln(y_{it}/y_{i,t-1}) = \ln(y_{it}) - \ln(y_{i,t-1})$. Whatever the concrete operationalization, a model of change has basically the same structure as a model in levels. The expected value of change is assumed to be some function of the explanatory variables X, Z and T. First, we will discuss a model that focuses on the change between two adjacent measurements of Y over time. As we will see, it is closely connected to a model in levels. Then we will discuss models of change that are due to certain events.

Consider the model in levels as it is specified in (3.11). For two arbitrary time points, t and $t-1$, it looks like this:

$$t : E(y_{it}) = \beta_0(t) + \beta_1 x_{1it} + \cdots + \beta_k x_{kit} + \gamma_1 z_{1i} + \cdots + \gamma_j z_{ji} \quad (3.15)$$

$$t-1 : E(y_{i,t-1}) = \beta_0(t-1) + \beta_1 x_{1i,t-1} + \cdots + \beta_k x_{ki,t-1} + \gamma_1 z_{1i} + \cdots + \gamma_j z_{ji} \quad (3.16)$$

If we compute the difference of both equations, we arrive at a model of absolute change:

$$E(y_{it} - y_{i,t-1}) = E(\Delta y_{it}) = \beta_0(t) - \beta_0(t-1) + \beta_1 \Delta x_{1it} + \cdots + \beta_k \Delta x_{kit} \quad (3.17)$$

Thus, our model in levels is easily transformed into a model of absolute change. Δy_{it} is also called a *change score* and the transformation is known as computing *first differences* (FD for short). Hence, as already mentioned, if models in levels $(X, Z \rightarrow Y)$ are true, models of change $(\Delta X \rightarrow \Delta Y)$ are true as well. This also corresponds to the usual interpretation of regression coefficients in (3.11). We usually say that a one unit *change* of a given independent variable X causes a *change* of β units of the dependent variable Y. But you should also note that when it comes to the estimation of both types of model, estimates of models in first differences may yield different estimates than models in levels. The most important reason for this difference is that, in order to estimate FD models, one exploits the over-time (within-unit) variation, which may provide other conclusions than the overall variation that consists of between- and within-unit variation.

Two other things are noteworthy, when computing first differences. (i) If the time trend is assumed to be linear, i.e. $\beta_0(t) = \beta_0 + \beta_1 t$, it will be eliminated as well because $(\beta_0 + \beta_1 t) - (\beta_0 + \beta_1(t-1)) = \beta_1$. (ii) The effects of time-constant variables Z drop out, because their differences are zero by definition. Hence, if model (3.11) is true, it implies that time-constant characteristics do not have an effect on the change of the dependent variable. Section 3.5.3.2, however, will discuss whether and how it is possible to include time-constant independent variables Z into models of ΔY.

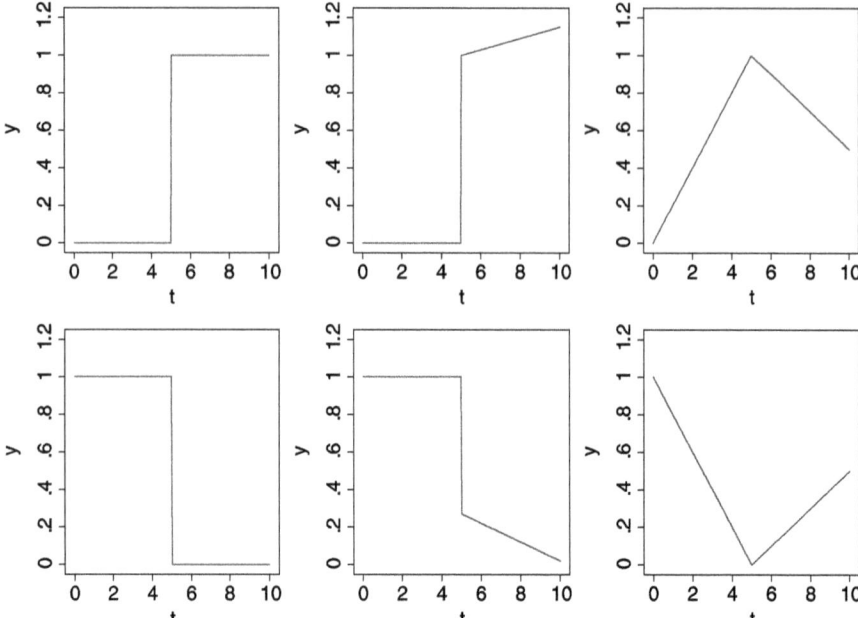

Fig. 3.4 Impact functions

Besides looking at continuous change using change scores, researchers are often interested in assessing the effect of events on the dependent variable Y. By an event, we mean a discontinuous change in some (mostly categorical) explanatory variable. Divorce (change in marital status), unemployment (change in employment status), or an economic crisis (change in the level of economic activity) are examples of such events. The assumption is that these events affect the level of the dependent variable Y. Figure 3.4 shows three possibilities for how an event happening, e.g., at $t = 5$ can have an impact on Y. (i) Y may have a significantly larger (or smaller) level after the event. (ii) The event may have an immediate increasing (or decreasing) impact on Y, and it may increase (decrease) with the passage of time. (iii) Y may increase (decrease) continuously up to the event and decrease (increase) continuously thereafter. In the first case, the time trajectory of Y is a simple step function, in the second case a step function with an increasing (decreasing) tail, and in the third case a linear function with a structural break. These and other kinds of *impact functions* can be modeled with the $\beta_0(t)$ term in a model of Y in levels (see (3.11)). For example, the simple step function in the first case is applied with a dummy variable $devent_{it}$, which equals 1 once the event has occurred (i.e., when $t \geq t_{event}$) and 0 otherwise:

$$E(y_{it}) = \beta_0 + \beta_1 devent_{it} + \beta_2 x_{2it} + \cdots + \beta_k x_{kit} + \gamma_1 z_{1i} + \cdots + \gamma_j z_{ji}$$

The effect of the event on the level of Y is estimated by the parameter β_1.

3.5 Modeling Panel Data

Note in passing that the dummy variable is not only a function of time T. It is also assumed to be different for each unit i. This is necessary whenever the timing of events is different for each unit (like, e.g., in the case of divorce). If the event in question affects each unit at the same point in time (e.g., an economic crisis in a specific year), this differentiation is not necessary and the impact function is only a function of T (like $\beta_0(t)$).

3.5.2.2 Categorical Dependent Variables

Models of change for categorical variables are less obvious, because it does not make sense to compute differences of a categorical variable. There are different options discussed in the literature. One of them uses the previous value of Y ($y_{i,t-1}$) as an independent variable in a model in levels. Thus, effects of X, Z and T are estimated, controlling for the former status of the unit. These kinds of methods have a close connection to *Markov modeling*, which tries to find a simple structure for the various transition matrices observed over time (see Table 3.2). For example, a simple assumption would be the hypothesis that—apart from random fluctuations—all yearly transition matrices are the same. The transition matrix at the bottom of Table 3.2 (see the row labeled "Total") would be an estimate of this simple process. This assumption is easily modeled with the following logistic model:

$$\Pr(y_{it} = q) = \frac{\exp(\beta_0 + \beta_1 y_{i,t-1})}{1 + \exp(\beta_0 + \beta_1 y_{i,t-1})} \quad (3.18)$$

Naturally, this model could also include independent variables X, Z, and a time trend $\beta_0(t)$, but to keep things simple, we have chosen not to do so. The model states that the probability of being observed in state $y_{it} = q$ depends only on the state being observed in the previous year $t - 1$, irrespective of whether q is observed in $t = 1981$, $t = 1982, \ldots$ or $t = 1987$. A slightly more complicated assumption posits that transition matrices are different for each year. This assumption could be modeled by interacting $y_{i,t-1}$ with a dummy for each year. In a similar way, other assumptions about the underlying process could be tested. For example, by using an impact function for $\beta_0(t)$, we could model the effect of certain events on the dependent categorical variable. Markov modeling is very popular in the context of categorical data analysis when all (dependent and independent) variables are categorical.

Another approach that aligns perfectly with our thinking consists of choosing one category of the dependent variable and modeling its change. More specifically, it models the conditional transition probability of making a change from category p to any other category of the dependent variable, given that the unit has been observed in category p at all previous points in time. In Sect. 3.3, we have defined this conditional transition probability as $\Pr(y_{it} \neq p | y_{i1} = \cdots = y_{i,t-1} = p)$. As a shortcut, we introduced $h_p(t) = \Pr(y_{it} \neq p | y_{i1} = \cdots = y_{i,t-1} = p)$, which is also termed the (discrete-time) hazard rate (of leaving state p). A model for the conditional transition probability resp. hazard rate would look like this (now with variables X and Z):

$$h_{ip}(t) = G\big(\beta_0(t) + \beta_1 x_{1it} + \cdots + \beta_k x_{kit} + \gamma_1 z_{1i} + \cdots + \gamma_j z_{ji}\big) \quad (3.19)$$

Again, $G(\cdot)$ is a suitable distribution function to make sure that the right-hand side of the equation provides values that are within the appropriate limits of probabilities (i.e., $0 \leq h_{ip}(t) \leq 1$). As will be discussed in Chap. 5, $G(\cdot)$ is either the logistic or the extreme value distribution function. Note also that the hazard rate now has an index i, because it is assumed to be different for each unit $i = 1, \ldots, n$. Equation (3.19) is a so-called (discrete-time) *hazard rate model*. Other terms for these kinds of analyses are *survival analysis*, because there is a close connection between the hazard rate and the survival probability, or *event history analysis*, because the (sudden) change from p to another category is also called an event. Survival analysis is the traditional term in biometrics and technometrics, where these methods originate from, while event history analysis is the more common name in the social sciences.

As (3.19) shows, one would usually prefer to model change (i.e., the conditional transition probability or hazard rate) and not the opposite (survival). For descriptive purposes, the survivor function is perfect (and sometimes easier to analyze than the conditional transition probability; see our discussion in Sect. 3.3). However, for explanatory purposes, one uses the conditional transition probability, because it measures the current rate of change for those who are still at risk, while the survival probability incorporates the history of all previous transitions and non-transitions. Thus, the conditional transition probability is a recent measure of change, which you need when you assume that the process itself changes over time (e.g., when you assume that the probability of union membership increases with membership duration, which implies a decreasing hazard rate). Moreover, it can be shown that the conditional transition probability is the fundamental parameter of the process, from which all the other parameters (survival probability, unconditional transition probabilities, etc.) derive (for the survival probability see (3.1)).

Although event history analysis is a convenient method to analyze the change in categorical variables, it should be stressed that it makes the strong assumption that the process of interest (in our case: union membership) has been observed from the very beginning. In our case, this assumption would imply that all employees had initiated their union membership in 1980, which may not be true, and, more importantly, which we often do not know, because the necessary information is not available from the panel data; this situation is also called (left) *censoring*. Markov modeling, on the other hand, often does not make such strong assumptions. Rather, it treats the yearly transition matrices as snapshots of the process and uses them to project the current distribution of the categorical dependent variable into the future.

3.5.3 Additional Models

In the preceding sections, we have discussed models for levels and change of the dependent variable, mostly focusing on relationships of type $(X, Z \rightarrow Y)$ and $(\Delta X \rightarrow \Delta Y)$, where all (independent and dependent) variables have either been included in levels or in changes (of course, excluding time-constant Z in the latter case, because they do not change by definition). Furthermore, we have assumed that the effects of the independent variables (β, γ) are fixed and do not change with some characteristic of either the units or the context. In this section, we want to generalize these assumptions and discuss other relationships that might be feasible (Sect. 3.5.3.1), as well as how to include time-constant Z in models of change (Sect. 3.5.3.2), and how to relax the assumption of fixed coefficients (Sect. 3.5.3.3).

3.5.3.1 Other Types of Relationship

Let us first discuss some variations of the types of relationship discussed so far. What about including X and Z in levels in a model of change (ΔY)? Or what about including the change of X (ΔX) in a model of Y in levels? Mathematically, these kinds of relations are possible, but from a substantive point of view, they do not always make sense. This is easily illustrated with some examples.

Relationships of the type $(X, Z \rightarrow \Delta Y)$ are typical of many change processes in the natural sciences. For example, radioactive decay (ΔY) is assumed to be related to the (time-varying) quantity of radioactive nuclei (X) that have not yet decayed. In our wage data, the amount of wage increases (ΔY) could be related to (time-constant) ethnicity (Z), because discrimination theories suggest that ethnic minorities are less often promoted than members of the majority population. Another example is the greater life expectancy of women compared to men. In this case, mortality (ΔY; changing from being alive to being dead) is related to the time-constant variable of gender (Z). From a purely computational point of view, no regression program will prevent you from adding a time-constant Z to the right-hand side of (3.17) that models ΔY. But from a substantive point of view, does that really make sense? How can something *constant* in time result in a *change* in something else?

When you rethink the two social science examples, you will probably realize that it is not ethnicity or gender as such that causes the change; rather, it is the fewer promotions or the more risky behavior that causes lower wage increases for minorities or higher mortality rates for males. In other words, it is the change ΔX of an independent variable X (job promotion, health-related behavior) that ultimately causes the change ΔY of the dependent variable Y. Thus, models of type $(\Delta X \rightarrow \Delta Y)$ would be much more useful, if we had information about the true causal factors. But very often we do not have access to these data and must use time-constant variables Z as indicators for them instead. Section 3.5.3.2 demonstrates how this is feasible, even if FD eliminates time-constant Z, as we have seen in Sect. 3.5.2.1. In the change model for categorical variables (3.19), they are included from the very beginning, because the model has not been derived by a linear transformation of the corresponding model in levels (3.12).

Relationships of the type $(\Delta X \rightarrow Y)$, where the level of the dependent variable is the result of changes in the independent variables, pose fewer problems. A good example is adaptive behavior. For example, subjective well-being (Y) is much less closely related to absolute income (X) than to changes in income (ΔX). This can be illustrated with research on the effects of unemployment. Becoming unemployed and experiencing an income loss is usually associated with significantly less subjective well-being. In the long run, however, even with lower incomes, people often return to their former levels of subjective well-being, because they adapt to the new financial situation. Hence, it is not the absolute level of income (X) that is effective, it is rather the change of income at the start of the unemployment spell that affects well-being. Hence, if you think that the change of some time-varying variable X affects the level of your dependent variable Y, include ΔX as an independent variable in (3.11) resp. (3.12).

Even more complex relationships result, if we extend the time dimension of the model. All models discussed up to now are, in a way, *static models*, because the current level (or change) of Y is a function of current levels (or change) of the independent variables X and Z. For this introductory textbook, this is perfectly alright, but you should know that even more complicated specifications are possible. The right-hand side of the equation may include past values of time-varying explanatory variables (e.g., $x_{i1,t-1}, x_{i1,t-2}, \ldots$). This would be called a *distributed lag model*, because the explanatory variables X affect the dependent variable Y with a time lag. Furthermore, the right-hand side may include lagged values of the dependent variable: $y_{i1,t-1}, y_{i1,t-2}, \ldots$. These specifications are called *dynamic models* and the Markov model (3.18) is an example of this kind.

Besides modeling the statistical properties of a change process (e.g., the Markov assumption), there are also substantive reasons for analyzing relationships of the type $(Y_{t-\tau} \rightarrow Y_t)$. As already noted in Sect. 3.4.2, there are instances of true state dependence, in which previous levels (τ years earlier) influence the present level of the dependent variable Y. Consider, for example, government spending: The present federal government's budget is fixed, in large part, by decisions that have been made in previous years. Thus, only a minor portion of the budget is available for new policies and large parts of present government spending can be explained by last year's budget. Or remember the union example: Being a member of a union may increase one's positive sentiments toward the union movement and, thus, may increase membership stability. Models of partial adjustment and adaptive expectations that have been popularized in economics also result in relationships, in which previous levels influence the present level of the dependent variable.

Finally, all the models discussed up to now posit that there is no feedback between X and Y, which may be an unrealistic assumption in some applications. Consider, for example, the relation between union membership and wages. Some economists ask whether there is a wage premium for union members. On the other hand, it is not unrealistic to assume that some employees become union members because they expect higher wages from being a union member. These kinds of reciprocal relationships $(X \leftrightarrow Y)$ are called (somewhat irritatingly) *non-recursive models*, while the former unidirectional relationships are termed *recursive*. Non-recursive and dynamic models are more difficult to handle and will not be covered in this introductory textbook.

3.5.3.2 How to Include Time-Constant Variables in Models of Change

When discussing models of change for continuous dependent variables in Sect. 3.5.2.1, we have seen that computing first differences eliminates time-constant variables Z from the model. However, as argued earlier, we may want to include these variables into our model, because we lack information about the true causal

3.5 Modeling Panel Data

factors that change over time. We also argued that, mathematically, you can compute first differences of your dependent variable and then regress ΔY on some time-constant variables Z. But how do the effects of the latter kinds of models relate to the effects of our former model in levels?

As an example, consider the assumption that wage growth differs with respect to ethnicity. We use a simple dummy variable for ethnicity (black = 1: Afro-American, black = 0: other) and if wage growth is assumed to be different among Afro-American and other employees, we have to create an interaction between ethnicity and the time trend. With a linear time trend and a main effect of ethnicity to control for different wage levels, a model in levels looks like this:

$$E(\ln(wage_{it})) = \beta_0 + \beta_1 \cdot t + \beta_2 \cdot black_i \cdot t + \gamma_1 \cdot black_i \quad (3.20)$$

If we compute first differences of (3.20), we arrive at the following model of change:

$$E(\Delta \ln(wage_{it})) = \beta_1 + \beta_2 \cdot black_i.$$

As expected, the main effect (γ_1) of ethnicity, a time-constant variable, cancels out. Thus, we get no numerical estimate of the wage gap for minorities (to this end, we should have estimated the model in levels (3.20)). But the interaction effect (β_2) is still part of the model on change. More generally speaking, whenever we interact a time-constant with a time-varying variable in a model on levels of Y, this time-constant variable is also part of a model on change of Y. In our simple model of change, both parameters have an interesting interpretation. $\hat{\beta}_1$ estimates the overall linear trend of (log) hourly wages for the non-black employees and $\hat{\beta}_2$ estimates how much this trend is different for Afro-American employees.

3.5.3.3 How to Relax the Assumption of Fixed Coefficients

A common feature of all models discussed so far is the assumption of *fixed* parameters. Fixed, in this context, means that each regression coefficient has a fixed value in the population and does not vary across the units of analysis.[14] This may be an unrealistic assumption. As Fig. 3.2 suggests, each unit seems to have its own intercept and slope in the wagepan data. Thus, each regression coefficient can be thought of as a random variable (with its own distribution function) and the specific values we observe for unit i can be interpreted as realizations of this random variable.

We illustrate this extension with a very simple example. Consider the data on hourly wages. A simple model posits an exponential growth of hourly wages (or equivalently, a linear growth of log hourly wages):

$$E(\ln(wage_{it})) = \beta_0 + \beta_1 t \quad (3.21)$$

This model is easily extended to include other independent variables (X, Z), but for our present purposes, it is sufficient to include time as the only "explanatory" variable. This model assumes an intercept β_0 and a slope β_1 that are the same for each unit. However, if we attach an index i to each parameter,

---

[14] In principle, the values of the regression coefficients can also change over time.

$$E(\ln(wage_{it})) = \beta_{0i} + \beta_{1i} t \qquad (3.22)$$

we obtain a much more general model with unit-specific intercepts and slopes. The unit-specific parameters β_{0i} and β_{1i} can now be modeled in separate regression functions. For example, the intercept could be partly random and partly a function of certain explanatory variables (e.g., years of schooling to control for different levels of human capital):

$$\beta_{0i} = \gamma_{00} + \gamma_{01} \cdot schooling_i + u_{0i} \qquad (3.23)$$

A random variable U_0 is included in the equation to control for the randomness of the intercepts. A common assumption is that u_{0i} is normally distributed with mean zero and variance $\sigma^2_{u_0}$. A similar model could be specified for the slope parameter:

$$\beta_{1i} = \gamma_{10} + \gamma_{11} \cdot ethnicity_i + u_{1i} \qquad (3.24)$$

Here we use another normally distributed random variable U_1 with mean zero and variance $\sigma^2_{u_1}$ to model the randomness of the slopes. Besides that, we use ethnicity as an explanatory variable, because we assume wage growth to differ with respect to ethnicity.

If we reinsert both equations into (3.22) and rearrange, we arrive at the following expression:

$$E(\ln(wage_{it})) = \underbrace{\gamma_{00} + \gamma_{01} \cdot schooling_i + \gamma_{10} \cdot t + \gamma_{11} \cdot ethnicity_i \cdot t}_{\text{fixed}} + \underbrace{u_{0i} + u_{1i} \cdot t}_{\text{random}} \qquad (3.25)$$

Compared to (3.21), this extended model includes fixed *and* random parameters. Therefore, these kinds of models are also known as *linear mixed models*. Note, also, that (3.22) operates at a different level than (3.23) and (3.24). To understand this statement, you have to remember the hierarchical nature of panel data including measurements within units (and sometimes within contexts). Equation (3.22) operates at the first (lowest) level: the level of measurements. β's denote parameters at this first level and have an index i attached to them. In contrast, (3.23) and (3.24) operate at the second level: the level of units. γ's denote parameters at this second level and their first index refers to the parameter at the first level to which they apply. This way of specifying the model is also known as *hierarchical linear* or *multi-level modeling*. But as (3.25) shows, it is quite easy to integrate these extended specifications into our modeling approach by using appropriate independent variables and interaction effects. Textbox 3.2 explains in greater detail how a panel regression model can be framed as a hierarchical linear model and how this motivated our extended notation.

Textbox 3.2 (Hierarchical linear models: Extended notation) As mentioned in Sect. 3.1, a panel data set has a hierarchical structure. This is obvious when data are organized in long format, i.e. measurements $t = 1, \ldots, T$ within units

3.5 Modeling Panel Data

$i = 1, \ldots, n$ (see Table 2.1). At the lowest level of this hierarchy (level 1), i.e., the level that changes the most quickly, we have the measurements over time, and at the higher level (level 2), i.e., the level changing less quickly, we have the units. If the data would also differentiate among different contexts (e.g., countries), we would even observe a third level. The dependent variable Y and the time-varying explanatory variables X are located at the lowest level, because they also change over time. The time-constant explanatory variables Z, on the other hand, are located at the second level, because they only vary between units.

Now, think about the parameters β and γ in our former regression models (e.g., model (3.11)): While the former measure effects of first level explanatory variables, the latter measure the effects of second level explanatory variables. When we talk about fixed and random effects in the context of panel models, we mean parameters that are either identical for all units or that may differ between units. In the latter case, the corresponding parameter is considered a random variable and the specific value of that parameter for one particular unit is thought of as a realization of that random variable. Since we are only assuming variation between units, only the β's can be random variables. Of course, part of the variation will be systematic and can be related to the characteristics of the units (Z). Hence, it is natural to model each parameter β_k as a function of some observed characteristics Z and a random component U. In line with our present notation, we indicate the effects of the Z with the Greek letter γ, and, to avoid confusion, we add another index to the γ that indicates the specific β-parameter to which they refer:

$$\beta_{ki} = \gamma_{k0} + \gamma_{k1}z_{1i} + \cdots + \gamma_{kj}z_{ji} + u_{ki}$$

Similarly, the random component U also has an index k attached to it.

Now let us think about our former regression. Take as an example model (3.11), assume no time trend, and reorder the equation like this:

$$E(y_{it}) = (\beta_0 + \gamma_1 z_{1i} + \cdots + \gamma_j z_{ji}) + \beta_1 x_{1it} + \cdots + \beta_k x_{kit}$$

If you think about the term in brackets as an "intercept" that is modified depending on the (time-constant) characteristics of the unit

$$E(y_{it}) = \beta_{0i} + \beta_1 x_{1it} + \cdots + \beta_k x_{kit}$$

then you are not too far away from a model with random coefficients. If you change β_0 and the γs to our new notation and add a random error term u_{0i}, you have specified a regression model with a random intercept:

$$E(y_{it}) = (\gamma_{00} + \gamma_{01}z_{1i} + \cdots + \gamma_{0j}z_{ji} + u_{0i}) + \beta_1 x_{1it} + \cdots + \beta_k x_{kit}$$

> This model is easily extended to a model that also includes random slopes by specifying similar regression models for $\beta_{1i}, \ldots, \beta_{ki}$, while all the models in the previous sections focused on regression functions with fixed intercepts and fixed slopes.

What is new here, is obviously the fact that the model now consists of fixed and random parts. In order to understand this fully, let us look ahead to the next section, where we discuss how to estimate our models. In that case, we have to replace the expected with the observed values of the dependent variable and acknowledge the fact that observed values may deviate randomly from our expectations. We include an error term e_{it} that measures how each measurement y_{it} for unit i deviates from its expected value:

$$\ln(wage_{it}) = \underbrace{\gamma_{00} + \gamma_{01} \cdot schooling_i + \gamma_{10} \cdot t + \gamma_{11} \cdot ethnicity_i \cdot t}_{\text{fixed}} + \underbrace{u_{0i} + u_{1i} \cdot t + e_{it}}_{\text{random}} \quad (3.26)$$

Usually one assumes that these deviations are due to a random measurement error, which is zero on average. Hence, e_{it} is assumed to be a realization of a normally distributed random variable E with zero mean and variance σ_e^2. Equation (3.26) is similar to a linear regression model as is well known from the analysis of cross-sectional data. It includes, however, a slightly more complicated error term. Basically, we try to decompose the error variance into different components, i.e., into random variation of (i) the measurements (σ_e^2), (ii) of the intercepts ($\sigma_{u_0}^2$), and (iii) of the slopes ($\sigma_{u_1}^2$). Therefore, another name for this approach is *error* or *variance components models*. Note, also, that if we specify a separate model for the slopes, the random part of (3.26) is, by definition, heteroscedastic. In other words, it varies with the explanatory variables of the model (in this case t). We will come back to the problem of heteroscedastic error terms in Sect. 3.6.2.

3.6 Estimating Models for Panel Data

Having discussed how we can put our hypotheses into a formal model, we can now turn to the question of how to find "good" estimates of the model parameters. When estimating a model with empirical data, our first goal is to reveal the "true" parameters in the population that were "really" operating when the values of the dependent variable were observed. The problem is that we make inferences about the world (the "population") with limited data (the "sample") that are more or less reliable (they are only "indicators"), without really knowing the process (the "model") that generated our data. Hence, there are many possibilities to misspecify our models: selective samples, measurement errors, omitted variables, wrong functional form, and more besides.

3.6 Estimating Models for Panel Data

If we misspecify the model, our estimation procedures will provide us with wrong (technically: *biased*) estimates of the model parameters that systematically deviate from the "true" parameters in the population. Sometimes, estimation procedures can be fortified against certain kinds of specification errors, but it is certainly better to specify the model correctly. Our estimation procedures shall also provide us with measures of the precision of the parameter estimates. A measure of each parameter's precision is its *standard error*, and, in some cases, parameter estimates are unbiased, but their standard errors are not. In that case, we can make point estimates about the "true" model parameters, but confidence intervals and all test statistics will be wrong.

Unbiasedness means that our estimates *on average* (across all possible samples from the population) are equal to the "true" population parameters. Focusing on the average can only be one goal of our estimations, since each single sample estimate may deviate from the "true" parameter due to sampling error (and in social sciences, collecting just one sample is difficult enough). Therefore, we prefer estimation methods that provide us with estimates that are as close as possible to the "true" parameters, i.e., that have smaller standard errors than other (less *efficient*) estimation methods. Hence, we want our estimation methods to be both unbiased and efficient. When the estimates of the standard errors are biased due to specification errors, we obtain an incorrect measurement of how the parameter estimates vary around their "true" population values. As it turns out, we can often also increase the efficiency of our parameter estimates by improving the estimates of our standard errors. Sections 7.2.1 and 7.2.2 explain these criteria in greater detail for the two estimation methods (ordinary least squares and maximum likelihood) that are used in this textbook. Both sections also provide a review of standard errors and test procedures that are available with both methods.

This overview of methods for panel data analysis is not the right place to discuss specification problems at length, but we want to offer at least a flavor of the most important ones and describe briefly how panel data can be used to deal with them. We start with a discussion of omitted variable bias. It is closely connected to assumptions about the stochastic part of the model, which we have not yet dealt with. Section 3.6.1 introduces a simple specification for the error term that allows us to control for certain kinds of omitted variables. Section 3.6.2 extends these ideas to more general types of error structure. Section 3.6.3 discusses problems resulting from measurement error. Finally, Sect. 3.6.4 provides a formal summary of the basic assumptions typical for panel data analysis. Overall, our discussion focuses mostly on models for continuous variables, but, of course, similar considerations also apply to models for categorical variables, as we will show with some examples.

3.6.1 Omitted Variable Bias (Unobserved Heterogeneity)

3.6.1.1 What Is the Problem?
Students of multivariate statistics know that neglecting important explanatory factors can seriously bias the estimates of the effects of the independent variables in a regression model. This is easily demonstrated in the context of cross-sectional data.

Example 3.2 (hetbias and nohetbias data) Human capital theory assumes that worker productivity increases with education and labor force experience. Usually education and labor market experience are positively correlated, e.g., because less educated individuals have less job stability due to unemployment. To simulate these assumptions, we have generated a data set named hetbias, which includes $n = 200$ employees with four different levels of education and varying degrees of labor force experience. The dependent variable is the (log) hourly wage, and wages have been generated in such a way that they correlate positively with experience ($r = 0.87$) and education ($r = 0.91$). Education and experience are also positively correlated ($r = 0.75$). We will use this data set to see what happens to the estimate of the experience effect, if we do not control for education. Since the possible bias of the experience effect depends on the correlation between experience and education, we have generated a second data set named nohetbias, which is identical to hetbias except for the fact that experience and education are not correlated.

The left panel of Fig. 3.5 shows the relationship between labor force experience and log hourly wages in the hetbias data. The solid regression line running from the lower-left to the upper-right corner of the plot area measures the effect of labor force experience in a bivariate regression model of (log) hourly wages on experience. As human capital theory assumes, wages are higher for employees with more experience in the labor market. But this simple bivariate regression model neglects the effect of education on labor income, and hence raises the question of whether the estimated positive effect of labor force experience is the true one.

In Fig. 3.5, the educational level of each employee is indicated by a number between 1 (low education) and 4 (high education). As explained, the hetbias data have been generated in such a way that education and experience are positively correlated. For example, employees with a great deal of labor force experience mostly have advanced educational degrees (the marker "4" prevails in the upper-right corner of the figure), while employees with little labor force experience mostly have low-level educational degrees (the marker "1" prevails in the lower-left corner).

Now what happens if we control for different levels of school education? This necessitates a multiple regression model with "labor force experience" as the independent variable and three dummies for the educational levels of the employees (low education is used as the reference category). Graphically, this is identical to computing four parallel regression lines, one for each educational level. As can be seen from the dashed level-specific regression lines in the left panel, the association between labor force experience and hourly wages is still positive for each educational level, but much smaller than in the bivariate regression model (compare the solid regression line). Obviously, the former bivariate effect overestimates the "true" effect of labor force experience, because without controlling for educational level,

3.6 Estimating Models for Panel Data

Fig. 3.5 Log hourly wages by labor force experience and education

labor force experience will also transport the effect of education, since education is positively correlated with labor force experience.

You should also remember that the problem of omitted variable bias does not exist when the omitted variables are statistically independent of the explanatory variables in the model. This is shown with the nohetbias data in the right panel of Fig. 3.5. If education and experience were independent of each other, the four scatter clouds corresponding to the four educational levels would be totally parallel (their centers would be on the some vertical line) and there would be no difference between the slope of the overall and the level-specific regression lines (compare the solid overall and the dashed level-specific regression lines in the right panel of Fig. 3.5). Hence, the effect of labor force experience on hourly wages would be the same, irrespective of whether we control for education or not.

3.6.1.2 How to Extend Panel Regression Models for Unobservables

Controlling for such "third" factors is an easy task, if we have information about them (as in the case of education, which is available in many surveys). We simply put them as another independent variable into our regression model. But what about those factors that are hard to measure ("ability" is a prominent example in the context of income analysis) or those which we have not yet considered? Should we not make provisions for having forgotten important determinants of our dependent variable?

As it turns out, panel data are an excellent tool to deal with these "third" factors that are *not* included in the model. Before showing that, we respecify our panel model to make explicit that we may have neglected certain important explanatory variables. This applies to both our time-constant and our time-varying explanatory variables. Instead of writing them all down (we do not know what they are anyway), we add two error terms to our model measuring the overall impact of all the unknown time-constant and time-varying factors, which influence Y:

$$y_{it} = \beta_0(t) + \beta_1 x_{1it} + \cdots + \beta_k x_{kit} + e_{it} + \gamma_1 z_{1i} + \cdots + \gamma_j z_{ji} + u_i \quad (3.27)$$

$u_i = \gamma_{j+1} z_{j+1,i} + \cdots + \gamma_{j+l} z_{j+l,i}$ summarizes the effect of the l unknown time-constant explanatory variables. $e_{it} = \beta_{k+1} x_{k+1,it} + \cdots + \beta_{k+m} x_{k+m,it} + v_{it}$ stands for the m unknown time-varying explanatory variables plus all random error (v_{it}) that affects the dependent variable (including measurement error). u_i, the error pertaining to the unit i, is termed *unobserved heterogeneity*, because it captures all the variation at the unit level that is not controlled for by the independent variables in the model. e_{it}, the error pertaining to each single measurement, is termed *idiosyncratic error*, because it captures all peculiarities that affect the dependent variable at each point in time for each unit besides the effects that are already controlled for in the model.

Note that we have dropped the expected value function on the left-hand side of the equation, because (3.27) specifies not only the systematic part of the dependent variable but also the stochastic part ($u_i + e_{it}$), which we have ignored in Sect. 3.5. Note also that models for categorical data are easily extended in a similar fashion (simply insert the right-hand side of (3.27) into $G(\cdot)$ in (3.12)). However, contrary to the discussion in Sect. 3.5.3.3 where we used a similar notation, here we make no distributional assumptions about the two error terms u_i and e_{it}. We think about them as some unknown parameters that differ between units and measurements.

Some people may view the inclusion of u_i and e_{it} only as a notational amendment, since it does not really let us off the hook with regard to the missing information. But it allows us to define more concretely the conditions that are necessary to achieve unbiased estimates of the parameters in the model. Since all effects are biased if the variables in the model are correlated with the neglected factors (compare the example in Fig. 3.5), we *must* assume—at least at the present state—that both X and Z are independent of U and E to ensure unbiasedness of the parameter estimates. Variables X and Z that meet these criteria are called *exogenous* in the econometric literature. We will come back to the issue of exogeneity in Sect. 3.6.4, when we know a little bit more about the other specification problems.

3.6.1.3 Why Are Panel Data Useful to Control for Unobserved Heterogeneity?

Obviously, it is difficult to prove whether this assumption is true, because unobserved heterogeneity and idiosyncratic errors by definition are unknown. But as we will shortly see, panel data allow us to control for unobserved heterogeneity at the unit level, even when this source of error (the u_i term) is correlated with the explanatory variables in the model. More specifically, panel data include repeated observations of the same units, which gives us the opportunity to control for all (observed and unobserved) characteristics of each unit that are constant over time. The necessary statistical tools are not entirely new. In fact, you will have already encountered the fundamental techniques during your basic training in statistics.

3.6 Estimating Models for Panel Data

Let us start with the example on (log) hourly wages. How do we test whether these wages increased significantly from 1981 to 1982? The increase from 1.513 in 1981 to 1.572 in 1982 (see Table 3.1) could be due to random error. As you will probably remember, the significance of this increase can be checked with a *T test for dependent observations*. The corresponding test statistic uses the difference of both numbers and the standard error of the difference:

$$t = \frac{\bar{y}_{.,82} - \bar{y}_{.,81}}{\hat{\sigma}_{(\bar{y}_{.,82} - \bar{y}_{.,81})}} \qquad (3.28)$$

For the wage data, the standard error is estimated to be 0.019. With an increase in log hourly wages by 0.059 points, the result is a highly significant test statistic: $t = 3.105$ and (one-sided) $p < 0.01$. We conclude that the geometric mean of hourly wages increased significantly from 1981 to 1982 by a factor of $\exp(0.059) = 1.061$ or roughly 6 %.

Now, consider a simple panel regression model for both years, with one time-constant Z and one time-varying explanatory variable X and a linear time trend T. We drop all observations from the other years (i.e., make the data a two-wave panel study) and write down the regression equation for each of the two remaining years:

$$t = 1981 \quad y_{i,81} = \beta_0 + \beta_1 \cdot 1981 + \beta_2 x_{2i,81} + e_{i,81} + \gamma_1 z_{1i} + u_i \qquad (3.29)$$

$$t = 1982 \quad y_{i,82} = \beta_0 + \beta_1 \cdot 1982 + \beta_2 x_{2i,82} + e_{i,82} + \gamma_1 z_{1i} + u_i \qquad (3.30)$$

and then compute their difference

$$y_{i,82} - y_{i,81} = \Delta y_{i,82} = \beta_1 + \beta_2 \Delta x_{2i,82} + \Delta e_{i,82} \qquad (3.31)$$

Due to the differencing, the error term u_i and the time-constant variable z_{1i} drop out of the equation. Thus, by computing a regression on the differenced data, we eliminate time-constant heterogeneity, whether it is observed (z_{1i}) or unobserved (u_i). Consequently, we do not need to bother about potential correlations between unobserved heterogeneity (u_i) and any explanatory variables. The transformation is known as computing first differences (see Sect. 3.5.2.1) and estimating a model with differenced data is termed *first differences (FD) estimation*. It controls for possible omitted variable bias caused by u_i, and the estimated regression coefficients ($\hat{\beta}$) are unbiased estimates of the "true" effects of the time-varying explanatory variables in the model. Of course, some other assumptions also need to be true to ensure unbiasedness (see Sect. 3.6.4), but at the moment, these details are not important in order to understand the main idea behind FD estimation. Some people see it as a disadvantage that FD estimation excludes *observed* time-constant variables Z. But this objection is not a real problem in many applications. It is true that FD estimation provides no numerical estimates of the effects of time-constant variables. Nevertheless, like the effects of unobserved heterogeneity u_i, their effects on the dependent variable are controlled for (see also the discussion in Sect. 4.1.2.3).

If the model includes only a linear time trend and no explanatory variables, (3.31) reduces to $\Delta y_{i,82} = \beta_1 + \Delta e_{i,82}$ and $\hat{\beta}_1 = 0.0588$ will equal the former difference in log hourly wages (with a standard error of $\hat{\sigma}_{\beta_1} = 0.0195$). The corresponding t

statistic is identical to the test statistic of the former T test. In other words, the T test for dependent observations, which you know from elementary statistics, is a special case of one important estimation method (FD estimation) for panel data.

Note also that the nominator of the test statistic (3.28) is nothing other than the arithmetic mean of all unit-specific differences: $(\sum_{i=1}^{n} \Delta y_{i,82})/n = [\sum_{i=1}^{n}(y_{i,82} - y_{i,81})]/n$. Instead of looking at the differences between (log) hourly wages 1982 and (log) hourly wages 1981, we can also compute an average log hourly wage for each unit, $\bar{y}_{i.}$, and then test whether the individual wage in 1982 differs significantly from this average. Let us call $\ddot{y}_{i,82} = y_{i,82} - \bar{y}_{i.}$ the demeaned log hourly wage of unit i in 1982. Instead of using (3.28), we would perform a T test for the demeaned data. This is simply a test of whether the mean (log) hourly wage in 1982 differs significantly from the overall mean, $\bar{y}_{..}$, of all 1981 and 1982 wages:

$$t = \frac{\bar{y}_{.,82} - \bar{y}_{..}}{\hat{\sigma}_{(\bar{y}_{.,82} - \bar{y}_{..})}}. \quad (3.32)$$

This T test will have the same result as the former T test for dependent observations. This is no surprise, because in the case of two measurements (a two-wave panel), the demeaned data are a simple transformation of the former differences: $\ddot{y}_{i,82} = \Delta y_{i,82}/2$. Computing a regression on the demeaned data is identical to another estimation method for panel data, called *fixed effects (FE) estimation*, which controls for unobserved heterogeneity in a similar way to FD estimation. In case of two-wave panels, FD and FE estimation yield identical results and, as shown, both can be traced back to familiar T tests. In the case of more than two waves, both estimation methods can have different results, but this will be discussed in greater detail in Chap. 4.

Both panel estimation methods can also be motivated from the viewpoint of experimental designs. The minimal design of an experiment is characterized by a sample of individuals that are randomly assigned to a treatment and a control group. Significant differences in the outcome variable between both groups indicate whether the treatment has been effective or not. Because of random assignment, this finding is easily defended against the objection that some other variable may have caused the difference. Even if other determinants of the outcome variable exist, with randomization, they should be randomly distributed across both groups. However, in many social science applications, randomization is not feasible. For example, if a job training program by law is open to everybody, we cannot randomly assign individuals to this program just because we want to assess its effects, e.g., on earnings. In this case, researchers often choose a much simpler design, the so-called one group pre-test post-test design. It focuses only on the treatment group and takes measures of the outcome variable before and after the treatment. The reasoning behind this design is the following: Even if the members of the treatment group have specific characteristics, if the treatment is effective there should be a significant difference between the post-test and the pre-test measurement. In other words, by looking at differences within the treatment group, the (time-constant) selective characteristics of their members are controlled for. This is essentially the same as what our former FD estimation does.

3.6 Estimating Models for Panel Data

Table 3.9 Union membership status 1985 by membership status 1984

	1984	1985		Total
		Member	No member	
	Member	100	37	137
		73.0 %	27.0 %	25.1 %
	No member	22	386	408
		5.4 %	94.6 %	74.9 %
	Total	122	423	545
		22.4 %	77.6 %	100.0 %

Source: wagepan data (see Example 3.1)

Now, let us turn to our categorical variable "union membership," and see how we can control for unobserved heterogeneity in that case. Much like we did in the wage example, we could ask: Is there a significant drop in union membership between 1984 and 1985? According to the data in Table 3.2, there has been a slight decrease in union membership of 2.7 percentage points, from 25.1 % in 1984 to 22.4 % in 1985. But this difference could also have resulted from random error.

Although it is not often discussed in elementary statistics courses, there is a test procedure, called *McNemar's test*, to check the difference of two proportions which are computed from dependent observations. Basically, it is a chi-square test of the off-diagonal cells in the transition matrix, i.e., on those individuals that change status. Table 3.9 shows the frequencies of the particular transition from 1984 to 1985. According to these data, 27.0 % or 37 of the $n = 137$ members left the union in 1985 and 5.4 % or 22 of the $n = 408$ non-members joined the union. If the overall percentage of union members is the same in 1985 as it was in 1984, these two frequencies should also be the same except for random error. Hence, under the null hypothesis of no change in the marginal percentages, the (estimated) expected frequencies in the two off-diagonal cells of the transition matrix should be $\widehat{F}_{12} = \widehat{F}_{21} = (37 + 22)/2 = 29.5$. The fact that the number of exits ($f_{12} = 37$) is a little larger and that the number of entries ($f_{21} = 22$) is a little smaller indicates that the overall percentage of union members has decreased slightly between 1984 and 1985. Whether these differences are significant or not can be tested with a chi-square test by comparing the observed and the expected number of individuals in both groups. Estimated expected frequencies are denoted by $\widehat{F}$, and observed frequencies by f. The indices i and j denote the corresponding row and column of the transition matrix:

$$X^2 = \sum_{i=1}^{2} \sum_{j=1}^{2} \frac{(f_{ij} - \widehat{F}_{ij})^2}{\widehat{F}_{ij}} = \frac{(37 - 29.5)^2}{29.5} + \frac{(22 - 29.5)^2}{29.5} = 3.8136 \quad \text{with } i \neq j$$

(3.33)

This test has one degree of freedom, and, compared to a χ^2-distribution with $df = 1$, the test statistic is not significant at the 5 % level ($p = 0.0508$). We conclude that the change in the overall percentage of union members (2.7 percentage points) could also result from random error.

If we reformulate the test problem within the context of panel models, we simply specify a logistic regression model for the probability of union membership that includes a linear time trend and the familiar unit-specific error term, u_i, controlling for the dependent observations in the two-wave panel including 1984 and 1985.[15] If y_{it} is a dummy for membership status ($1 =$ member, $0 =$ no member), the model resembles the following equation:

$$\Pr(y_{it} = 1) = \frac{\exp(\beta_0 + \beta_1 t + u_i)}{1 + \exp(\beta_0 + \beta_1 t + u_i)} \quad (3.34)$$

For logistic regression models of this kind, there are similar estimation methods to those for continuous variables, which also control for unobserved heterogeneity u_i. They are called *conditional maximum likelihood estimation (CML)*, and we will discuss these methods in greater detail in Chap. 5. In the case of two-wave panels, they are similar to first differences. In our example, CML estimation uses only data from those individuals who change their membership status (i.e., individuals in the off-diagonal cells of the transition matrix).[16] It provides an estimate of the trend parameter which is negative ($\hat{\beta}_1 = -0.5199$), indicating that the probability of union membership has slightly decreased. However, if we take its estimated standard error into account ($\hat{\sigma}_{\beta_1} = 0.2692$) and perform a Z test, we see again that it is not statistically different from zero ($p = 0.053$). In Chap. 5, we will also show how to interpret the parameters of logistic regression models. Usually, one computes the anti-logarithm of the regression coefficient, in our case $\exp(\hat{\beta}_1) = \exp(-0.5199) = 0.5946$, and interprets it as the multiplicative change in the odds of union membership. In our case, this means that—controlling for unobserved heterogeneity—the odds of union membership have decreased by a factor of 0.5946 or roughly 41.5 %. This sounds like a lot, but taking into account the standard error of this odds ratio, the 95 % confidence interval (0.3508, 1.0078) includes the factor 1 (no change) indicating that there is no significant change. Interestingly, this odds ratio equals the ratio of entries into and exits from a union between 1984 and 1985 ($22/37 = 0.5946$). We see, again, that there is a close connection between simple techniques for dependent observations (McNemar's test) and more advanced methods of panel data analysis.

Before we proceed with a discussion of the two error terms, let us summarize the main conclusions of this section. At the beginning, we asked "Why are panel data useful to control for unobserved heterogeneity?" The answer is: Because panel data include repeated observations for each unit of analysis, and this allows us to base our estimations on the within-unit variation, which is unaffected by time-constant

---

[15] You may wonder why this model does not include an idiosyncratic error term. We want to keep the model as simple as possible and, thus, ignore e_{it}, because the dependent variable (a probability) is stochastic by definition (for more details see Sect. 5.1.2).

[16] Notice the similarities to first differencing. If membership status is indicated by a dummy variable ($1 =$ member, $0 =$ no member), differencing this variable will yield change scores of 1 (change from non-member to member), -1 (change from member to non-member), or 0 (no change). McNemar's test will use only those units with change scores $\Delta y_{i,85} \neq 0$.

characteristics of the units (both the known and the unknown ones). Each unit, so to speak, is used as its own control.[17]

We showed this for both continuous and categorical dependent variables Y with a model specified in *levels*, although the estimation was done using only *change* over time. Therefore, some purists think that panel regression models should be specified as change models from the very beginning. But it is important to separate problems of model specification from problems of parameter estimation. It is one question, whether a model should be specified in levels or in changes; and it is quite another, how to achieve unbiased estimates of its parameters and what kind of information to use for it.

3.6.2 Serially Correlated and Heteroscedastic Errors

The preceding discussion showed that unobserved heterogeneity, u_i, can be dealt with when we have access to panel data. But what about the idiosyncratic error e_{it}? A simple starting point is to treat e_{it} like the error term in regression models for cross-sectional data. In these kinds of models, it is usually assumed that the error e_{it} for each unit i is independent of the variables in the model (exogeneity assumption), has constant variance (homoscedasticity assumption), and is independent of the error influencing Y for any other unit $j \neq i$ (no autocorrelation assumption). If all these assumptions are met, ordinary least squares (OLS) estimation will provide unbiased standard errors and will be the most efficient estimation method (for more details see Textbox 4.1).

There is no easy way to eliminate e_{it}, as in the case of unobserved heterogeneity u_i (e.g., by using FD or FE estimation). Hence, the exogeneity assumption is crucial to ensure unbiasedness of the parameter estimates. If the other two assumptions are not met, parameter estimates are still unbiased, but standard errors are not, because their formulas are based on the assumption of homoscedasticity and no autocorrelation. That is why they are sometimes called *theoretical standard errors*. In this section, we want to discuss the latter two assumptions. Obviously, they are vital for unbiased standard errors and efficient estimation. We discuss, first, why they often make sense for cross-section data, and then consider why they present a problem for panel data.

Consider, for example, a cross-sectional survey on household incomes and the possibility of measurement error, which is part of the error term. Does the fact that Mr. Schulz underreports his income influence the probability that Mrs. Mayer, whom he does not know, underreports her income as well? No! Thus, it is a plausible starting point for cross-sectional data to assume that errors across units (respondents) are independent of one another (i.e., not autocorrelated). You may object, however, that underreporting income is typical for certain social groups (e.g., self-employed individuals), and that if Schulz and Mayer belong to the same group, it

---

[17] However, it is important to keep in mind that not all unknown determinants of Y are controlled for. Unknown determinants, e_{it}, that vary over time are still effective (see Sect. 3.6.2).

is nevertheless possible that the measurement error is similar for both respondents (and for all the other individuals in the same social group). But this only appears to be a problem of autocorrelation. When incomes for a certain group are systematically underestimated, it is rather a problem of bias (in this case, due to an omitted variable). You should control for the group characteristic by including a suitable indicator into your regression model (e.g., occupational status). Moreover, if you are not controlling for this variable (let us call it O), you have additional specification problems, if the variables in the model (X) correlate with variable O that predicts underreporting incomes. Necessarily, the variables in the model (X) will correlate with the error term (via the omitted variable O), and this violates the exogeneity assumption of independent error terms. All estimated effects of X will be biased due to an omitted variable.

All in all, the assumption of uncorrelated error terms is often a reasonable one for cross-sectional data, because the units of analysis are sampled independently of one another and rarely share any characteristics that cannot be controlled for.[18] However, the example about incomes from self-employment may be used to highlight the problems of heteroscedasticity. As you know, it is much more difficult for self-employed individuals to report exact monthly incomes, because they do not receive a regular monthly payment similar to, e.g., white collar workers. Hence, it is a plausible assumption that the incomes of the self-employed include more measurement error (i.e., have greater error variance) than incomes of other occupational groups with regular monthly payments. This would invalidate the homoscedasticity assumption of constant error variance across all units of analysis. Generally speaking, heteroscedasticity is more of a problem for cross-sectional data than autocorrelation, because—even when units in a cross-section are sampled *independently* of one another—the risk of observing greater error with respect to certain units remains.

The situation is less straightforward when analyzing panel data, because one observes the *same* units repeatedly over time. Generally speaking, sources of error (i.e., omitted variables and measurement error) that are constant over time are less of a problem than errors that change over time. For example, if it is true that self-employed individuals underreport their incomes, then it is likely that they do it to the same extent, more or less, in each consecutive panel wave. Because of these time-constant impacts on the dependent variable Y, measurements over time will be serially correlated.[19] This kind of (measurement) error is easily controlled for by the error term u_i, which measures all unobserved characteristics that are time-constant for the units. The estimation procedures that we discussed in the previous section (FD and FE estimation) will take care of all serial correlations in the data that are

---

[18] The no autocorrelation assumption may be at stake, however, when the cross-sectional data result from a cluster sample with units from the same cluster sharing similar and often hard-to-control characteristics.

[19] This is the basis of the intra-class correlation coefficient (3.5) that attributes all serial correlations in the data to time-constant characteristics of the higher-level units (see Textbox 3.1).

generated by such time-constant unobserved variables at the unit level (and, in doing so, will also take care of the omitted variable bias due to these variables).

Yet, if the source of the error changes over time, then it should be modeled with the error term e_{it}. Idiosyncratic errors, e_{it}, may be serially correlated as well. As an example, consider the economic situation of the companies that have employed the respondents included in the wagepan data (see Example 3.1). Certainly, the economic situation of these companies will have influenced what they were able to pay their employees. Information about the economic situation at the enterprise level is not easy to obtain for the social scientist and—since it often changes over time—is a possible candidate for an unknown time-varying variable that should be included in e_{it}. If the economic situation of some employers has been extraordinarily positive in one year, say t, then it will have resulted in above average incomes for some wagepan respondents, and this positive income effect—though at a falling rate— will have carried over to some of the following years $t+1, t+2, \ldots$. Hence, e_{it} and $e_{i,t+1}, e_{i,t+2}, \ldots$ will be correlated. Therefore, with panel data, it is the assumption of uncorrelated idiosyncratic error terms e_{it} that is at greater risk.

As mentioned in the beginning, OLS estimation is not efficient in the case of correlated error terms. This is also the case for FD and FE estimation, when the idiosyncratic errors e_{it} are correlated. Estimated standard errors will be wrong (mostly underestimated), resulting in wrong decisions in tests of significance (mostly in favor of the alternative hypothesis). In brief, correlated error terms increase the risk of reporting significant results when they do not exist. There are several solutions to the problem. One solution is to extend the model to account for correlated error terms. Another is to estimate so-called *robust standard errors* instead of theoretical standard errors (see Sect. 7.2.1 for a discussion of robust standard errors). Finally, generalized estimating equations (GEE) provide a flexible environment to deal with all kinds of correlation structures among the unobservables.

A similar problem arises if the variance of the error term, σ_e^2, is not constant. If the assumption of *homoscedastic* error terms is violated, OLS, FD, and FE estimation are inefficient too. Again, estimated standard errors are biased (but the direction of the bias is less clear than in the case of correlated error terms), and significance tests may produce incorrect conclusions. However, compared to correlated error terms, heteroscedasticity is not a specific problem for panel data. As already mentioned, it is also quite frequent in cross-sectional data. Heteroscedastic error terms are easily dealt with by using robust standard errors that control for heteroscedasticity. Of course, if you know the source of the misspecification that caused the heteroscedasticity, it is always a better strategy to revise the model itself. For example, the assumption of random slopes results in a model, where the error term is also a function of the variables in the model (see (3.26)). In this case, the source of the heteroscedasticity is known and can be explicitly modeled.

3.6.3 Measurement Error Bias

In the previous section, we discussed measurement error as a source of heteroscedasticity and correlated error terms. But measurement error can have additional negative effects to which you should pay attention, especially in the case of panel data. Unfortunately, measurement error is treated very poorly in many introductory econometrics textbooks. Many economists think that economic data include only few measurement errors. Furthermore, they argue that measurement error in the dependent variable (v_{it}) is already captured in the error term e_{it}. To learn more about the treatment of unreliable measurements, of both the dependent and the independent variables, you should have a look at more advanced econometrics textbooks or consult the methodological literature from psychology and sociology, where models controlling for measurement error have a long history.

An important distinction in this context is that between latent and manifest variables. Our theories focus on the "true" error-free relationships between variables that represent our theoretical constructs (e.g., human capital, productivity, union commitment). These are called *latent variables*. The possibly unreliable indicators of these theoretical constructs that are provided by social science data (e.g., years of education, earned income, union membership) are called *manifest variables*. A simple measurement model assumes that the values of a manifest variable, say Y, are a linear function of the underlying latent variable, Y^*, plus some random measurement error:

$$y_{it} = \lambda y_{it}^* + v_{it} \tag{3.35}$$

Equation (3.35) assumes that all variables have been standardized, and thus includes no regression constant. The standardized regression coefficient λ is also called the *reliability* of the manifest variable. In this bivariate regression model, it equals the correlation between the indicator (the manifest variable) and its underlying construct (the latent variable). In the case of categorical variables, measurement error can be modeled as a process of misclassification, in which only a certain percentage of each category of the latent variable is assigned correctly to the corresponding category of the manifest variable, while the rest is erroneously assigned to other categories. The categories of the latent variable are also called *latent classes*.

In the following, we want to illustrate some of the problems when analyzing unreliable panel data. We start with the seemingly simple case of an unreliable dependent variable, and then proceed to the more difficult case of an unreliable independent variable. Since we do not have the space in this textbook to treat measurement models for panel data more thoroughly, our examples should alert you to those situations in which the basic panel regression models of this textbook have to be extended.

3.6.3.1 Measurement Error in the Dependent Variable

Measurement error in the dependent variable is no problem for cross-sectional data, if it is independent of the variables in the model. In that case, it is indeed sufficiently treated in the error term of the model. But for panel data, it remains a problem. It can be shown that in dynamic models, estimates of true state dependence will be attenuated, i.e., biased towards zero. Instead of providing a formal proof, we will present a simple example with a categorical dependent variable.

Consider, again, the process of becoming a union member. Let us assume that the "true" change in membership status between two years, say 1980 and 1981, is measured correctly in the `wagepan` data. The corresponding transition probabilities

3.6 Estimating Models for Panel Data

Table 3.10 Union membership status 1981 by membership status 1980

True transition matrix

1980	1981	
	No member	Member
No member	88.97 %	11.03 %
Member	33.58 %	66.42 %

Transition with error

1980	1981	
	No member	Member
No member	84.21 %	15.79 %
Member	41.98 %	58.02 %

Source: wagepan data (see Example 3.1 observed with 5 % misclassification)

from Table 3.2 are reproduced in the upper part of Table 3.10. Hence, in the population, the probability of becoming a union member is assumed to be about 11 %, while the probability of leaving the union amounts to 33.6 %. Assume, furthermore, that we observe this process with a two-year panel survey, which, unfortunately, is less reliable than the survey from which we obtained the wagepan data. As a consequence, 5 % of the union members were classified erroneously as not belonging to a union and, similarly, 5 % of the non-members were recorded erroneously as union members (this is just an example; errors of classification could be different for members and non-members). With these assumptions, union membership reveals a much higher rate of membership turn-over between 1980 and 1981 than is the case in reality, simply because of the many misclassified members and non-members. More specifically, if the upper part of Table 3.10 describes the "true" turn-over process, and the probability of misclassification is 5 % (irrespective of origin status), we expect 15.8 % (and not 11.0 %) of the non-members to join a union in 1981, while 42.0 % (and not 33.6 %) of the members should leave their union in 1981 (see the lower part of Table 3.10).[20] In other words, the stability of union membership and non-membership is much lower than in reality, and corresponding regression models will underestimate an effect of true state dependence.

---

[20] The observed (erroneous) transition probabilities were computed by multiplying and adding the corresponding probabilities. For example, the respondents observed as union members in both years have one of the following characteristics (in brackets: probability of observing that characteristic): (i) being correctly classified as a member in both years ($p_{22} \cdot 0.95 \cdot 0.95$), (ii) being a member in 1980 and a non-member in 1981 who is misclassified in 1981 ($p_{21} \cdot 0.95 \cdot 0.05$), (iii) being a non-member in 1980 and a member in 1981 who is misclassified in 1980 ($p_{12} \cdot 0.05 \cdot 0.95$), or (iv) being a non-member in both years who is misclassified in both years ($p_{11} \cdot 0.05 \cdot 0.05$). The p_{jk} are the "true" (unconditional) probabilities (not the transition probabilities) of belonging to the respective group. By adding the four probabilities (i) to (iv), we arrive at the observed transition probability of 58.02 %.

Fig. 3.6 Path diagram of hourly wages

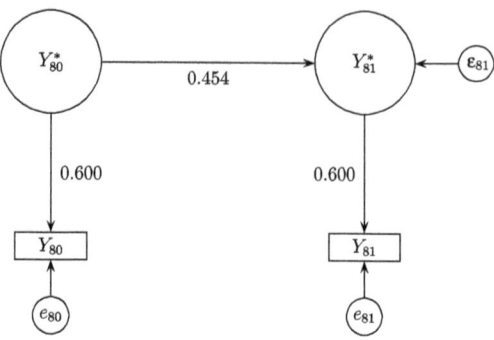

A similar example can be constructed for our (continuous) wage data. Assume that the first-order serial correlation for 1981 describes the "true" process at the level of the latent variable, i.e., Corr($y^*_{i,81}, y^*_{i,80}$) = 0.454 (see Table 3.1). Now, imagine again the less reliable panel survey. Let us assume that "true" wage can only be measured with reliability $\lambda = 0.6$. Figure 3.6 illustrates this example with a path diagram (a similar diagram could have been drawn for the union example). We adopt the convention that latent variables should be symbolized by circles, while manifest variables should be represented with rectangles. Additionally, the path diagram includes symbols for the error terms. In our case, the arrows indicate the true effect of last year's wage on the current wage and the dependence of the yearly measurements on the underlying latent variables (all other effects on current "true" wages and on the yearly measurements are captured by the error terms). Path analysis uses standardized regression coefficients to indicate the strength of the corresponding relationship. In a bivariate regression model, as in the case of regressing Y^*_{81} on Y^*_{80}, the standardized regression coefficient is identical to the correlation coefficient (thus, we have used Corr($y^*_{i,81}, y^*_{i,80}$)). Both regression models for the observed variables Y_{81} and Y_{80} are already in standard form (see (3.35)). With these assumptions, it can be shown that the serial correlation of the unreliable income measurements equals Corr($y_{i,81}, y_{i,80}$) = 0.6 · 0.454 · 0.6 = 0.163 (see Sect. 7.1). Hence, while the relationship is quite strong (0.454) in reality, it appears rather weak (0.163) with the (unreliable) observed data. Again, corresponding regression models will underestimate the effect of "true" state dependence and, similar to the union example, the cause of this bias is random noise introduced by the measurement process.

However, measurement error in the dependent variable is not only a problem for our substantive models. When you remember that specialized estimation procedures for panel data focus on changes over time (FD estimation) or on deviations from unit-specific means (FE estimation), you can imagine that it also affects the quality of various estimation methods for panel data. In the following chapters, we present the assumptions and features of these methods in greater detail. At that point, we will also discuss how FD and FE estimation are negatively affected by unreliable change scores and unreliable demeaned data.

3.6.3.2 Measurement Error in the Explanatory Variables

Measurement error in the explanatory variables is already a problem for cross-sectional data. Its effect is that regression coefficients of unreliably measured independent variables are biased towards zero (called *measurement attenuation bias* in the literature). The situation is identical to the one depicted in Fig. 3.6. Simply replace Y_{80}^* and Y_{80} by any explanatory variable from 1981, say X_{81}^* and X_{81}, and the diagram shows a bivariate cross-sectional regression model that is estimated with a cross-section from 1981. Naturally, if measurement error in the explanatory variables is a problem for cross-sectional data, it is just as much of a problem for panel data. Again, you can use Fig. 3.6 as an illustration. Replace Y_{81}^* and Y_{81} by Y_t^* and Y_t and also Y_{80}^* and Y_{80} by X_t^* and X_t. Now the path diagram represents a bivariate panel regression model that is estimated from $t = 1, \ldots, T$ waves of a panel survey.

3.6.3.3 Structural Equation Models

In sum, models for panel data are equally plagued by measurement error, and the distinction between latent and manifest variables provides a methodology for dealing with this problem. Instead of specifying a single regression equation, one would specify a *system* of regression equations representing both the process of measurement and the structural relations of the underlying theory (i.e., the relations between the latent variables). This methodology is known as *structural equation modeling* (SEM). If the underlying latent variables are assumed to be continuous, *factor analysis* is used for the measurement model. If both the latent and the manifest variables are assumed to be categorical, *latent class analysis* is used for the measurement model.

As already mentioned in Sect. 1.1.1.6, panel data provide valuable information to assess the reliability of the measurements (the manifest variables). With cross-sectional data, we are forced to use *different* indicators of the same construct in order to estimate each indicator's reliability (a typical example is an attitude measured by several items on a Likert scale). With panel data, on the other hand, we have repeated measurements of the *same* construct over time that allow us to assess its reliability. This is identical to test-retest reliability, one of the classical methods of reliability estimation. However, if we need some of the repeated observations to assess the reliability of the measurements, we have fewer opportunities to estimate the parameters of the underlying process. Hence, questions of identification are always prevalent with these kinds of measurement models.

To understand this statement, have a look again at Fig. 3.6. We made assumptions about the reliabilities and the effect Y_{80}^* on Y_{81}^* (the stability of the latent variable). Using these assumptions, we predicted the correlation (a standardized covariance) of the observed (manifest) variables Y_{80} on Y_{81}. Since we assumed standardized data, we implicitly made an assumption about their variances too ($\text{Var}(y_{80}) = \text{Var}(y_{81}) = 1$). In other words: Given the assumptions specified in the path diagram, we can derive a *model-implied* variance-covariance matrix of the observed variables, which in this simple example includes three numbers (one covariance and two variances).

Now let us think about the more realistic case in which we have an *observed* variance-covariance matrix and from these "data" want to estimate the parameters of the path diagram (i.e., the model that describes the population). How many parameters are there?—One stability coefficient, two reliabilities, and two variances of the latent variables, altogether five parameters. Trying to estimate five unknowns from only three data is not feasible. Even when we assume both reliabilities to be the same and hence, have only four different parameters to estimate, this is not possible either. In other words, the path model is not identified given the few data. However, if we would have three measurements and altogether six data (three covariances and three variances), the corresponding three-wave path diagram would be identified, at least when we assume that the (now) two stability coefficients are the same, as are the three reliabilities (resulting—together with the three variances—in five parameters to be estimated).

Although we do not have the space to explain how this estimation is actually done, the former reasoning already shows the principle. Given the assumptions specified in the path diagram, one can derive the variances and covariances of the manifest variables implied by the model. By minimizing the deviations between the model-implied and the observed variance-covariance matrix, one can estimate the regression coefficients that most likely have generated the statistical relationships between the observed variables. As the term "likely" suggests, estimation is done using maximum likelihood (ML).

3.6.4 A Formal Summary of the Main Estimation Assumptions

At the end of this discussion of specification errors, it is helpful to summarize the main conclusions. We have discussed various assumptions that are essential to obtaining "good" estimates of our model parameters and their standard errors, and we have illustrated—mostly by examples from elementary statistics—how panel data can be used to relax some of these assumptions. In order to obtain unbiased and efficient parameter estimates, we have to assume the following.

1. *Exogeneity.* This means that our explanatory variables are independent of all the factors that we cannot control, either because they are unknown to us or because we have no data concerning them. This includes unknown factors at the unit level (u_i), as well as unknown factors at the level of measurements (e_{it}). Furthermore, unobserved heterogeneity (u_i) and idiosyncratic errors (e_{it}) should be independent of one another. More specifically, we assume:

$$E(u_i | x_{1i1}, \ldots, x_{1iT}, \ldots, x_{ki1}, \ldots, x_{kiT}, z_{1i}, \ldots, z_{ji}, e_{it}) = 0$$
$$E(e_{it} | x_{1i1}, \ldots, x_{1iT}, \ldots, x_{ki1}, \ldots, x_{kiT}, z_{1i}, \ldots, z_{ji}, u_i) = 0$$
(A.1)

2. *Homoscedasticity.* This means that both error terms, u_i and e_{it}, have constant variance. In formal terms:

$$\text{Var}(u_i | x_{1i1}, \ldots, x_{1iT}, \ldots, x_{ki1}, \ldots, x_{kiT}, z_{1i}, \ldots, z_{ji}) = \sigma_u^2$$
$$\text{Var}(e_{it} | x_{1i1}, \ldots, x_{1iT}, \ldots, x_{ki1}, \ldots, x_{kiT}, z_{1i}, \ldots, z_{ji}) = \sigma_e^2$$
(A.2)

3.6 Estimating Models for Panel Data

3. *No serial correlation.* This means that the (time-varying) idiosyncratic errors are independent of one another. In formal terms:

$$\text{Corr}(e_{it}, e_{is} | x_{1i1}, \ldots, x_{1iT}, \ldots, x_{ki1}, \ldots, x_{kiT}, z_{1i}, \ldots, z_{ji}) = 0, \quad t \neq s \quad (A.3)$$

4. *No measurement error.* This means that the observed values are identical to the true values of Y, X and Z:

$$y_{it} = y_{it}^*, \quad x_{1it} = x_{1it}^*, \quad \ldots, \quad x_{kit} = x_{kit}^*, \quad z_{1i} = z_{1i}^*, \quad \ldots, \quad z_{ji} = z_{ji}^* \quad (A.4)$$

Furthermore, there are some other, more technical assumptions, which will be discussed in the following chapters.

Among the four assumptions, the most important one concerns the exogeneity, because it ensures unbiasedness of the parameter estimates (besides the fourth assumption of perfectly reliable measurements). What (A.1) says is that each error term is zero on average, once we control for all the other variables in the model.[21] Homoscedastic and uncorrelated error terms (assumptions (A.2) and (A.3)) guarantee efficient estimates and unbiased standard errors.

Why does it make sense to specify all the assumptions in terms of conditional expected values, variances, and correlations? As you may remember from our former discussion, biased estimates, heteroscedastic error terms, and serially correlated errors may result from misspecifications of the model. The effect of labor force experience was overestimated, when we did not control for education. Serially correlated errors in an analysis of labor income may have resulted from temporary changes in the employers' business situation. These problems disappear if we specify our models correctly (i.e., if we include an education variable or an indicator of the business cycle at the firm level). Hence, our assumptions are conditional on the model we are estimating.

You may also wonder, why these conditions mention all the T measurements of each time-varying explanatory variable (e.g., $x_{1it}, \ldots, x_{1iT}$). This is because we are dealing with panel data. With one cross-section of data, it is not necessary to consider what happens in other time periods. However, with panel data, past and future values of the explanatory variables can affect the dependent variable and, thus, the error terms. This more comprehensive definition of (A.1) is also known as the assumption of *strict exogeneity* (as opposed to contemporaneous exogeneity in case of cross-section data). Strict exogeneity is at stake when we are analyzing models with lagged dependent variables and feedback processes (non-recursive models).

Finally, as we have seen in Sect. 3.6.1, the assumption concerning u_i in (A.1) is not a real problem with panel data, because repeated observations of the same units over time allow us to control for unobserved heterogeneity, u_i, even if it is correlated

---

[21] Using the expected value function is a more general form of saying that the error terms and the variables in the model are not related. $\text{Corr}(x, u) = 0$, for example, would only imply that they are not *linearly* related.

with the variables in the model. This is one of the great advantages of panel data over cross-sectional data. We have learned that there are special estimation methods for this problem (FD, FE) and in the following chapters we will study these methods in greater detail. Furthermore, if both parts of assumption (A.1) hold, panel data give us the opportunity to estimate the effects of the explanatory variables much more efficiently than with cross-sectional data. However, it is interesting to consider what happens with first differences, fixed effects, and other panel estimation methods, when assumptions (A.2)–(A.4) do not hold. This discussion is deferred to subsequent chapters.

3.7 Overview of Subsequent Chapters

In this chapter, we showed how to describe and model panel data. Throughout the text, we draw the distinction between continuous and categorical dependent variables and indicated different strategies for analyzing these kinds of data. We also introduced possible estimation methods for the panel regression models. All of them make use of repeated observations for each unit of analysis due to the panel design. We motivated these techniques by referring to simple statistical tests for dependent observations you hopefully remember from your undergraduate courses in statistics. Finally, we discussed possible specification errors, among them biased estimates due to omitted variables, which can be nicely controlled with panel data. The following two chapters discuss panel regression models in greater detail. Chapter 4 focuses on the analysis of continuous dependent variables, and Chap. 5 on the analysis of categorical dependent variables.

Both chapters can be read independently of each other. They start with simple cross-sectional regression models applied to panel data, which are then fortified against the hierarchical clustering of data by computing robust standard errors. After that, we will introduce models analyzing the level of the dependent variable. These models account for the specific nature of panel data, and, in doing so, they try to improve both the estimates and the standard errors. Finally, both chapters end with models analyzing the change in the dependent variable. For continuous variables this discussion will be rather short, while for categorical variables it is necessary to introduce a whole new methodology (event history analysis).

Because this is an introductory textbook, we will *not* focus on systems of regression equations (which are used to model non-recursive relationships $X \leftrightarrow Y$), and we will not discuss models that distinguish between latent and manifest variables (which are used to model measurement error). However, the following chapters will give you references to the more advanced literature that discusses these kinds of models. Throughout this text, we will also ignore the fact that panel data provide only an incomplete picture of what is happening in reality. First of all, the units of analysis are observed at discrete points in time, although change happens continuously. Second, they enter the panel at a certain point in time, but the process of interest may have already begun before the start of the panel. Only in Chap. 5, when we discuss (event history) models for the change of categorical dependent variables, we will take up the issue of censored observations and models in discrete and continuous time. But it should be stressed that problems of incomplete observations and discrete- versus continuous-time modeling are challenges for all types of model discussed in this textbook.

Panel Analysis of Continuous Variables

4

This chapter deals with linear models for continuous dependent variables Y. In the first part of this chapter, we will discuss models focusing on the level of Y. As discussed in Chap. 3, models for the level of continuous variables focus on the expected value of Y. More specifically, in Sect. 3.5.1.1, we introduced the following regression function, in which $E(y_{it})$ is regressed on a set of independent variables, which may be either time-constant (Z) or time-varying (X):

$$E(y_{it}) = \beta_0(t) + \beta_1 x_{1it} + \cdots + \beta_k x_{kit} + \gamma_1 z_{1i} + \cdots + \gamma_j z_{ji} \quad (4.1)$$

Subscript i refers to the $i = 1, \ldots, n$ units, which have been observed at $t = 1, \ldots, T$ equidistant points in time. Typical continuous variables would be measures of individual earnings and attitudes, capital investments of industrial enterprises, or data concerning government spending. As mentioned in Chap. 3, the units can be individuals, firms, nations, or other objects of analysis. Most of the following examples will focus on individuals, and in various places we will use the terms "units" and "individuals" interchangeably. In (4.1), y_{it} denotes the value of the continuous dependent variable for unit i at time point t. It is modeled as a linear-additive function of the values of j time-constant $z_{1i}, \ldots, z_{ji}$ and k time-varying independent variables $x_{1it}, \ldots, x_{kit}$. $\gamma_1, \ldots, \gamma_j$ and $\beta_1, \ldots, \beta_k$ denote the corresponding regression coefficients. The term $\beta_0(t)$ determines the overall level of the dependent variable. Since its level may change over time, $\beta_0(t)$ can be any function of time to control for possible time trends. If there is no time trend, it reduces to the familiar regression constant $\beta_0(t) = \beta_0$. In the more general case, $\beta_0(t)$ can be a linear, quadratic, exponential or non-parametric function of time t (see Sect. 3.5.1.1). This model is the *classical linear regression model* and is extensively discussed in Sect. 4.1. Returning to the examples above, typical research questions would be: Do certain employees earn higher wages than others? How does the amount of capital investment vary with the characteristics of firms? Does globalization influence the amount of government spending?

In the second part of this chapter, we will focus on linear models for the change of Y. In Chap. 3, we discussed different ways in which to model ΔY (see Sect. 3.5.2.1). We will begin our discussion with models that focus on the change in Y between $t-1$ and t as a function of changing time-varying explanatory variables X during the same period:

$$E(y_{it} - y_{i,t-1}) = E(\Delta y_{it}) = \beta_0(t) - \beta_0(t-1) + \beta_1 \Delta x_{1it} + \cdots + \beta_k \Delta x_{kit} \quad (4.2)$$

This specification of change is termed *first differences (FD) regression*, because it can be derived by looking at the differences at the level of Y at time points t and $t-1$ (see Sect. 3.5.2.1). The FD regression model and other specifications of change are extensively discussed in Sect. 4.2. Typical research questions would be: How do household incomes change after certain life events (e.g., divorce)? How do capital investments at the firm level increase with overall economic growth? Does government spending decrease with increasing globalization? The chapter concludes with some suggestions for further reading (Sect. 4.3).

4.1 Modeling the Level of Y

This section illustrates how to estimate linear regression models that focus on the level of the dependent variable. As an illustrative example, we use the wagepan data introduced in Sect. 3.2 (see Example 3.1). In their analysis of these data, Vella and Verbeek (1998) use the natural logarithm of hourly wages as the dependent variable. The authors try to estimate the wage premium that union members may gain, controlling for the fact that union membership itself may be a purposive decision. This is a much more complicated research question than we are able to solve at the moment. But if we ignore for the moment the potential endogeneity[1] of union membership, we can use Vella and Verbeek's wage equation to illustrate panel models for continuous dependent variables. The question of whether union members earn higher wages than non-members on average is a typical analysis of the levels of Y. Vella and Verbeek's wage equation includes the usual variables from human capital theory; namely, years of schooling and labor force experience (a linear and a quadratic term). Additionally, it controls for ethnicity, health and family status, place of residence, and characteristics of current job (occupation, industry). The main variable of interest is a dummy for union membership. Most of these variables are time-varying, except ethnicity and years of schooling (see the descriptive analysis in Sect. 3.4.1).

We begin our analysis with classical cross-sectional models, and ask whether they can easily be applied to panel data (Sect. 4.1.1). An obvious choice is to pool the observations from all panel waves and analyze them with ordinary least squares (OLS), as if they came from one large cross-section (Sect. 4.1.1.1). However, as mentioned in the previous chapter (Sect. 3.6.2), standard errors are underestimated if the data do not include independent observations. As a remedy, we will introduce

---

[1] An independent variable is called endogenous, when it is not only an explanatory variable of Y, but at the same time causally determined by Y itself. For the wagepan data, one could hypothesize that the explanatory variable union is also determined by the dependent variable lwage, because higher income groups may be less inclined to become union members.

4.1 Modeling the Level of Y

robust standard errors that control for the hierarchical (observations within units) nature of the data (Sect. 4.1.1.2). In doing so, we arrive at more conservative test results of our parameter estimates, reflecting the fact that we do not have as many independent data as the $N = n \cdot T$ panel observations suggest.

Nevertheless, parameter estimates can also be improved upon, if we are willing to model the panel structure of the data. This is the task of Sect. 4.1.2 and it is based on the assumption that the OLS error term consists of two independent components: one operating at the level of units, and the other operating at the level of individual measurements over time. This assumption has two advantages. Firstly, by using estimation methods that take advantage of the specific features of panel data, we are able to control for omitted variable bias at the unit level, something which is not possible with cross-section data (see Sect. 3.6.1). Secondly, such models allow us to compute more efficient estimates of the parameters and their standard errors than in pooled OLS regression models, even with robust standard errors.

4.1.1 Ignoring the Panel Structure

4.1.1.1 Pooled Ordinary Least Squares

A first step in the analysis of the wagepan data could be to pool the information from all $t = 1, \ldots, 8$ panel waves for all $i = 1, \ldots, 545$ individuals and treat them as though they represented independent information for $n = 8 \cdot 545 = 4{,}360$ individuals. To make the following explanation as simple as possible, we ignore possible time trends in the data and specify only a regression constant $\beta_0(t) = \beta_0$.[2] In order to estimate the parameters of the model, we have to make an assumption regarding how each observed value y_{it} relates to its expected value $E(y_{it})$:

$$y_{it} = E(y_{it}) + \varepsilon_{it} = \beta_0 + \beta_1 x_{1it} + \cdots + \beta_k x_{kit} + \gamma_1 z_{1i} + \cdots + \gamma_j z_{ji} + \varepsilon_{it} \quad (4.3)$$

Assuming that the residual ε_{it} behaves like the OLS error term (see the following Textbox 4.1), we can apply OLS to the pooled data and arrive at the same estimates that Vella and Verbeek (1998) present in their Table III. Table 4.1 shows these estimates in the column named "pooled OLS". According to these data, union membership—ceteris paribus—increases hourly wages by a factor of 1.1566 ($= \exp(0.1455)$) or roughly 16 %. The effect is highly significant ($t = 8.63$, $p < 0.01$).

OLS is easily applied to cross-section data. If certain assumptions are met, statistical theory shows that, in this case, OLS estimates are unbiased and efficient. But what happens if we apply this technique to data that are, in fact, panel data? In order to understand the potential problems, it is helpful to review the main OLS assumptions (see Textbox 4.1).

---

[2] It should be stressed, however, that the assumption of $\beta_0(t) = \beta_0$ is only made for ease of exposition. Since most y trend over time, it is usually necessary to model the time trend. Otherwise, the estimates of the other explanatory variables are biased. The analysis of trends is deferred to Sect. 4.2.

Table 4.1 Determinants of log hourly wages: Pooled OLS (1980–1987)

Variable	Pooled OLS		Robust pooled OLS	
	Estimate	Std. Err.	Estimate	Std. Err.
Union membership (yes = 1)	0.1455	0.0169	0.1455	0.0263
Education (years)	0.0905	0.0046	0.0905	0.0086
Experience (years)	0.0759	0.0097	0.0759	0.0114
Experience squared	−0.0022	0.0007	−0.0022	0.0008
Hispanic (yes = 1)	−0.0585	0.0219	−0.0585	0.0404
Afro-American (yes = 1)	−0.1545	0.0230	−0.1545	0.0465
Lives in rural area (yes = 1)	−0.1314	0.0185	−0.1314	0.0316
Married (yes = 1)	0.1100	0.0153	0.1100	0.0241
Poor health (yes = 1)	−0.0580	0.0539	−0.0580	0.0670
Lives in North-East (yes = 1)	0.0197	0.0233	0.0197	0.0412
Lives in South (yes = 1)	−0.0784	0.0210	−0.0784	0.0389
Lives in North-Center (yes = 1)	−0.1057	0.0226	−0.1057	0.0400
Constant	0.2237	0.0780	0.2237	0.1345
R^2	0.2636		0.2636	
$F(X)$	67.49		34.34	
df_1, df_2	23	4,336	23	544
N	4,360		4,360	
n			545	
T			8	

Note: Models control for industry (11 dummies, regression coefficients not shown)
Source: wagepan data (see Example 3.1)

Textbox 4.1 (OLS assumptions) Consider the following linear model for a cross-section of $i = 1, \ldots, n$ units: $y_i = \beta_0 + \beta_1 x_{1i} + \cdots + \beta_k x_{ki} + \varepsilon_i$. y_i is a continuous dependent variable and $x_{1i}, \ldots, x_{ki}$ can be either continuous or categorical explanatory variables. ε_i captures measurement error in the dependent variable and all unknown explanatory variables of Y that have not been controlled for in the model. OLS estimates of the parameters $\beta_0, \beta_1, \ldots, \beta_k$ are found by minimizing the sum of squared residuals $\sum_i (y_i - \hat{y}_i)^2$. The residual is defined as the difference between observed (y_i) and predicted values ($\hat{y}_i = \hat{\beta}_0 + \hat{\beta}_1 x_{1i} + \cdots + \hat{\beta}_k x_{ki}$) of the dependent variable. The statistical properties of OLS estimation rest on the following assumptions:
1. The data are a simple random sample of a well-defined population.
2. The model is linear in its parameters $\beta_0, \beta_1, \ldots, \beta_k$.
3. Each explanatory variable is neither a constant nor a linear function of the other explanatory variables.
4. The error term is independent of the variables in the model: $E(\varepsilon_i | x_{1i}, x_{2i}, \ldots, x_{ki}) = 0$.

4.1 Modeling the Level of Y

5. The error has constant variance given any value of the explanatory variables: $\text{Var}(\varepsilon_i | x_{1i}, x_{2i}, \ldots, x_{ki}) = \sigma^2$.
6. The error is uncorrelated between any two units i and j ($i \neq j$), given any value of the explanatory variables: $\text{Corr}(\varepsilon_i, \varepsilon_j | x_{1i}, x_{2i}, \ldots, x_{ki}) = 0$.
7. The error is normally distributed with mean 0 and variance σ^2: $\varepsilon_i \sim \text{Normal}(0, \sigma^2)$.

The first assumption is necessary for making statistical inferences about the population using the sampled data. The second assumption allows us to use least squares estimation. If the model were not linear in its parameters, we would have to use other estimation techniques. The third assumption guarantees that a numerical value exists for each regression coefficient (technically, the model is identified). The fourth assumption is the most important one, because it makes sure that our regression estimates $\hat{\beta}_0, \hat{\beta}_1, \ldots, \hat{\beta}_k$ are *unbiased*: $E(\hat{\beta}_0) = \beta_0$, $E(\hat{\beta}_1) = \beta_1$, ..., $E(\hat{\beta}_k) = \beta_k$. Literally speaking, unbiasedness means that our estimates are correct; not necessarily in a single sample, but on average; i.e., they are identical, across a multitude of samples, to the "true" population parameters $\beta_0, \beta_1, \ldots, \beta_k$. If either measurement error, or the unknown explanatory variables of Y, correlates with the explanatory variables $x_{1i}, x_{2i}, \ldots, x_{ki}$ in the model, assumption 4 does not hold and OLS estimates will be biased. If assumptions 5 and 6 are true, OLS estimates are not only identical on average to the true population parameters, but the variance of the estimates across different samples is also smaller than for any other estimation method. This characteristic is known as *efficiency*. In sum, if assumptions 4 to 6 are true, OLS estimates are the best linear unbiased estimates (BLUE) of $\beta_0, \beta_1, \ldots, \beta_k$. Finally, assumption 7 guarantees that even in small samples, we can use standard test procedures and confidence intervals based on the normal distribution. If we have a large sample, this assumption is not necessary. In this case, normality can be assumed by referring to the central limit theorem. Discussions of these OLS assumptions and proofs of the statistical properties of OLS can be found in any modern textbook on regression analysis.

With panel data, at least one of these assumptions (assumption 6: no serial correlation) is at stake. This can be illustrated by estimating the residuals $\hat{\varepsilon}_{it}$ from the former (pooled OLS) regression model:

$$\hat{\varepsilon}_{it} = \ln(wage_{it}) - (0.2237 + 0.1455 \cdot union_{it}$$
$$+ 0.0905 \cdot educ_i + \cdots - 0.1617 \cdot industry11_{it}) \qquad (4.4)$$

If we now correlate the residuals from the first time point ($t = 1$) with the residuals from the second time point ($t = 2$) for all $i = 1, \ldots, 545$ individuals, we obtain a significant positive correlation: $\text{Corr}(\hat{\varepsilon}_{i1}, \hat{\varepsilon}_{i2}) = 0.3374$. The same is true for all other pairs of time points (see the upper part of Table 4.2).

Table 4.2 Serial correlations in pooled OLS model and in raw data

Year	1980	1981	1982	1983	1984	1985	1986	1987
	Correlation of residuals from the pooled OLS model in Table 4.1							
1980	1.0000							
1981	0.3374	1.0000						
1982	0.3198	0.4689	1.0000					
1983	0.2714	0.4469	0.5701	1.0000				
1984	0.1643	0.3751	0.4939	0.5413	1.0000			
1985	0.2187	0.3573	0.4526	0.4925	0.5268	1.0000		
1986	0.1839	0.2597	0.3815	0.3853	0.4224	0.5083	1.0000	
1987	0.1882	0.3586	0.3429	0.4100	0.4549	0.5475	0.5573	1.0000
Mean	$\bar{r}_{OLS} = 0.3942$							

Year	1980	1981	1982	1983	1984	1985	1986	1987
	Correlation of residuals from an empty pooled OLS model							
1980	1.000							
1981	0.454	1.000						
1982	0.432	0.611	1.000					
1983	0.408	0.582	0.690	1.000				
1984	0.316	0.506	0.626	0.675	1.000			
1985	0.356	0.469	0.588	0.625	0.664	1.000		
1986	0.297	0.407	0.523	0.549	0.565	0.632	1.000	
1987	0.310	0.480	0.498	0.563	0.588	0.672	0.693	1.000
Mean	$\bar{r}_y = 0.5277$							

Source: wagepan data (see Example 3.1)

What can explain this pattern? We have already discussed the problem of dependent observations in the preceding chapter. In Sect. 3.4.2, we mentioned three causes of serial correlation of the dependent variable: (i) time-constant explanatory variables that cause Y to be persistently above (or below) the average, (ii) serially correlated time-varying explanatory variables, and (iii) true state dependence of the dependent variable itself. In Sect. 3.2, we showed that the amount of serial correlation in the raw wagepan data is quite high (on average, $\bar{r}_y = 0.5277$, see Table 3.1). In order to understand how our pooled OLS model controls for this serial dependence in the data, let us replicate the serial correlation coefficients of the raw data with a simple regression model. This is easily done by specifying only a regression constant and no explanatory variables in our pooled OLS model (the so-called empty model). The regression constant is estimated as $\hat{\beta}_0 = 1.6491$ and equals the overall average of (log) hourly wages across units and time (see Table 3.6). Computing and correlating the residuals of this simple model reproduces the former correlation matrix (see the lower part of Table 4.2, and compare Table 3.1).

Now, if we compare the lower and upper part of Table 4.2, we see that the pooled OLS estimates from Vella and Verbeek control for some of the causes of serial cor-

4.1 Modeling the Level of Y

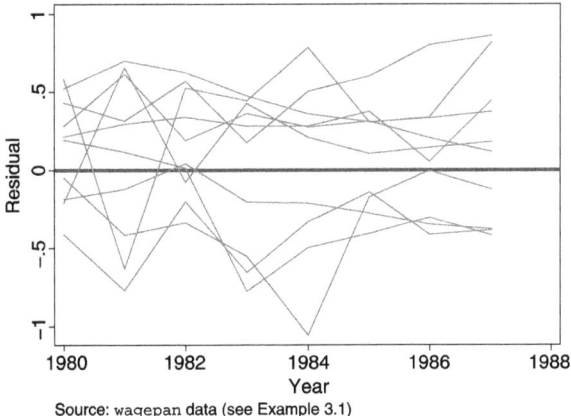

Fig. 4.1 Residuals (connected by *lines*) for each of ten individuals

Source: wagepan data (see Example 3.1)

relation. More specifically, they control for two time-constant explanatory variables; namely, ethnicity and years of schooling, and for several time-varying explanatory variables. Consequently, the remaining serial correlation in this more refined model is much lower (on average, $\bar{r}_{OLS} = 0.3942$; see Table 4.2). But still, it is not zero as the OLS assumptions require.

The remaining serial correlation is due to explanatory variables that are not included in Vella and Verbeek's model, either because there are no data about them or because they are theoretically unknown. Worker productivity is a typical example of a theoretically important but hard-to-measure explanatory factor. Economic theory postulates that more productive employees receive higher wages on average than less productive employees. Of course, productivity should be correlated with schooling, but perhaps not perfectly. In other words, there are individual characteristics (such as productivity) that are not controlled for by the independent variables in the model. If they are constant over time, this unobserved heterogeneity causes some individuals to have disproportionately higher (or lower) wages in *all* years than could be expected from the independent variables in the model. This situation is illustrated in Fig. 4.1 for ten individuals from the wagepan dataset. The lines connect the residuals from the pooled OLS model for each individual across the eight time points. In most cases, each line runs either consistently above or consistently below zero, indicating that there is something specific (and time-constant) about each individual that has not been accounted for by the independent variables in the model. Hence, if we correlate these residuals, they will show positive association.[3]

4.1.1.2 Robust Standard Errors

Pooled OLS is only unbiased, if we are ready to assume that this unobserved heterogeneity (e.g., differences with respect to productivity) is independent of the ex-

---

[3] It should be noted that, in most cases, serial correlation is positive (see also footnote 5 on p. 69). Negative serial correlation would imply that negative residuals, at one point in time, are associated with mostly positive residuals at the next point in time (and vice versa). Such an oscillatory pattern of residuals is hardly observed with panel data.

planatory variables in the model (see assumption 4 in Textbox 4.1). However, even if assumption 4 is true, estimated standard errors are biased. This is because assumption 6 (no serial correlation) is most likely violated. Error terms at different time points will certainly correlate with one another, if there is unobserved unit-specific heterogeneity that is constant over time, even when it is uncorrelated with the variables in the model (see Sect. 3.6.2). A simple remedy for correlated error terms is to compute robust standard errors. They are called *empirical standard errors*, because they use the distribution of estimated residuals instead of the theoretical formulas derived from the classical OLS assumptions (for a detailed discussion of empirical and theoretical standard errors, see Sect. 7.2.1). In our case, we need standard errors that are robust against the fact that each unit i is represented with T observations in the data set, which therefore have something in common. Each of these T measurements constitutes a *cluster* of observations and the corresponding standard errors have to be robust against this clustering in the data. More precisely, cluster-robust standard errors assume that observations are independent across clusters (i.e., units), but not necessarily within clusters.

Table 4.1 shows the pooled OLS estimates using cluster-robust standard errors. Estimated regression coefficients remain the same, but standard errors are larger than in the former pooled OLS model, indicating that neglecting the serial dependence in panel data provides seemingly precise estimates and often seemingly significant test statistics.[4] The effect of union membership remains significant nonetheless ($t = 5.53$, $p < 0.01$). But, for example, the effect of Hispanic ethnicity is no more significant (pooled OLS: $t = -2.67$, $p < 0.01$; robust pooled OLS: $t = -1.45$, $p = 0.147$).

It should be stressed that computing robust standard errors does not change the estimates of the regression coefficients. Hence, unbiasedness of robust pooled OLS estimates hinges on the same assumptions as "simple" pooled OLS estimates, among them the assumption of independent unobserved heterogeneity. Moreover, cluster-robust standard errors control for any kind of serial dependence within clusters. They are, so to speak, a kind of broad-spectrum antibiotic. But if we have a specific model of the source of serial correlation, we can develop more refined estimation techniques (a focused antibiotic) that provide more efficient estimates of both the regression coefficients and the standard errors. This is the topic of the following section.

4.1.2 Modeling the Panel Structure

Apparently, pooled OLS makes unrealistic assumptions about panel data. However, as shown in the previous chapter (see Sect. 3.6.1.2), the model is easily extended to account for unobserved heterogeneity at the unit level:

---

[4]Since we are using cluster-robust standard errors, the degrees of freedom, $df_2 = n - 1$, of the overall F test depend on the number of clusters ($n = 545$ units), and not on the number of observations ($N = 4,360$) in the data set.

4.1 Modeling the Level of Y

$$y_{it} = \beta_0 + \beta_1 x_{1it} + \cdots + \beta_k x_{kit} + \gamma_1 z_{1i} + \cdots + \gamma_j z_{ji} + u_i + e_{it} \quad (4.5)$$

The stochastic part of the model, $\varepsilon_{it} = u_i + e_{it}$, now distinguishes between two components: (i) u_i: unobserved predictors of Y that are specific to the unit and therefore time-constant, (ii) e_{it}: unobserved predictors of Y that are specific to the time point *and* the unit (including measurement errors). Unmeasured productivity would be an example of the first type, while unmeasured economic performance of the individual's employer would be an example of the second type. Given a sample of adults and a comparatively short observation period of eight years, productivity can safely be assumed to be time-constant in the wagepan data and, as already discussed, productivity is positively associated with wages. Depending on their economic performance, some employers may be able to increase the wages of their employees above the average level in certain years, while other employers would be unable to pay even the average wage. Possibly, economic performance changes each year in the observation period, which may explain why certain individuals show particularly high (or low) wages and correspondingly extraordinary high residuals in some years but not in all. In many applications, these time-varying errors, e_{it}, will show a very unsystematic pattern.

Depending on our assumptions about these two error terms, different estimation procedures are available. A simple starting point is the assumption that the time-varying error, e_{it}, has the same properties as the error term in OLS estimation (for more details, see the following sections). In other words, e_{it} is assumed to be purely random "white noise". Therefore, some authors call it *idiosyncratic* error. Yet, the main discussion revolves around the unit-specific error, u_i. As already mentioned, u_i is a measure of unobserved heterogeneity at the unit level, and different estimation strategies exist depending on our assumptions about this heterogeneity. Again, a simple starting point is the assumption that unobserved heterogeneity is uncorrelated with the variables in the model. In other words, u_i is a sort of random disturbance at the individual level. This is called a *random effect* in the literature, and *random effects (RE) estimation* is used to assess the effects of the explanatory variables in the model. For many applications, however, assuming uncorrelated heterogeneity is not a very realistic assumption. Remember our example, where the unobserved variable productivity can be positively correlated both with time-constant and time-varying explanatory variables (schooling increases productivity and more productive employees need not be unionized to achieve higher wages). With correlated heterogeneity, we have to use other techniques that have become known as *fixed effects (FE) estimation*. In that context, the unit-specific error, u_i, is termed a *fixed effect*, stressing the fact that it is typical for unit i and is fixed over time.

The distinction between random and fixed effects also has to do with the fact that in the first case, u_i is assumed to be a random draw from the universe of all possible values of a random variable having a certain distribution (e.g., the normal distribution), while in the second case, u_i is assumed to be a parameter that is to be estimated from the data of the sampled unit i (and hence, may be different in another sample). Therefore, some scholars argue that statistical inference about the population is only possible with random effects, while fixed effects would always be sample-specific and thus ill-suited to statistical inference. As we will see later, the practical relevance of this argument is rather limited.

In the following, we start with the more realistic assumption of correlated heterogeneity and introduce FE estimation first (Sect. 4.1.2.1). We then proceed with explaining RE estimation, also because it is more cumbersome to compute and because it partly builds on results from FE estimation (Sect. 4.1.2.2). After the reader has developed a solid understanding of both estimation techniques, we will then discuss their connections and present testing procedures for deciding which of them fits the data better (Sect. 4.1.2.3). At the end of this rather long introduction to modeling the panel structure of continuous panel data, we conclude by summarizing what the different models actually mean in applied research (Sect. 4.1.2.4).

4.1.2.1 Correlated Heterogeneity: Fixed Effects Estimation

This is a rather lengthy section on FE estimation. The technique is most easily understood within the familiar OLS context when using a dummy for each unit (the following section on p. 128). However, with large data sets, the dummy variable approach is not a very practical technique. Therefore, FE estimates are usually obtained from demeaned data (section on p. 133). Once we know how to compute FE estimates, we can discuss when and how they deviate from pooled OLS estimates (section on p. 140) and how FE estimation controls for the serial dependence in the data (section on p. 144).

FE Estimation Using Dummy Variables If we think about u_i as something typical of unit i that is unfortunately unknown to us, we simply could estimate this unit-specific heterogeneity by including a dummy for each unit. This approach has become known as *least squares dummy variables* (LSDV) regression. Certainly, this is hard to realize with Example 3.1, because in that case we would have to estimate the effects of $n = 545$ different dummies (except one, the reference category). But in other applications with few units of analysis, this is a very practical approach. LSDV is also interesting from a didactic point of view, because it shows what is going on with FE estimation. So, let us switch to an example from political science, in which the units of analysis are 18 OECD countries. If we now speak about unobserved unit-specific heterogeneity, we mean unobserved heterogeneity with respect to countries.

Example 4.1 (garmit data) Garrett and Mitchell (2001) assess the impact of globalization on welfare state effort in the OECD countries. Public spending (as a percentage of gross domestic product—GDP) is one of their dependent variables. Globalization is defined in terms of total trade, imports from low wage countries, and foreign direct investments. Two conflicting hypotheses can be found in the literature about the effect of globalization on public spending. The efficiency hypothesis states that globalization induces a downward pressure on public spending, while the compensation hypothesis claims that globalization is associated with higher demand for social security,

4.1 Modeling the Level of Y

which in turn increases public spending. To test these hypotheses, the authors collect data from OECD statistics for 18 countries over a period of 33 years (1961–1993). Thus, the full data matrix consists of $n = 18$ units observed over $T = 33$ time points.

All in all, the dataset includes $N = n \cdot T = 594$ data records, but due to missing data, the number of valid cases depends on which variables we are using in our regression model. The data are a typical example of an *unbalanced* panel data set, where the number of available measurements, T_i, varies between units $i = 1, \ldots, n$. The analysis of unbalanced panels is a rather straightforward extension of the following methods (see Sect. 4.1.3.3), but for the following discussion, it is easier to have a balanced panel. Therefore, we exclude New Zealand, Norway, and Switzerland from the analysis and test only two globalization indicators (lowwage: imports from low wage countries as a percentage of total imports, trade: total trade as percentage of GDP). All other variables have non-missing values for the remaining 15 countries, resulting in a reduced dataset of $N = n \cdot T = 15 \cdot 33 = 495$ data records. For simplicity, we call it the garmit data, although it includes only a subsample of Garrett and Mitchell's analysis.

The time dimension in this example ($T = 33$) is much larger than in the previous example (with $T = 8$) and even exceeds the number of units in the data set ($n = 15$). Thus, as described in Sect. 3.1, the data is a typical example of what we have called a macro panel (where $n < T$). Certainly, for a full analysis of the time dimension of this macro panel, we need more refined methods that cannot be covered in this textbook, but see Kittel and Winner (2005) for an extensive analysis of the garmit data. We are focusing on micro panels, hence this example is only used to illustrate the LSDV approach, which is very practical when the number of units is small.

One of the key variables in Garrett and Mitchell's analysis is imports from low wage countries. If the compensation hypothesis were true in the observation period, we would expect a positive sign of this variable, because cheap imports from low wage countries put pressure on local labor markets and thus increase public spending, because of increased payments to the unemployed. Applying pooled OLS to the garmit data, however, results in a highly significant negative effect ($\hat{\beta}_{lowwage} = -0.2868$, $t = -6.37$, $p < 0.01$; see Table 4.3), which would support the opposite efficiency hypothesis, assuming a downward pressure on public spending.

But as we know from the previous section, pooled OLS is not a very useful estimation procedure, because it ignores the panel structure of the data. Observations belonging to the same country are not independent of one another and have something in common that has to be controlled for in the model. Garrett and Mitchell's model controls for each country's policy orientation (left: proportion of cabinet portfolios held by social democratic/labor parties, cdem: same for Christian democratic parties), business cycle (growthpc: growth of per capita GDP) and welfare dependency (unemp: unemployment rate, depratio: dependency ratio). But

Table 4.3 Determinants of public spending: Pooled OLS (1961–1993, 15 OECD countries)

Variable	Estimate	Std. Err.
Low wage imports	−0.2868	0.0450
Trade	0.1430	0.0132
Unemployment	0.5016	0.0958
GDP/capita growth	−1.1495	0.1209
Dependency ratio	−0.8997	0.1156
Left cabinet portfolios	0.0415	0.0084
Christian Democrat portfolios	0.0161	0.0119
Constant	70.6190	4.7626
R^2	0.6654	
$F(X)$	138.37	
df_1, df_2	7	487
N	495	

Source: garmit data (see Example 4.1)

these five controls probably do not capture all country-specific heterogeneity. Therefore, a more realistic model should include u_i to control for unmeasured country-specific heterogeneity. This is equivalent to including a dummy for each of the countries. However, including a dummy for each country results in a model that is not identified. Either one excludes one country dummy (for the reference country) or one has to specify a model without a constant. Two options are available for coding: *Dummy coding* (i.e., dummy variables coded 1 and 0) is the most popular one and results in effects that measure differences from the reference country. They are sometimes called *cornered effects*. *Effect coding* (i.e., dummy variables coded +1 and −1) is not always applicable,[5] but results in effects measured as differences from the overall mean, which are easy to interpret. This is why they are also called *centered effects*.

Choosing the United States (country No. 15) as the reference category, and using dummy coding, the LSDV model looks like this (again ignoring a time trend and specifying $\beta_0(t) = \beta_0$):

$$spend_{it} = \beta_0 + \sum_{j=1}^{14} \gamma_j \cdot country_{ji} + \sum_{k=1}^{2} \beta_k \cdot global_{kit} + \sum_{l=1}^{5} \beta_l \cdot controls_{lit} + e_{it} \quad (4.6)$$

All in all, the model includes two globalization indicators, five controls and 14 1/0-coded country dummies. Except the dummies, all variables change over time. It is easy to see that this model is equivalent to (4.5), if we write down the equation for

---

[5] The effects of effect coded dummy variables will measure differences from the overall mean, if (and only if) the different categories of the corresponding categorical variable (in this case: country) have identical frequencies. In our case, we have reduced the data to a balanced panel, hence each country is observed the same number of times.

4.1 Modeling the Level of Y

each country separately (taking into account that the corresponding country-specific dummy equals one, while all other country dummies are zero):

$$i = 1 \text{ (Australia)}: \quad spend_{1,t} = \beta_0 + \gamma_1 \cdot 1 + \sum_{k=1}^{2} \beta_k \cdot global_{k,1,t}$$
$$+ \sum_{l=1}^{5} \beta_l \cdot controls_{l,1,t} + e_{1,t}$$

$$i = 2 \text{ (Austria)}: \quad spend_{2,t} = \beta_0 + \gamma_2 \cdot 1 + \sum_{k=1}^{2} \beta_k \cdot global_{k,2,t}$$
$$+ \sum_{l=1}^{5} \beta_l \cdot controls_{l,2,t} + e_{2,t}$$

$$\vdots \qquad \vdots$$

$$i = 15 \text{ (USA)}: \quad spend_{15,t} = \beta_0 + 0 + \sum_{k=1}^{2} \beta_k \cdot global_{k,15,t}$$
$$+ \sum_{l=1}^{5} \beta_l \cdot controls_{l,15,t} + e_{15,t}$$

Setting $\beta_0 + \gamma_i$ equal to u_i and β_0 equal to u_{15} results in (4.5) including country-specific heterogeneity.

$$i = 1 \text{ (Australia)}: \quad spend_{1,t} = \sum_{k=1}^{2} \beta_k \cdot global_{k,1,t} + \sum_{l=1}^{5} \beta_l \cdot controls_{l,1,t}$$
$$+ u_1 + e_{1,t}$$

$$i = 2 \text{ (Austria)}: \quad spend_{2,t} = \sum_{k=1}^{2} \beta_k \cdot global_{k,2,t} + \sum_{l=1}^{5} \beta_l \cdot controls_{l,2,t}$$
$$+ u_2 + e_{2,t}$$

$$\vdots \qquad \vdots$$

$$i = 15 \text{ (USA)}: \quad spend_{15,t} = \sum_{k=1}^{2} \beta_k \cdot global_{k,15,t} + \sum_{l=1}^{5} \beta_l \cdot controls_{l,15,t}$$
$$+ u_{15} + e_{15,t}$$

We also see that it is not possible to estimate a separate constant β_0, as suggested by (4.5). Within the LSDV approach and dummy coding, this constant equals the country-specific heterogeneity for the reference country (in our case: USA).

Table 4.4 shows the results of the LSDV model. In this more refined model, imports from low wage countries have only a small negative—and insignificant—effect on government spending ($\hat{\beta}_{lowwage} = -0.0909$, $t = -1.95$, $p = 0.052$), and hence support neither the compensation nor the efficiency hypothesis. Why is this the case? Apparently, the variable "low wage imports" conveys some of the unmeasured country-specific heterogeneity, and once this is controlled for, its effect is much smaller. In the section on p. 140 we will discuss this so-called omitted variable bias in greater detail. For the moment, it is enough to see that effects may change dramatically, once we control for the longitudinal structure of our data.

Table 4.4 Determinants of public spending: LSDV (1961–1993, 15 OECD countries)

Variable	Dummy coding		Effect coding		No constant	
	Estimate	Std. Err.	Estimate	Std. Err.	Estimate	Std. Err.
Low wage imports	−0.0909	0.0466	−0.0909	0.0466	−0.0909	0.0466
Trade	0.1999	0.0274	0.1999	0.0274	0.1999	0.0274
Unemployment	1.1575	0.1076	1.1575	0.1076	1.1575	0.1076
GDP/capita growth	−0.8543	0.0878	−0.8543	0.0878	−0.8543	0.0878
Dependency ratio	−0.2203	0.1386	−0.2203	0.1386	−0.2203	0.1386
Left cabinet portfolios	−0.0162	0.0075	−0.0162	0.0075	−0.0162	0.0075
Christian Democrat portfolios	−0.0531	0.0161	−0.0531	0.0161	−0.0531	0.0161
Australia	−3.8218	1.3209	−7.0117	1.0091	31.3287	5.8706
Austria	8.9904	2.1694	5.8005	0.9883	44.1409	6.1833
Belgium	−0.7248	3.1779	−3.9147	1.9397	34.4257	7.0038
Canada	−2.7451	1.4908	−5.9349	0.9760	32.4055	5.8921
Denmark	5.8155	1.8110	2.6256	0.8050	40.9660	5.9435
Finland	1.2328	1.5458	−1.9571	0.8637	36.3833	5.8593
France	8.2138	1.2929	5.0239	0.9160	43.3643	6.0354
Germany	9.9746	1.7765	6.7847	0.9183	45.1251	5.8133
Ireland	−9.4742	2.4700	−12.6641	1.5849	25.6763	7.2893
Italy	6.7647	1.7176	3.5748	1.1471	41.9152	5.9426
Japan	−0.2513	1.3936	−3.4412	1.4326	34.8992	5.4223
Netherlands	5.2266	2.6759	2.0367	1.4331	40.3771	6.5668
Sweden	15.8134	1.8709	12.6235	0.8835	50.9639	5.8519
United Kingdom	2.8337	1.4183	−0.3562	0.8486	37.9842	6.1813
Constant	35.1505	6.0612	38.3404	6.0287	35.1505	6.0612
R^2	0.8456		0.8456		0.9908	
$F(X)$	171.92		171.92		171.92	
df_1, df_2	7	473	7	473	7	473
F (dummies)	39.42		39.42		67.63	
df_1, df_2	14	473	14	473	15	473
N	495		495		495	
n	15		15		15	
T	33		33		33	

Source: `garmit` data (see Example 4.1)

We do not comment on the other explanatory variables in the model, but it is important to remember that none of them is constant over time (such as, for example, schooling and ethnicity in Example 3.1). If they were, it would not be possible to estimate their effects, because they are linearly related to the dummies. Each dummy captures all the characteristics of the corresponding unit, hence it is impossible to estimate on top of that the effect of a variable that is constant for the unit.

Finally, it is interesting to take a look at the estimates of the country dummies. Not all of them are significant, but Sweden, for instance, directs a significantly greater proportion of its GDP toward public expenditure. Government spending is on average 15.8 percentage points higher than in the US, which channels 35.2 % of its GDP (see the constant) toward public expenditure.[6] On the other side of the continuum, we find Ireland, where public spending is 9.5 percentage points lower than in the US, on average. However, as long as we do not know what the driving forces are behind these differences, we are unable to draw any theoretical conclusions. In other words, LSDV only controls for unobserved heterogeneity without providing information about the unobserved explanatory variables.

Table 4.4 also shows the effects of the two other LSDV specifications. With a balanced panel, it is possible to use dummy variables coded $+1$ and -1 and to measure centered effects (see footnote 5 on p. 130). The corresponding estimates in the column "effect coding" show that the overall average of all country effects amounts to 38.3 % (see the constant) and that the amount of government spending in Ireland lies about 12.7 percentage points below that average. Finally, specifying no regression constant and including *all* country dummies shows the absolute amount of government spending in each country, when all other explanatory variables are zero (see the column "no constant"). For example, the US directs 35.2 % of its GDP toward public expenditure (see the row labeled "constant"), while Ireland spends only 25.7 %. Subtracting these two percentages results in the estimate (9.5) for the Irish dummy in the LSDV model with dummy coding. Calculating the average of all country dummies (31.3, 44.1, ..., 35.2) results in the constant (38.3) of the LSDV model with effect coding. In other words, all three LSDV specifications are reparametrizations of the same country structure in the data. It should also be noted that if we had used other countries in our analysis, all effects would be different, irrespective of the type of coding. More generally speaking, FE estimation of country-specific (or more generally: unit-specific) effects always depends on the specific sample of countries (units) used in the analysis. No one would think about these effects as being realizations of a larger population of countries. This differs notably from the random effects that we will discuss in Sect. 4.1.2.2.

FE Estimation Using Time-Demeaned Data Now, let us consider what these dummy variables add to the picture. Take, as an example, the effects from the LSDV model with dummy coding. They measure how the observations in the respective country deviate *on average* from the reference country *while controlling* for the other variables in the model. Intuitively, the country effect u_i is nothing other than the difference of the adjusted mean government spending in country i minus the adjusted mean government spending in the reference country (the term "adjusted"

---

[6]More specifically, government spending in the US amounts to 35.2 %, if all independent variables are zero (i.e., `lowwage = trade = unemp = growthpc = depratio = left = cdem = 0`). This is not a very realistic situation, and it is perhaps a better strategy to center these variables around their US mean before putting them into the regression model. A similar caveat applies to all the other LSDV specifications.

should indicate that all other variables in the model have been controlled for). Put differently, all country dummies would be exactly zero if we were to center all measurements of public spending on their country-specific means. This observation suggests another way to control for country-specific (or more generally: unit-specific) heterogeneity. Some simple mathematical transformations will show why this is the case.

As an example, we use (4.5) with one time-constant and one time-varying independent variable:

$$y_{it} = \beta_0 + \beta_1 x_{1it} + \gamma_1 z_{1i} + u_i + e_{it} \tag{4.7}$$

Let us further assume that the data include only three measurements per unit of analysis. Thus, the data for an arbitrary unit i look like this:

$$t = 1: \quad y_{i1} = \beta_0 + \beta_1 x_{1i1} + \gamma_1 z_{1i} + u_i + e_{i1}$$
$$t = 2: \quad y_{i2} = \beta_0 + \beta_1 x_{1i2} + \gamma_1 z_{1i} + u_i + e_{i2}$$
$$t = 3: \quad y_{i3} = \beta_0 + \beta_1 x_{1i3} + \gamma_1 z_{1i} + u_i + e_{i3}$$

Now, what do we mean by "demeaning" variables? It sounds complicated, but mathematically, this transformation is very easy. First of all, we add the equations of unit i for all three time points and then divide the sum by three. This results in the average of all equations for unit i:

$$\bar{y}_{i.} = \beta_0 + \beta_1 \bar{x}_{1i.} + \gamma_1 z_{1i} + u_i + \bar{e}_{i.} \tag{4.8}$$

$\bar{y}_{i.}$, for example, is the average of all values of the dependent variable for unit i: $\bar{y}_{i.} = (y_{i1} + y_{i2} + y_{i3})/3$. $\bar{x}_{1i.}$ and $\bar{e}_{i.}$ are the averages of the time-varying explanatory variable X and the idiosyncratic error E. Since both z_{1i} and u_i are time-constant, taking the average of each of them results once more in their original values, z_{1i} and u_i. Finally, we subtract (4.8) from each of the time-point-specific equations for unit i:

$$t = 1: \quad (y_{i1} - \bar{y}_{i.}) = (\beta_0 + \beta_1 x_{1i1} + \gamma_1 z_{1i} + u_i + e_{i1})$$
$$- (\beta_0 + \beta_1 \bar{x}_{1i.} + \gamma_1 z_{1i} + u_i + \bar{e}_{i.})$$
$$= \beta_1 (x_{1i1} - \bar{x}_{1i.}) + (e_{i1} - \bar{e}_{i.})$$
$$t = 2: \quad (y_{i2} - \bar{y}_{i.}) = (\beta_0 + \beta_1 x_{1i2} + \gamma_1 z_{1i} + u_i + e_{i2})$$
$$- (\beta_0 + \beta_1 \bar{x}_{1i.} + \gamma_1 z_{1i} + u_i + \bar{e}_{i.})$$
$$= \beta_1 (x_{1i2} - \bar{x}_{1i.}) + (e_{i2} - \bar{e}_{i.})$$
$$t = 3: \quad (y_{i3} - \bar{y}_{i.}) = (\beta_0 + \beta_1 x_{1i3} + \gamma_1 z_{1i} + u_i + e_{i3})$$
$$- (\beta_0 + \beta_1 \bar{x}_{1i.} + \gamma_1 z_{1i} + u_i + \bar{e}_{i.})$$
$$= \beta_1 (x_{1i3} - \bar{x}_{1i.}) + (e_{i3} - \bar{e}_{i.})$$

$\ddot{y}_{i3} = (y_{i3} - \bar{y}_{i.})$, $\ddot{x}_{1i3} = (x_{1i3} - \bar{x}_{1i.})$, and $\ddot{e}_{i3} = (e_{i3} - \bar{e}_{i.})$ are the "demeaned" values of Y, X, and the idiosyncratic error E for unit i at time point $t = 3$. More specifically, we talk about *time-demeaning*, because for each variable, this transformation implies that we subtract from each original value the average of that variable for the

4.1 Modeling the Level of Y

corresponding unit *over time*. Or to put it differently: We center all measurements on their unit-specific means.

If we now consider the general case with t time points and several time-constant and time-varying independent variables, time-demeaning equation (4.5) results in the following general model:

$$(y_{it} - \bar{y}_{i.}) = \beta_1(x_{1it} - \bar{x}_{1i.}) + \cdots + \beta_k(x_{kit} - \bar{x}_{ki.}) + (e_{it} - \bar{e}_{i.})$$
$$\ddot{y}_{it} = \beta_1 \ddot{x}_{1it} + \cdots + \beta_k \ddot{x}_{kit} + \ddot{e}_{it}$$
(4.9)

The important thing about this transformation is that unobserved unit-specific heterogeneity, u_i, has disappeared. It should be noted that all time-constant independent variables Z have disappeared as well. Apparently, time-demeaning eliminates all—observed *and* unobserved—time-constant unit-specific heterogeneity.[7] However, without u_i in the equation for the transformed data, and given that the idiosyncratic errors e_{it} still have nice properties, we can safely apply OLS to the pooled and time-demeaned data and obtain estimates of the effects of the X that are unaffected by omitted variable bias at the unit level. A pooled OLS estimator that is based on the time-demeaned variables is called the *fixed effects estimator*. Textbox 4.2 summarizes the main features and assumptions of FE estimation.

Textbox 4.2 (FE assumptions) FE estimation controls for unit-specific heterogeneity by eliminating (demeaning) all time-constant information for each unit i from the data. As a side effect, it is not possible to estimate the effects of the time-constant explanatory variables. In other words, FE estimation builds on a linear model for panel data that puts all observed and unobserved time-constant explanatory variables into the error term u_i: $y_{it} = \beta_0 + \beta_1 x_{1it} + \cdots + \beta_k x_{kit} + u_i + e_{it}$. The statistical properties of FE estimation rest on assumptions that are very similar to those of OLS estimation:

1. The units $i = 1, \ldots, n$ in the panel data set are a simple random sample from a cross-section of a well-defined population.
2. The model is linear in its parameters $\beta_0, \beta_1, \ldots, \beta_k$.
3. Each independent variable $x_{1it}, \ldots, x_{kit}$ changes over time and is not a linear function of the other independent variables.
4. Idiosyncratic error is independent of the variables in the model and independent of unit-specific unobserved heterogeneity: $E(e_{it}|x_{1i1}, \ldots, x_{1iT}, \ldots, x_{ki1}, \ldots, x_{kit}, u_i) = 0$. It should be noted that the assumption expects independence from *all* measurements of each variable over time, a characteristic that has become known as *strict exogeneity*.

---

[7] We did not observe this in our LSDV analysis of the `garmit` data, because the model did not include any time-constant variables. As mentioned, if there had been any time-constant explanatory variables, it would have been impossible to estimate their effects when specifying country dummies.

5. Idiosyncratic error has constant variance, given any value of the independent variables and unit-specific effects: $\text{Var}(e_{it}|x_{1i1}, \ldots, x_{1iT}, \ldots, x_{ki1}, \ldots, x_{kit}, u_i) = \sigma_e^2$.
6. Idiosyncratic error is uncorrelated between any two observations t and s ($t \neq s$) of unit i, given any value of the independent variables and unit-specific effects: $\text{Corr}(e_{it}, e_{is}|x_{1i1}, \ldots, x_{1iT}, \ldots, x_{ki1}, \ldots, x_{kit}, u_i) = 0$.
7. Idiosyncratic error is normally distributed with mean 0 and variance σ_e^2: $e_{it} \sim Normal(0, \sigma_e^2)$, given any value of the independent variables and unit-specific effects.

The first assumption is necessary for making statistical inferences regarding the population using the sampled data. The second assumption ensures that we can use least squares estimation to estimate the parameters of the model. If the model were not linear in its parameters, we would have to use other estimation techniques. The third assumption guarantees that a numerical value exists for each regression coefficient. The fourth assumption is the most important one, because it makes sure that FE estimates $\hat{\beta}_0, \hat{\beta}_1, \ldots, \hat{\beta}_k$ are *unbiased*: $E(\hat{\beta}_0) = \beta_0$, $E(\hat{\beta}_1) = \beta_1, \ldots, E(\hat{\beta}_k) = \beta_k$. Assumption 4, however, is far more demanding than assumption 4 for OLS with cross-section data (see Textbox 4.1). Strict exogeneity (as opposed to contemporaneous exogeneity) assumes that there are no feedback mechanisms caused by unobserved effects over time. Furthermore, if assumptions 5 and 6 are true, FE estimates are the best linear unbiased estimates (BLUE) of $\beta_0, \beta_1, \ldots, \beta_k$. Finally, assumption 7 guarantees that we can use standard test procedures and confidence intervals based on the normal distribution. Otherwise, we have to rely on asymptotic approximations, which are only true in large samples. Discussions of these FE assumptions and proofs of the statistical properties of FE can be found in Wooldridge (2010, 265).

FE estimates parameters using only variation around the unit-specific means. This is why FE is also called the *within estimator*. This is illustrated in Fig. 4.2 for a subset of data (Sweden and Ireland) from our Example 4.1. The left panel shows the original data on government spending for both countries plotted along the time axis. A simple LSDV model would include a dummy for Sweden, measuring unit-specific heterogeneity, and perhaps a non-linear time trend. The effect of the dummy captures all between-country variance, and the time trend is estimated controlling for observed *and* unobserved country heterogeneity. The right panel shows the time-demeaned data, in which the differences between Sweden and Ireland are hardly visible, because all between variance has been eliminated. In this case, it is no longer necessary to include a country dummy in the model.[8] However, estimates for the

---

[8]Of course, we would get exactly the same picture, if we would graph the *residuals* from a regression including only a dummy for Sweden.

4.1 Modeling the Level of Y

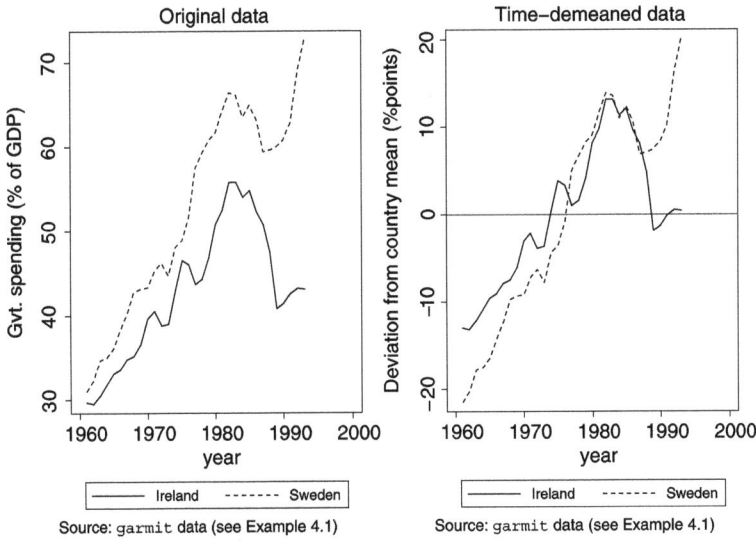

Fig. 4.2 Government spending in Ireland and Sweden (1961–1993)

non-linear time trend will be the same as in the LSDV model and hence will be computed net of country heterogeneity.

Having introduced FE estimation, we can now return to Example 4.1 and reestimate the government spending equation with time-demeaned data. We get exactly the same estimates as the ones from the LDSV model (see Table 4.5, column "FE"). Instead of demeaning the data by hand, we have used a readily programmed FE estimation routine, which is available in many statistical software packages. The advantage of using a programmed algorithm for FE estimation is the correct computation of degrees of freedom. A simple OLS algorithm would not know that we demeaned the data beforehand. This can be seen in the last column of Table 4.5 labeled "pooled OLS, time-demeaned data". Parameter estimates are the same, while standard errors and F statistics differ from the corresponding FE test statistics. The following Textbox 4.3 describes how to correct the pooled OLS test statistics.

Textbox 4.3 (Degrees of freedom for time-demeaned data) If we are transforming the data manually and using a simple OLS routine to compute FE estimates, it is important to tell the program to estimate a model *without* a constant, because (4.9) does not include an intercept (time-demeaning does eliminate the constant β_0 as well). Without a constant term, OLS computes $n \cdot T - k$ degrees of freedom (and not $df = n \cdot T - k - 1$) based on k independent variables and a total of $N = n \cdot T$ pooled observations. However, since we subtracted $i = 1, \ldots, n$ unit-specific means from the original observations, we have lost another n degrees of freedom. Thus, if no programmed algorithm for FE estimation is available, standard errors and test statistics have to be recomputed with the correct degrees of freedom, which equal $df = n \cdot T - k - n = n \cdot (T - 1) - k$. More specifically, OLS standard errors have to be multiplied with the square root of $(n \cdot T - k)/(n \cdot (T - 1) - k)$, while the F statistic has to be multiplied with $(n \cdot (T - 1) - k)/(n \cdot T - k)$ (readers should check this with the data in Table 4.5).

There is, however, a peculiarity with many programmed FE routines that we need to understand. Surprisingly, they provide an estimate of a regression constant (see column "FE" of Table 4.5), although time-demeaning eliminates the original intercept of the model (see (4.9)). The estimate of this constant (38.3404) is equal to the average of all country effects (compare the constant in the LSDV model with effect coding). This happens because some FE routines use a slightly different transformation than (4.9). They do not subtract the unit-specific means from all variables; rather, they subtract how much those unit-specific means deviate from the corresponding overall mean (across i and t). In that case, the regression model should include a constant term β_0:

$$\left(y_{it} - (\bar{y}_{i.} - \bar{\bar{y}}_{..})\right) = \beta_0 + \beta_1\left(x_{1it} - (\bar{x}_{1i.} - \bar{\bar{x}}_{1..})\right) + \cdots + \beta_k\left(x_{kit} - (\bar{x}_{ki.} - \bar{\bar{x}}_{k..})\right) \\ + \left(e_{it} - (\bar{e}_{i.} - \bar{\bar{e}}_{..})\right) \quad (4.10)$$

$\bar{\bar{y}}_{..}, \bar{\bar{x}}_{1..}, \ldots, \bar{\bar{x}}_{k..}$ are the overall averages (across i and t) of the variables in the model and $\bar{\bar{e}}_{..}$ is the overall average of the idiosyncratic errors. Using this slightly more complicated transformation, one estimates the regression constant as the average of all estimated unit-specific effects: $\hat{\beta}_0 = \sum_{i=1}^{n} \hat{u}_i/n = \bar{u}_{.}$.

Finally, it is interesting to have a look at the overall fit measures and test statistics. The results for the LSDV model without a constant are a little peculiar, so we ignore them for a moment.[9] The other LSDV models show R^2 values of 0.8456 (see Table 4.4), while the FE model has an R^2 of 0.7179 (see Table 4.5). Why are these R^2 statistics different? FE estimation uses time-demeaned data and the corresponding R^2 measures the explained portion of the within variance. This is why it is also called the "within" R^2. LSDV estimation, on the other hand, uses the original data; therefore, the corresponding R^2 measures the explained proportion of the overall

---

[9] Its R^2 and the F statistic for the country effects are not comparable to the other models and have no substantive meaning. Basically, the F statistic tests whether country effects *and* the constant are zero. Correspondingly, the computation of R^2 assumes that the average of Y is zero and uses $\sum(y_{it} - 0)^2$ and not $\sum(y_{it} - \bar{\bar{y}}_{..})^2$ in its denominator.

4.1 Modeling the Level of Y

Table 4.5 Determinants of public spending: FE (1961–1993, 15 OECD countries)

Variable	FE		Pooled OLS, time-demeaned data	
	Estimate	Std. Err.	Estimate	Std. Err.
Low wage imports	−0.0909	0.0466	−0.0909	0.0459
Trade	0.1999	0.0274	0.1999	0.0269
Unemployment	1.1575	0.1076	1.1575	0.1059
GDP/capita growth	−0.8543	0.0878	−0.8543	0.0865
Dependency ratio	−0.2203	0.1386	−0.2203	0.1364
Left cabinet portfolios	−0.0162	0.0075	−0.0162	0.0073
Christian Democrat portfolios	−0.0531	0.0161	−0.0531	0.0158
Constant	38.3404	6.0287		
R^2		0.7179		0.7179
$F(X)$		171.92		177.37
df_1, df_2	7	473	7	488
F (dummies)		39.42		
df_1, df_2	14	473		
N		495		495
n		15		
T		33		

Source: `garmit` data (see Example 4.1)

variance.[10] Unless the within variance is very large compared to the between variance, LSDV R^2 values are usually very high, because including a dummy for each unit explains the between variance perfectly.

One can also test for the joint significance of either all independent variables or all unit-specific effects by using appropriate linear restrictions on the model parameters (see Sect. 7.2.1). The corresponding F statistic is computed from the residual sum of squares of the restricted (SSR_r) and the unrestricted model (SSR_{ur}):

$$f = \frac{(SSR_r - SSR_{ur})/q}{SSR_{ur}/(n \cdot (T-1) - k)}, \quad f \sim F(q, n \cdot (T-1) - k) \quad (4.11)$$

q equals the number of restrictions and k the number of independent variables in the unrestricted model. For example, in our case, testing for the joint significance of lowwage, trade, unemp, growthpc, depratio, left, and cdem implies comparing two models: the unrestricted model in Table 4.4 (e.g., the LSDV model with dummy coding), and a restricted model, where the $q = 7$ parameters of these variables have been set to zero (i.e., a LSDV model, in which these variables have

---

[10] Some FE routines also report an "overall" R^2, measuring the proportion of the overall variance explained by all independent variables *except* the unit-specific effects. Therefore, its value is not identical to the R^2 from LSDV estimation.

been excluded). With $SSR_r = 30{,}485.27$ and $SSR_{ur} = 8{,}601.41$ we compute the F statistic as follows:

$$f = \frac{(30{,}485.27 - 8{,}601.41)/7}{8{,}601.41/(15 \cdot (33-1) - 7)} = 171.92$$

The resulting value $f = 171.92$ has to be compared to a table of the F distribution with $df_1 = q = 7$ and $df_2 = n \cdot (T-1) - k = 473$ degrees of freedom. It will show that 171.92 is highly significant ($p < 0.01$). Hence, the hypothesis that both the two globalization indicators and the five control variables have no effect on government spending has to be rejected.

Similarly, we can test for the joint significance of all country-specific effects by comparing the same unrestricted model with a LSDV model, in which all $q = n - 1 = 14$ country dummies have been excluded:[11]

$$f = \frac{(18{,}636.46 - 8{,}601.41)/14}{8{,}601.41/(15 \cdot (33-1) - 7)} = 39.42$$

Again, the resulting f value is highly significant ($p < 0.01$) telling us that there are significant country differences after lowwage, trade, unemp, growthpc, depratio, left, and cdem have been controlled for. Usually, it is not necessary to specify the corresponding restricted model, because most FE programs have internal algorithms to compute the F statistic for the hypothesis that all unit-specific effects are zero (see Table 4.5). However, for didactic reasons, testing the joint significance of all unit-specific effects is most easily understood with the LSDV model, because it simply means dropping the unit dummies from the regression equation.

When and How Do FE Estimates Deviate from Pooled OLS Estimates? Having introduced FE estimation with time-demeaned data, we can now return to Example 3.1 about the union effect on wages. As already mentioned, the LSDV approach is not very useful for the wagepan data, because these data include a large number of units (545 individuals).[12] Therefore, FE estimation using time-demeaned data is a practical alternative to control for unobserved heterogeneity at the unit level. Table 4.6 shows the results in the column labeled "FE". According to the F-Test, unobserved heterogeneity is highly significant, i.e., all individual-specific effects, u_i, are jointly significantly different from zero ($f = 6.73$, $p < 0.01$). Now let us take a look at the FE estimates for the variables in the wage equation. We are unable to estimate the effects of the time-constant variables schooling and ethnicity, because the FE estimator uses only the within variation, which is zero by definition for time-constant variables. With respect to Vella and Verbeek's research question, the most interesting effect is that of union membership. Compared to the pooled OLS estimate from Sect. 4.1.1.1, the union effect is now only half as large, but still

---

[11] Remember: There are only $n - 1$ country dummies in the unrestricted model, because one dummy had to be excluded for model identification.

[12] Nevertheless, a model including these many dummies is feasible on today's computers.

4.1 Modeling the Level of Y

Table 4.6 Determinants of log hourly wages: Pooled OLS and FE (1980–1987)

Variable	Logarithm of hourly wage				Unit-specific heterogeneity	
	Pooled OLS		FE		Pooled OLS	
	Estimate	Std. Err.	Estimate	Std. Err.	Estimate	Std. Err.
Union membership (yes = 1)	0.1455	0.0169	0.0793	0.0194	0.0662	0.0118
Education (years)	0.0905	0.0046	(dropped)		0.0905	0.0033
Experience (years)	0.0759	0.0097	0.1125	0.0085	−0.0365	0.0068
Experience squared	−0.0022	0.0007	−0.0041	0.0006	0.0020	0.0005
Hispanic (yes = 1)	−0.0585	0.0219	(dropped)		−0.0585	0.0154
Afro-American (yes = 1)	−0.1545	0.0230	(dropped)		−0.1545	0.0161
Lives in rural area (yes = 1)	−0.1314	0.0185	0.0501	0.0290	−0.1815	0.0130
Married (yes = 1)	0.1100	0.0153	0.0398	0.0183	0.0703	0.0107
Poor health (yes = 1)	−0.0580	0.0539	−0.0167	0.0471	−0.0413	0.0378
Lives in North-East (yes = 1)	0.0197	0.0233	0.0775	0.0802	−0.0579	0.0163
Lives in South (yes = 1)	−0.0784	0.0210	0.1079	0.0623	−0.1863	0.0147
Lives in North-Center (yes = 1)	−0.1057	0.0226	−0.0222	0.0584	−0.0835	0.0158
Constant	0.2237	0.0780	1.0645	0.0659	−0.8408	0.0546
R^2	0.2636		0.1898		0.3535	
$F(X)$	67.49		44.46		103.06	
df_1, df_2	23	4,336	20	3,795	23	4,336
$F(U, Z)$			6.73			
df_1, df_2			544	3,795		
r_v			−0.1277			
$\hat{\sigma}_v$			0.3986			
$\hat{\sigma}_e$			0.3495			
$\hat{\rho}_{FE}$			0.5654			
N	4,360		4,360		4,360	
n			545			
T			8			

Note: Models control for industry (regression coefficients not shown)
Source: wagepan data (see Example 3.1)

significant ($t = 4.08$, $p < 0.01$). The wage premium of union membership amounts to roughly 8 % ($\exp(0.0793) = 1.0825$). Evidently, controlling for unobserved individual heterogeneity via FE estimation decreases the union effect. Why is this the case? Generally speaking, this happens when we have omitted an important Z that correlates with the X variables in the model. In the following, we will use the wagepan data to learn a little bit more about possible omitted variable bias.

Remember our discussion from Sect. 4.1.1.1, when we used the unobserved variable "productivity" to explain why panel data are serially correlated. If productivity or other unobserved characteristics are correlated with union membership, the estimated union effect will be biased, if this unobserved individual heterogeneity is not controlled for. If, for example, more productive employees are less often union members than less productive employees (e.g., because they prefer individual over collective action), then union membership will carry some of the productivity effects on wages, if productivity is omitted from the regression equation. Therefore, the pooled OLS estimate of union membership is higher than the FE estimate.

This can be shown more formally. Consider, again, the pooled OLS estimation that ignores unit-specific heterogeneity u_i. u_i is sort of an omitted variable. Let us call the estimate of the union effect in this misspecified model $\hat{\beta}_{union}^{OLS}$, while the estimate in the FE model is denoted $\hat{\beta}_{union}^{FE}$. It is easy to show that the bias (i.e., the difference between both estimates) equals the following product:

$$\hat{\beta}_{union}^{OLS} - \hat{\beta}_{union}^{FE} = \hat{\gamma}_v \cdot \hat{\delta}_{union} = 1 \cdot \hat{\delta}_{union} \quad (4.12)$$

It states that the bias is related (i) to the effect of all time-constant unit characteristics on the dependent variable and (ii) to how the variable in question (union) is related to these time-constant unit characteristics and to all other variables in the model. The first aspect is measured by $\hat{\gamma}_v$, the second by $\hat{\delta}_{union}$.

To understand this formula, we define v_i as the sum of the effects of all observed and unobserved characteristics of unit i that are constant over time: $v_i = \gamma_1 z_{1i} + \cdots + \gamma_j z_{ji} + u_i$. Since v_i includes both observed *and* unobserved characteristics, we call it unit-specific heterogeneity (as opposed to unobserved heterogeneity u_i). By definition, this sum has an effect of $\hat{\gamma}_v = 1$ on Y ((4.5) can be written as $y_{it} = \beta_0 + \beta_1 x_{1it} + \cdots + \beta_k x_{kit} + 1 \cdot (\gamma_1 z_{1i} + \cdots + \gamma_j z_{ji} + u_i) + e_{it})$. $\hat{\delta}_{union}$, on the other hand, is a measure of the association between union membership and unit-specific heterogeneity v_i, while controlling for all other variables in the model. More specifically, it is the estimate of the union effect in an auxiliary model, in which v_i is regressed on all independent variables of the misspecified pooled OLS model (including union membership):

$$v_i = \delta_0 + \delta_{union} \cdot union_{it} + \delta_{educ} \cdot educ_i + \cdots + \delta_{ind11} \cdot industry11_{it} + \varepsilon_{it} \quad (4.13)$$

In order to compute this regression model, we need an estimate of unit-specific heterogeneity v_i. A straightforward estimate is derived from the averaged equation (4.8) that we used for time-demeaning. We extend it to more than one X and Z variable:

$$\bar{y}_{i.} = \beta_0 + \sum_{l=1}^{k} \beta_l \bar{x}_{li.} + \sum_{m=1}^{j} \gamma_m z_{mi} + u_i + \bar{e}_{i.}$$

Given that idiosyncratic error is zero on average ($\bar{e}_{i.} = 0$), an estimate of all observed ($\sum_j \gamma_j z_{ji}$) and unobserved (u_i) unit-specific heterogeneity is as follows:

$$\hat{v}_i = \sum_{m=1}^{j} \hat{\gamma}_m z_{mi} + \hat{u}_i = \bar{y}_{i.} - \hat{\beta}_0 - \sum_{l=1}^{k} \hat{\beta}_l \bar{x}_{li.} \quad (4.14)$$

4.1 Modeling the Level of Y

Using the FE estimates of β_0[13] and $\beta_1, \ldots, \beta_k$ and the corresponding unit-specific means $\bar{x}_{li.}$ of all $l = 1, \ldots, k$ time-varying explanatory variables, we arrive at estimates $\hat{v}_i$ which can be regressed on all independent variables of the misspecified pooled OLS model (see (4.13)).[14]

The coefficients of this auxiliary regression model are shown in the last column, labeled "unit-specific heterogeneity," in Table 4.6. If we now apply (4.12), we see that the pooled OLS estimate of union membership is $\hat{\gamma}_v \cdot \hat{\delta}_{union} = 1 \cdot 0.0662$ units larger than the corresponding FE estimate (a positive bias), while the pooled OLS estimate of experience is 0.0365 units smaller (a negative bias).[15] The general lesson from (4.12) is that there is always bias in the pooled OLS estimates, once the observed and unobserved time-constant characteristics are correlated with the explanatory variables in the model (i.e., when $\delta \neq 0$). As we can also see from the effect of experience, omitted variable bias is not always positive. In some cases, FE estimates will be larger than pooled OLS estimates, and it is a fallacy to believe that FE estimation always attenuates pooled OLS estimates (as might be expected from the union effect or the effect of low wage imports in Example 4.1).

As an overall measure of association, some FE programs compute the correlation between the estimates $\hat{v}_i$ and all (time-varying) variables in the model. The aggregate effect of all k time-varying explanatory variables is estimated as

$$\sum_{l=1}^{k} \hat{\beta}_l \bar{x}_{li.} = 0.0793 \cdot \overline{union}_{i.} + 0.1125 \cdot \overline{exper}_{i.} + \cdots - 0.1617 \cdot \overline{industry11}_{i.}$$

and in case of the wagepan data, the aggregate correlates negatively with unit-specific heterogeneity ($r_v = -0.1277$, see Table 4.6). It should be noted, however, that we are using estimates, $\hat{v}_i$, of unit-specific heterogeneity, and that these estimates are only as good as the FE model from which they are derived. If this model is misspecified (even though it controls for unobserved heterogeneity, there may be other reasons for misspecification), then these estimates $\hat{v}_i$ will be worthless. Some people think that the estimates $\hat{v}_i$ can be used to detect which variable has been omitted from the model. According to the auxiliary regression model in Table 4.6, unit-specific heterogeneity is positively related to union membership and schooling, negatively related to labor force experience, negatively related to Afro-American and Hispanic ethnicity, and so forth. Does this tell us something about the omitted

---

[13] At this point, it becomes obvious why some FE programs estimate a regression constant. Otherwise, estimates of the unobserved individual-specific effect, u_i, would have to control for the overall level of the dependent variable.

[14] Many programs provide estimates of v_i. Thus, it is not necessary to perform the computations manually. Unfortunately, some programs label $\hat{v}_i$ with the letter u, giving the wrong impression that $\hat{v}_i$ is identical to an estimate of *unobserved* heterogeneity u_i. But, as already mentioned, $\hat{v}_i$ includes unobserved *and observed* heterogeneity.

[15] The effects of the time-constant independent variables (e.g., schooling) have exactly the same estimates as those in the pooled OLS regression model, because the estimates $\hat{v}_i$ include the effects of $z_{1i}, \ldots, z_{ji}$ (see (4.14)).

variable(s)? No! This information is pretty useless, since $\hat{v}_i$ not only includes unobserved heterogeneity but also the effects of all the time-constant Z in the model. Hence, it is no surprise that it correlates, for example, positively with schooling (one of the Z variables). Even the fact that it correlates in a certain way with the time-varying X in the model (e.g., positively with union membership) is not very informative, because it is unknown how much of this correlation is due to u_i in $\hat{v}_i$ and how much is due to the effects of the Z.

How Does FE Estimation Control for Serial Dependence? Finally, we can ask how FE estimation controls for serial dependence in the data. We will answer this question in two steps. Since FE estimation uses demeaned data to control for unobserved heterogeneity, we will first discuss serial dependencies in the demeaned data. In the second step, we will discuss which assumptions FE estimation makes about the serial dependencies in the original data.

For the first step, we need estimates of the FE residuals (i.e., the idiosyncratic errors). To this end we use the estimates of unit-specific heterogeneity, $\hat{v}_i$, and the aggregate effect of all k time-varying explanatory variables, $\sum_l^k \hat{\beta}_l x_{lit}$:

$$\hat{e}_{it} = y_{it} - \hat{v}_i - \sum_{l=1}^{k} \hat{\beta}_l x_{lit} \qquad (4.15)$$

These are the residuals of a linear regression model of the demeaned data controlling for all time-varying explanatory variables in the model. The upper half of Table 4.7 shows how these residuals correlate over time. The first-order serial correlations of the FE model (the correlations below the main diagonal) are lower than the first-order serial correlations of the raw demeaned data, which we already discussed in the previous chapter (see Sect. 3.4.2 and Table 3.7). The latter correlations can be replicated with a simple FE model that includes no independent variables. Using this trick, the lower half of Table 4.7 shows the first- and higher-order serial correlations of the raw demeaned data.[16] Most of them are larger than the corresponding correlations from the FE model, including all the independent variables. However, the differences are not very large. On average, the correlations of the raw demeaned data amount to $\bar{r}_{\ddot{y}} = 0.1776$, while the correlations of the residuals of the FE model amount to $\bar{r}_{FE} = 0.1634$.[17] While demeaning already controls for all (observed and unobserved) heterogeneity at the unit level, FE estimation additionally controls for the effects of time-varying explanatory variables and, obviously, their serial dependence also adds to the serial correlation of log hourly wages. Otherwise, $\bar{r}_{FE}$ would not be smaller than $\bar{r}_{\ddot{y}}$.

---

[16] You should check that the first-order serial correlations from Table 3.7 are identical to the numbers below the main diagonal in the lower part of Table 4.7.

[17] Since, in both cases, some of the correlations are positive and some are negative, we computed the average of their absolute values.

4.1 Modeling the Level of Y

Table 4.7 Correlation of the residuals from the FE model

	FE model							
	1980	1981	1982	1983	1984	1985	1986	1987
1980	1.0000							
1981	0.0030	1.0000						
1982	−0.1231	0.0067	1.0000					
1983	−0.2032	−0.0866	0.0478	1.0000				
1984	−0.3529	−0.2159	−0.0736	0.0216	1.0000			
1985	−0.2801	−0.3115	−0.1865	−0.1338	0.0127	1.0000		
1986	−0.2541	−0.3392	−0.2166	−0.2167	−0.1257	0.0123	1.0000	
1987	−0.2421	−0.2033	−0.3412	−0.2135	−0.1056	0.0559	0.1913	1.0000
Mean	$\bar{r}_{FE} = 0.1634$							

	Raw demeaned data							
	1980	1981	1982	1983	1984	1985	1986	1987
1980	1.0000							
1981	0.0251	1.0000						
1982	−0.1086	0.0315	1.0000					
1983	−0.1874	−0.0781	0.0620	1.0000				
1984	−0.3496	−0.2223	−0.0717	0.0275	1.0000			
1985	−0.2890	−0.3406	−0.2081	−0.1502	0.0335	1.0000		
1986	−0.2909	−0.3509	−0.2373	−0.2132	−0.1175	0.0346	1.0000	
1987	−0.2840	−0.2228	−0.3576	−0.2268	−0.1090	0.0921	0.2514	1.0000
Mean	$\bar{r}_{\ddot{y}} = 0.1776$							

Source: wagepan data (see Example 3.1)

Surprisingly, higher-order serial correlations in the upper half of Table 4.7 are still quite high. For example, the residuals of the FE model for $t = 1980$ and $t = 1982$ correlate with $r = −0.1231$. Most of them are negative and increase with the time lag between measurements. This increasing pattern of negative serial correlations should not bother you, because it is a necessary consequence of demeaning. If all observations within each unit are centered around their mean (see the example in Fig. 4.2), then residuals within each unit cannot be all positive or all negative. A positive residual must be outweighed by a negative one, and vice versa. This pattern produces the negative serial correlations, and if there is a trend in the data, these serial correlations will increase with the time lag between the observations.

Now, let us focus on the second step, and discuss the serial dependencies in the original data. Apart from the dependencies due to serially correlated X, FE estimation assumes that all remaining dependencies in the original data are due to time-constant characteristics of the units (i.e., observed time-constant Z and unobserved heterogeneity U). Let us assume for a moment that all heterogeneity at the unit level is unobserved and that, as a result, Z does not exist (or can be thought of as part of U). FE distinguishes between unobserved heterogeneity at the unit level (u_i)

and idiosyncratic errors at the level of single measurements (e_{it}). Furthermore, it assumes that idiosyncratic errors are independent of all X, Z, and unobserved heterogeneity U, have constant variance, and are not serially correlated (see Textbox 4.2). Given these assumptions, it is easy to derive variances and covariances of the composite error $\varepsilon_{it} = u_i + e_{it}$ and finally derive a formula for the expected degree of serial correlation if these assumptions are true (for a derivation of this formula see Textbox 3.1):

$$\text{Corr}(\varepsilon_{it}, \varepsilon_{is}) = \frac{\text{Cov}(\varepsilon_{it}, \varepsilon_{is})}{\sqrt{\text{Var}(\varepsilon_{it})} \cdot \sqrt{\text{Var}(\varepsilon_{is})}} = \frac{\sigma_u^2}{\sigma_u^2 + \sigma_e^2}, \quad t \neq s \quad (4.16)$$

According to this formula, the expected degree of serial correlation is a function of the variances of both error components. σ_u^2 is the variance of unobserved heterogeneity and σ_e^2 is the variance of the idiosyncratic errors. Since FE estimation is unable to distinguish between observed and unobserved heterogeneity, we can replace σ_u^2 with $\hat{\sigma}_v^2$. Based on (4.14) and (4.15), both variances can be estimated as follows:

$$\hat{\sigma}_v^2 = \frac{\sum_{i=1}^n \hat{v}_i^2}{n-1} \quad (4.17)$$

$$\hat{\sigma}_e^2 = \frac{\sum_{i=1}^n \sum_{t=1}^T \hat{e}_{it}^2}{n \cdot (T-1) - k} \quad (4.18)$$

k equals the number of time-varying explanatory variables X in the model. $\hat{\sigma}_v$ and $\hat{\sigma}_e$ are listed in Table 4.6 in the column labeled "FE". Based on these values, the correlation (4.16) is estimated as $\hat{\rho}_{FE} = 0.5654$. It is similar, however not identical, to the intra-class correlation coefficient (ICC) that we introduced in Sect. 3.3. More specifically, the ICC assumes that unobserved heterogeneity U is a random variable (see Textbox 3.1), while FE estimation treats the u_i as fixed constants. But the FE correlation ρ_{FE} has the same underlying concept: It measures the "closeness" of measurements of the same unit relative to the "closeness" of measurements between different units.

If we compute ρ_{FE} for an empty model (i.e., a model without any explanatory variables), we arrive at an estimate of $\hat{\rho}_{FE} = 0.5045$,[18] which is a bit smaller than the former estimate for the full model ($\hat{\rho}_{FE} = 0.5654$). This happens because the full model controls for the effects of the time-varying explanatory variables X, which the empty model does not. Hence, σ_e^2 is smaller in the full model (as discussed above, part of the within variance is attributed to the X) and correspondingly, ρ_{FE} in the full model is larger. In other words: Once we control for the extra variation that is brought about by time-varying determinants of Y, the "closeness" of measurements

---

[18] This estimate is easily derived from the between-unit variance ($\hat{\sigma}_b^2 = 0.3907^2$) and within-unit ($\hat{\sigma}_w^2 = 0.3872^2$) variance of (log) hourly wages (see Table 3.6). In the FE model, $\hat{\sigma}_v^2 = \hat{\sigma}_b^2$ and $\hat{\sigma}_e^2 = \hat{\sigma}_w^2$ (see Textbox 3.1), hence $\hat{\rho}_{FE} = 0.3907^2/(0.3907^2 + 0.3872^2) = 0.5045$. In other words: in the case of an empty FE model ($k = 0$) (4.17) and (4.18) reduce to (3.2) and (3.3).

4.1 Modeling the Level of Y

within units is even larger. But the more important point is that the "closeness" is the same, irrespective of the time lag between t and s. This *equal correlation assumption* of the FE model may not be plausible for some applications (see Sect. 4.1.3.2).

4.1.2.2 Uncorrelated Heterogeneity: Random Effects Estimation

As we have seen in the last section, FE estimation is unable to estimate the effect of time-constant independent variables. In our last example, this has not been a major problem, because we have not been interested in achieving a numerical estimate of the effect of schooling on wages. Our main interest concerned the wage premium of union membership, and we only wanted to make sure that all time-constant characteristics of the sample members were controlled for (i.e., observed ones like schooling, and unobserved ones like productivity). This is essentially what FE estimation does; it provides no numerical estimates of their effects, but it controls for them. However, there may be applications where the effect of time-constant variables Z on the dependent variable Y is one of the main research questions. A study on the effects of marital disruption on psychological distress is an example.

> *Example 4.2* (johnson-wu data) Johnson and Wu (2002) analyze a sample of 2,033 individuals over a 13-year period from 1980 to 1992. During this period, survey respondents have been interviewed four times (1980, 1983, 1988, 1992). Obviously, the four panels are not equally spaced, which should be controlled for when modeling possible time trends in the data. However, for didactic purposes, we have deferred the analysis of trends to a later section. Panel attrition has been quite substantial over the course of the study, and the authors confine their analysis to those individuals that answered the survey in two or more of the waves. 269 participated in two waves, 158 in three waves, and the majority (1,166) in all four waves. In other words, their data represent a typical *unbalanced panel*.
>
> Many studies had already found that divorced individuals show higher levels of psychological distress than married individuals. But it is unclear, as Johnson and Wu argue, whether distress can be attributed causally to the preceding marital disruption, because most studies do not rule out the possibility that individuals with poor mental health have a higher probability of getting divorced. Even if this second explanation turns out to be untrue, it is still an open question whether higher levels of psychological distress are only a temporary or a permanent consequence. Hence, their analysis tries to test three theoretical explanations: (a) social role theory maintaining that the role of being divorced is inherently more stressful than that of being married, (b) crisis theory attributing higher stress to role transitions and transient stressors of the disruption process, and (c) social selection theory claiming that higher stress levels among the divorced result from the selection of those with poor mental health into divorce. Social role theory assumes that distress is causally related to divorce and is a permanent phenomenon as long as the person re-

mains divorced. Crisis theory also attributes distress causally to divorce, but sees it as a temporary phenomenon. Finally, social selection theory denies the causal effect of divorce and explains higher distress with antecedent factors that increase the probability of divorce itself.

For each explanation, Johnson and Wu use specific indicators. Current family status (three dummies for being divorced (divorce), widowed (widow1), or cohabiting (cohab1) with married as the reference category) is used as an indicator of the social role of the respondent. Whether divorce results in a temporary or permanent crisis is measured by two time variables (time from the current interview to the divorce: todiv, time since last divorce: frmdiv). Finally, social selection is measured by a dummy (socsel) that equals 1 if the respondent already experienced marital disruption before the beginning of the study. Furthermore, the authors control for schooling (educr), age (ager), and gender of their respondents (sexr).

Unfortunately, the authors had no independent measure of mental health before the start of the study, which would have allowed them to identify much more precisely those individuals with poor mental health. Instead, they use former divorce experiences as an indicator of poor mental health. This operationalization is far from optimal, because it assumes that all former divorces are only an effect of poor mental health, while divorces observed during the course of the study are assumed to have a negative effect on psychological distress. Since we are using the johnson-wu data only for illustrative purposes, we are ignoring these measurement problems.

Their dependent variable psydis is a summary index derived from five survey items measuring mental distress, subjective health status, global happiness, and life satisfaction (two items). All five items have been standardized and the index psydis is defined as the average of the five z-scores. For example, an individual having a value of 1 on the index is one standard deviation more distressed than the average respondent in the sample.

In order to make our following methodological discussion as simple as possible, we use a slightly different model than Johnson and Wu and include only those $n = 1,166$ individuals in our analysis that participated in all $T = 4$ panel waves. This constitutes a *balanced panel* with $N = n \cdot T = 4,664$ data records. However, because of this selection, the following estimates do not exactly replicate Johnson and Wu's results, but the substantive findings remain the same.

To make things simple, the following analysis ignores the question of whether the consequences of divorce are temporary or permanent (but see Sect. 4.2.2). We will only focus on the question of whether the social role of being divorced creates distress when taking into account that individuals with poor mental health are more likely to get divorced (i.e., have a higher probability of being selected into divorce). The social selection indicator socsel is a key variable in this comparison, because

4.1 Modeling the Level of Y

it is hypothesized to measure all pre-divorce disposition to psychological distress, which, when significant, would challenge the causal explanation of the divorce effect. Hence, we want to know whether there is still a negative effect of the respondent's current social role of being divorced (`divorce`) on psychological distress, even if `socsel` is controlled for.

If the comparison between social role and social selection explanations is the research question, we have to decide how to model the dependent variable and how to estimate the parameters of the model.

1. First, it is not quite clear whether this research question implies an analysis of the level of, or the change in, psychological distress. If one only wishes to show that divorce (a change in marital status)—net of social selection—has a positive effect on distress (i.e., changes distress levels in the positive direction), an analysis of *change* in a sample of individuals at risk for divorce (i.e., a sample of married individuals) would be the preferred research design. Within this research design, all the relevant information comes from those individuals who change their marital status. However, if one wishes to show that—net of social selection—the average distress levels of divorced individuals are higher than the average distress levels of married individuals, an analysis of distress *levels* in a sample of divorced and married individuals would be the preferred research design. In this section, we will use this second research design and analyze the level of psychological distress. However, in Sect. 4.2.1 on change models, we will also apply the first research design. Having seen both research designs, we can then understand that the differences between both research designs are not that significant, and that the essential question is how to control for social selection. As it will turn out, this is most convincingly done in both research designs by comparing both marital statuses *within* the partnership biographies of individuals over time. Naturally, this is only feasible for those individuals who have been married at least once and divorced at least once during the observation period.

2. Having decided on an analysis of levels, the next question is how to estimate the parameters of the model. If the task is to estimate the divorce effect net of selection, this problem is easily tackled with FE estimation, because it controls for all time-constant characteristics of the sample members and hence also for any pre-divorce disposition to psychological distress. However, if we want to compare the importance of the two theoretical explanations (social selection, social role), we need an estimate of their respective size. This is not feasible with FE, because it does not provide estimates for the time-constant variables in the regression equation (among them `socsel`). Therefore, we would be looking for other panel estimation techniques.

As a reference point, we start again with pooled OLS. Table 4.8 (column "pooled OLS") shows a significant positive effect of social selection on psychological distress ($t = 5.86$, $p < 0.01$). Individuals with higher divorce probabilities, as indicated by their divorce experiences in the past (`socsel = 1`), show distress levels that are 0.1925 standard deviations higher than the distress level of the average respondent in the sample. However, as we already know, pooled OLS is not very useful for panel data. If we look at the correlations of the OLS residuals from different

Table 4.8 Determinants of psychological distress: Pooled OLS, FE, and RE (1980–1992)

Variable	Pooled OLS		FE		RE	
	Estimate	Std. Err.	Estimate	Std. Err.	Estimate	Std. Err.
Social selection (yes = 1)	0.1925	0.0329	(dropped)		0.2221	0.0455
Divorced (yes = 1)	0.3020	0.0575	0.1696	0.0492	0.2215	0.0472
Widowed (yes = 1)	0.1918	0.0837	0.1021	0.0877	0.1462	0.0790
Cohabiting (yes = 1)	−0.4935	0.1202	−0.3475	0.1040	−0.3912	0.0998
Age (years)	0.0012	0.0010	0.0087	0.0017	0.0042	0.0012
Female (yes = 1)	−0.0300	0.0196	(dropped)		−0.0227	0.0302
Education (years)	−0.0365	0.0036	−0.0234	0.0096	−0.0329	0.0049
Constant	0.3613	0.0676	−0.1195	0.1415	0.1825	0.0882
R^2	0.0466		0.0162		0.0441	
$F(U)$ or X_1^2			4.43		1,464.19	
df_1, df_2			1,165	3,493	1	
$F(X, Z)$ or X_2^2	32.49		11.51		127.10	
df_1, df_2	7	4,656	5	3,493	7	
r_u			−0.0623		0.0000	
$\hat{\sigma}_u$			0.5062		0.4325	
$\hat{\sigma}_e$			0.4678		0.4678	
$\hat{\rho}_{FE}, \hat{\rho}_{RE}$			0.5394		0.4609	
N	4,664		4,664		4,664	
n			1,166		1,166	
T			4		4	

Source: johnson-wu data (see Example 4.2)

panel waves, we observe a fair degree of serial correlation in the data. According to the figures in Table 4.9, the average of these serial correlation coefficients equals $\bar{r}_{OLS} = 0.4662$.

Given our discussion in Sect. 4.1.1.1, this large degree of serial correlation comes as no surprise. Because we analyze a panel, i.e., repeated measurements of the same units over time, observations for each unit are not independent of one another. Unobserved individual heterogeneity, as we have said, is one of the causes for OLS residuals to be correlated over time and for OLS estimates to be inefficient and possibly biased. A safer approach would be to use FE estimation. The estimates in Table 4.8 (column "FE") show us that there is a significant causal effect of divorce, net of social selection and all the other controls in the model ($t = 3.45$, $p < 0.01$). Individuals getting divorced during the course of the study show distress levels that are 0.1696 standard deviations higher than the distress level of the average respondent in the sample. Unfortunately, FE estimation conceals the effects of the time-constant variables (among them socsel), which Johnson and Wu would like to compare with the divorce effect.

4.1 Modeling the Level of Y

Table 4.9 Correlation of the residuals from the pooled OLS model

	Wave 1	Wave 2	Wave 3	Wave 4
Wave 1	1.0000			
Wave 2	0.4985	1.0000		
Wave 3	0.4201	0.4918	1.0000	
Wave 4	0.3936	0.4505	0.5425	1.0000
Mean	$\bar{r}_{OLS} = 0.4662$			

Source: johnson-wu data (see Example 4.2)

Now, let us assume for a moment that unobserved heterogeneity at the unit level is uncorrelated with all explanatory variables in the model. In this case, we have no problem with omitted variable bias, and it seems like we could use the pooled OLS estimate that would provide us with an unbiased estimate of the divorce effect. This is easily seen from (4.12): If social selection is independent of the variables in the model, the association between divorce and unit-specific heterogeneity $\hat{v}_i$ will be essentially zero ($\hat{\delta}_{divorce} = 0$), when controlling for all the other variables in the model. In that case, no bias will be observed: $\hat{\beta}_{divorce}^{OLS} - \hat{\beta}_{divorce}^{FE} = 1 \cdot \hat{\delta}_{divorce} = 0$. However, even if unobserved heterogeneity is uncorrelated with the variables in the model, it still exists, because not all relevant unit characteristics are likely to be included in the model. As a consequence, measurements over time will be correlated for each unit (see Sect. 3.6.2) and the no-autocorrelation assumption of OLS estimation is at stake (see Textbox 4.1).

This can be tested more formally with a Lagrange multiplier (LM) test, proposed by Breusch and Pagan (1980). It tests the hypothesis $H_0: \sigma_u^2 = 0$ (or Corr($\varepsilon_{it}, \varepsilon_{is}$) = 0, $t \neq s$) versus $H_1: \sigma_u^2 \neq 0$. If the null hypothesis is true, pooled OLS can be used to estimate the regression coefficients. Lagrange multiplier tests play a role within the context of maximum likelihood estimation, which is not a topic that we will address in this chapter. Therefore, we will not discuss the Breusch–Pagan test in greater detail. The popularity of the test comes from the fact that the test statistic itself is easily computed from the residuals ($\hat{\varepsilon}_{it} = y_{it} - \hat{y}_{it}^{OLS}$) of the pooled OLS model:[19]

$$X_1^2 = \frac{n \cdot T}{2 \cdot (T-1)} \left(\frac{\sum_{i=1}^{n}(\sum_{t=1}^{T}\hat{\varepsilon}_{it})^2}{\sum_{i=1}^{n}\sum_{t=1}^{T}\hat{\varepsilon}_{it}^2} - 1 \right)^2 \qquad X_1^2 \sim \chi^2(1) \qquad (4.19)$$

It is also readily available in many statistical program packages. The LM statistic X_1^2 is distributed as χ^2 with $df = 1$ degree of freedom, and, in our case, it amounts to $X_1^2 = 1{,}464.19$, which is highly significant ($p < 0.01$) (see Table 4.8, column "RE"). We conclude that the OLS assumption of uncorrelated error terms is not valid for Johnson and Wu's data, as could be expected from the serial correlations in Table 4.9. The test suggests giving up pooled OLS estimation.

---

[19] We use X^2 to symbolize test statistics that are distributed as χ^2. However, since there are several of them, we distinguish them by indices (X_1^2, X_2^2, etc.).

In this situation, one option is to use robust standard errors (see Sect. 4.1.1.2). The other is to use our knowledge of the nature of the serial correlations and generalize OLS estimation in such a way that it can deal with them. If our model of the serial correlations and the assumption of independent unobserved heterogeneity are true, this generalized ordinary least squares (GLS) estimation will provide us with more efficient estimates of both the regression coefficients and the standard errors. This technique is also known as random effects (RE) estimation.

We start with the pooled OLS model (4.3) including an error term, ε_{it}, which is now assumed to be independent of the systematic part of the model. We generalize this model by assuming that the error term consists of two components u_i and e_{it} measuring unobserved heterogeneity at the unit level and idiosyncratic errors at the level of measurements. We have

$$y_{it} = \beta_0 + \beta_1 x_{1it} + \cdots + \beta_k x_{kit} + \gamma_1 z_{1i} + \cdots + \gamma_j z_{ji} + \varepsilon_{it} \quad \text{with } \varepsilon_{it} = u_i + e_{it} \quad (4.20)$$

Unobserved heterogeneity (u_i) is assumed to be one of the causes of serially correlated measurements, but contrary to the fixed effects approach in the previous section, the u_is are now seen as one component of the stochastic part of the model. Similar to the traditional OLS error term, they are treated as realizations of a random variable, which gives the whole procedure its name (*random effects estimation*). Because the error term ε_{it} is split into two components u_i and e_{it}, another name is the *error* or *variance components model*. Textbox 4.4 summarizes the main features of RE estimation.

Textbox 4.4 (RE assumptions)   RE estimation builds on the linear model (4.20) for panel data that includes both time-dependent and time-constant independent variables and an error term consisting of two components: unobserved heterogeneity at the unit level, u_i, and idiosyncratic errors, e_{it}, at the level of measurements over time. The assumptions for RE estimation are very similar to the FE assumptions, except that u_i is now assumed to be a random variable with specific characteristics:

1. The units $i = 1, \ldots, n$ in the panel data set are a simple random sample from a cross-section of a well-defined population.
2. The model is linear in its parameters $\beta_0, \beta_1, \ldots, \beta_k$ and $\gamma_1, \ldots, \gamma_j$.
3. The independent variables must not necessarily change over time. Time-constant variables $z_{1i}, \ldots, z_{ji}$ are allowed in model (4.20). But no independent variable, whether time-constant or time-varying, is allowed to be a constant or a linear function of the other independent variables.
4. Idiosyncratic error is independent of the variables in the model and independent of unit-specific unobserved heterogeneity: $E(e_{it}|x_{1i1}, \ldots, x_{1iT}, \ldots, x_{ki1}, \ldots, x_{kit}, z_{1i}, \ldots, z_{ji}, u_i) = 0$. Additionally, unit-specific unobserved heterogeneity on average may be different from zero. The average level of U is captured by the constant β_0 in RE models. But unit-

specific heterogeneity must be independent of the variables in the model: $E(u_i|x_{1i1},\ldots,x_{1iT},\ldots,x_{ki1},\ldots,x_{kit},z_{1i},\ldots,z_{ji}) = \beta_0$.
5. Idiosyncratic error has constant variance, given any value of the independent variables and unit-specific effects: $\text{Var}(e_{it}|x_{1i1},\ldots,x_{1iT},\ldots,x_{ki1},\ldots,x_{kit},z_{1i},\ldots,z_{ji},u_i) = \sigma_e^2$. The same is true for unit-specific error: $\text{Var}(u_i|x_{1i1},\ldots,x_{1iT},\ldots,x_{ki1},\ldots,x_{kit},z_{1i},\ldots,z_{ji}) = \sigma_u^2$.
6. Idiosyncratic error is uncorrelated between any two observations t and s ($t \neq s$) of unit i, given any value of the independent variables and unit-specific effects: $\text{Corr}(e_{it},e_{is}|x_{1i1},\ldots,x_{1iT},\ldots,x_{ki1},\ldots,x_{kit},z_{1i},\ldots,z_{ji},u_i) = 0$.
7. Unobserved heterogeneity is normally distributed, $u_i \sim Normal(0,\sigma_u^2)$, given any value of the independent variables. The same is assumed for idiosyncratic errors: $e_{it} \sim Normal(0,\sigma_e^2)$.

The first assumption is necessary for making statistical inferences about the population using the sampled data. The second assumption ensures that we can use least squares estimation to estimate the parameters of the model. If the model were not linear in its parameters, we would have to use other estimation techniques. The third assumption guarantees that a numerical value exists for each regression coefficient. The fourth assumption ensures unbiasedness of the RE estimates: $E(\hat{\beta}_0) = \beta_0$, $E(\hat{\beta}_1) = \beta_1,\ldots, E(\hat{\beta}_k) = \beta_k$ and $E(\hat{\gamma}_1) = \gamma_1,\ldots, E(\hat{\gamma}_j) = \gamma_j$. It should be noted that strict exogeneity is also necessary with RE estimation. Assumptions 5 and 6 guarantee the efficiency of RE estimation. If assumptions 4 to 6 are true, RE estimates are consistent and asymptotically normally distributed. Assumption 7 is not really necessary, but if we make that assumption, we can also apply maximum likelihood estimation, which is another estimation technique for RE models. When assumptions 4 to 6 hold, it can also be shown that RE estimation is more efficient than pooled OLS or FE estimation. Discussions of these RE assumptions and proofs of the statistical properties of RE can be found in Wooldridge (2010, 257).

Basically, RE estimation assumes that the two error components u_i and e_{it} (i) are uncorrelated with the explanatory variables in the model, (ii) have constant variance σ_u^2 and σ_e^2, and (iii) are independent of each other and across different units. Finally, (iv) e_{it} is assumed not to be serially correlated over time. Given these assumptions, it is easy to derive variances and covariances of the composite error $\varepsilon_{it} = u_i + e_{it}$ and finally a formula for the expected amount of serial correlation, if these assumptions are true (for a derivation of this formula see Textbox 3.1):

$$\text{Corr}(\varepsilon_{it},\varepsilon_{is}) = \frac{\text{Cov}(\varepsilon_{it},\varepsilon_{is})}{\sqrt{\text{Var}(\varepsilon_{it})} \cdot \sqrt{\text{Var}(\varepsilon_{is})}} = \frac{\sigma_u^2}{\sigma_u^2 + \sigma_e^2}, \quad t \neq s \quad (4.21)$$

The expected degree of serial correlation is a function of the variances of both error components.

As discussed in Sect. 3.4.2, this model of serial dependence assumes an equal correlation structure, where the amount of serial correlation is the same irrespective of the time lag $(t - s)$ between observations. Of course, it is easy to imagine more complicated correlation structures than the ones implied by the RE assumptions. For example, instead of the equal correlation structure, it may be more useful to assume either correlations that are different for each specific combination of t and s, or that decline with the time lag between both measurements (for the latter see Table 3.8). On the other hand, it is also easy to think of even simpler processes. Pooled OLS, for example, assumes that observations are not serially correlated at all and that the error term has constant variance (see Textbox 4.1). In a way, the usual OLS assumptions are only a special case of the other, more complicated, correlation structures. With this idea in mind, we can now consider how to generalize OLS in order to deal with these complexities. Unfortunately, these derivations need a basic understanding of matrix algebra, which we did not presuppose for this textbook. Basically, these matrix operations transform the data (Y, X, Z) in a way to get rid of the more complicated correlation structures. After this transformation, the data are supposed to behave according to the OLS assumptions, and pooled OLS can be applied to the transformed data. The necessary mathematical derivations for this claim are a little involved (see Greene (2008, 202) for more details), but estimating the following model with pooled OLS provides us with the RE estimates that control for the equal correlation structure of the error term:

$$y_{it} - \theta \bar{y}_{i.} = \beta_0(1-\theta) + \beta_1(x_{1it} - \theta \bar{x}_{1i.}) + \cdots + \beta_j(x_{kit} - \theta \bar{x}_{ki.}) + \gamma_1(z_{1i} - \theta z_{1i})$$
$$+ \cdots + \gamma_k(z_{ji} - \theta z_{ji}) + (u_i - \theta u_i) + (e_{it} - \theta \bar{e}_{i.}) \quad (4.22)$$

Obviously, before applying pooled OLS, we have to transform the data by subtracting from each variable Y, X, and Z a fraction θ of its mean.[20] The fraction θ depends on the variance of both error components and on the number of measurements over time:

$$\theta = 1 - \sqrt{\frac{\sigma_e^2}{\sigma_e^2 + T\sigma_u^2}} \quad (4.23)$$

Since we subtract only part of each variable's mean, this operation is also called *quasi-demeaning* (as opposed to demeaning in FE estimation when $\theta = 1$). We call θ the *demeaning parameter*. Compared to FE estimation, time-constant variables remain in the model, because we are subtracting only a fraction of each variable's mean.

---

[20] Note that the mean of a time-constant explanatory variable Z equals the variable itself (e.g., $\bar{z}_{1i} = z_{1i}$). Note, also, that you have to generate a new "variable" constant that equals $(1-\theta)$ to obtain an estimate of the regression constant. Include constant in your model as an additional variable and force your regression program not to estimate a regression constant. The RE constant is estimated by the regression coefficient of the variable constant.

4.1 Modeling the Level of Y

But how are quasi-demeaning and RE estimation feasible, if we do not know the population variances σ_u^2 and σ_e^2? In order to apply (4.22), we need estimates for both variances. The estimated standard deviations of both error components are shown in the lower part of Table 4.8 (column "RE"). As you can see, the estimate of σ_e ($\hat{\sigma}_e = 0.4678$) is identical to the corresponding estimate of the FE model (see column "FE" and (4.18)). The estimate of σ_u ($\hat{\sigma}_u = 0.4325$) is a function of the sum of squared residuals of the corresponding between effects model and the estimate of σ_e:

$$\hat{\sigma}_u^2 = \max\left(0, \frac{\sum_{i=1}^n (\bar{y}_{i.} - \hat{y}_{i.}^{BE})^2}{n-k-j-1} - \frac{\hat{\sigma}_e^2}{T}\right) \quad (4.24)$$

k equals the number of time-varying and j the number of time-constant explanatory variables in the model. $\hat{y}_{i.}^{BE}$ are the predicted values of a between effects model that we discuss later (see Textbox 4.5 on p. 158).

Using (4.21) and the two estimates of σ_u^2 and σ_e^2, the serial correlation of the composite error is estimated as $\hat{\rho}_{RE} = 0.4609$ (see Table 4.8, column "RE"), which comes pretty close to the mean of all serial correlations of the residuals in the pooled OLS model ($\bar{r}_{OLS} = 0.4662$, see Table 4.9). In case of an empty RE model including no explanatory variables, (4.18) and (4.24) reduce to (3.8) and (3.10) and $\hat{\rho}_{RE}$ equals the ICC coefficient (see Textbox 3.1). For the johnson-wu data, it is estimated as $\hat{\rho} = 0.4788$, indicating that the $(k+j=7)$ explanatory variables of the full RE model explain only very few of the serial dependencies in the data.

Having estimates of both variance components, we can now proceed and find estimates of the RE regression coefficients. Applying (4.22) with $\hat{\theta} = 1 - \sqrt{\hat{\sigma}_e^2/(\hat{\sigma}_e^2 + T\hat{\sigma}_u^2)}$ based on estimates of σ_u^2 and σ_e^2 has become known as *feasible generalized least squares* (FGLS) estimation. Of course, we can perform all these transformations by hand, but there is no need to, because readily programmed algorithms for RE estimation exist in many statistical software packages. Reestimating Johnson and Wu's model with pooled OLS and the transformed data results in estimates that are shown in column "RE" of Table 4.8. Using the estimates of both variance components we can compute

$$\hat{\theta} = 1 - \sqrt{\frac{\hat{\sigma}_e^2}{\hat{\sigma}_e^2 + T\hat{\sigma}_u^2}} = 1 - \sqrt{\frac{0.4678^2}{0.4678^2 + 4 \cdot 0.4325^2}} = 0.5243$$

In other words, about 52.4 % of each variable's mean has been subtracted from the original values. We can also illustrate that this transformation controls for the dependencies over time in the johnson-wu data. According to the statistical theory, residuals from a RE model should be statistically independent. The serial correlations of the residuals from our RE model are shown in Table 4.10. On average, they are much smaller ($\bar{r}_{RE} = 0.0075$) than the serial correlations for the pooled OLS residuals ($\bar{r}_{OLS} = 0.4662$, see Table 4.9). Of course, they are not exactly zero, because we are dealing with one specific sample, while the statistical argument applies to the expected value (the "average") of all samples that could be selected from the

Table 4.10 Correlation of the residuals from the RE model

	Wave 1	Wave 2	Wave 3	Wave 4
Wave 1	1.0000			
Wave 2	0.0741	1.0000		
Wave 3	−0.0683	0.0207	1.0000	
Wave 4	−0.0909	−0.0312	0.1406	1.0000
Mean	$\bar{r}_{RE} = 0.0075$			

Source: johnson-wu data (see Example 4.2)

population. Note, however, that two of the six serial correlations are not significantly different from zero.

After controlling for the panel structure of our data with RE estimation, we can now assess the effect of social selection on psychological distress more efficiently than with pooled OLS. The effect of the variable socsel remains significant ($z = 4.88$, $p < 0.01$) and increases slightly to a value of 0.2221. Compared to the pooled OLS parameter (0.1925), this is a more efficient estimate of Johnson and Wu's hypothesis that part of the greater distress among divorced individuals is caused by social selection (i.e., individuals with poor mental health being selected into divorce). A similar strong effect is found for *current* family status, as indicated by the significant regression coefficient of the variable divorce ($z = 4.70$, $p < 0.01$). Currently divorced individuals (divorce = 1) show distress levels that are 0.2215 standard deviations higher than the distress level of the average respondent in the sample. Obviously, the social role of being divorced increases distress levels significantly and to a similar degree as social selection does. However, we shall refrain from making a decision about the relative importance of the three theoretical explanations offered by Johnson and Wu (2002), because we have not yet controlled for possible time trends in the data, and thus cannot distinguish between the temporary and permanent effects of divorce. This will be done in Sect. 4.2.2 (see Table 4.15).

Before comparing RE and FE estimations, we will take a final look at the test statistics and fit measures of the RE model. First of all, it is not obvious how to compute R^2. The quasi-demeaned data are only a technical instrument to arrive at the GLS estimates without using complicated matrix algebra. Therefore, we are only interested in the estimates (including their variances and covariances), but not in the proportion of explained variance of the transformed dependent variable. To compute R^2, we use the RE estimates and the *original* (untransformed) data to estimate a predicted value of the dependent variable $\hat{y}_{it}^{RE} = 0.1825 + 0.2221 \cdot socsel_{it} + 0.2215 \cdot divorce_{it} + \cdots - 0.0329 \cdot educ_{it}$. The square of the correlation of predicted and observed psychological distress, $R^2 = (\text{Corr}(y, \hat{y}^{RE}))^2$, is the correct measure of explained variance we are looking for. In our case, it is distressingly low: Only 4.4 % of the variance in distress levels is explained by selection and role indicators, plus three additional controls (see Table 4.8, column: "RE").

With respect to test statistics, we have to keep in mind that the statistical properties of GLS estimators (consistency, efficiency) are only valid where large samples are used. Therefore, instead of T and F tests, we have to use the normal and the chi-square distribution. Testing single parameters is possible using the Z statistic, which

4.1 Modeling the Level of Y

is normally distributed. Linear restrictions on a set of parameters can be tested with a so-called Wald statistic X_2^2, which is χ^2 distributed. The Z statistic is computed exactly like the T statistic, only the test distribution is the normal instead of the T distribution. To compute the Wald statistic manually, you need matrix algebra. Basically, the Wald statistic is a weighted sum of all the differences between the tested parameters and the corresponding parameters assumed in the null hypothesis. These differences are weighted with the estimated variances and covariances of the estimated parameters (see Sect. 7.2.1). Statistical programs for RE estimation will automatically compute this statistic for you.

But you can also use a relation between the Wald and the F statistic that holds for linear models and OLS estimation. In that case, the Wald statistic equals q times the F statistic (q measuring the number of restrictions tested). For instance, if we want to test the joint hypothesis that all our $q = 7$ independent variables in the RE model (except the constant) have zero effects, we quasi-demean the data and use pooled OLS to compute the RE estimates manually with the transformed data (see (4.22)). The combined test of all $q = 7$ restrictions is based on comparing the residual sum of squares of the restricted (SSR_r) and the unrestricted model (SSR_{ur}, see Sect. 7.2.1). The corresponding F statistic is computed as follows:

$$f = \frac{(SSR_r - SSR_{ur})/q}{SSR_{ur}/(n \cdot T - k - 1)}, \qquad f \sim F(q, n \cdot T - k - 1) \qquad (4.25)$$

For Johnson and Wu's data, we arrive at

$$f = \frac{(1{,}050.64 - 1{,}022.72)/7}{1{,}022.72/(1{,}166 \cdot 4 - 7 - 1)} = 18.16$$

multiplied by $q = 7$ results in the Wald statistic ($X_2^2 = 127.10$) shown in Table 4.8. Compared to a chi-square distribution with $df = q = 7$ degrees of freedom, this value is highly significant ($p < 0.01$).

4.1.2.3 Combining Fixed and Random Effects Estimation: A Hybrid Model

Having introduced three different estimation procedures for panel data—pooled OLS, FE, and RE estimation—we can now discuss two further questions. How do these methods relate to one another, and can we combine their virtues? We start with the relationships between the methods (the following section on p. 157) and then briefly summarize their virtues (first section on p. 163). As it turns out, there is sort of a hybrid model that combines both the FE and the RE model (second section on p. 164). We conclude with a discussion of tests showing whether using FE or RE estimation makes a difference (section on p. 166).

Relationship Between Pooled OLS, FE, and RE As an introduction to the first question, let us consider the effect of current marital status in the former RE model. The estimate of the divorce effect (0.2215) represents a comparison between the distress levels of currently divorced and currently married individuals, while controlling for the values of the other independent variables. This is neither a pure between-respondent comparison (because current family status changes between panel waves for 138 individuals) nor a pure within-respondent comparison (because 1,028 respondents never divorce during the observation period). A within estimator like FE ($\hat{\beta}^{FE}$) would only use the variance within the time dimension, while a between estimator ($\hat{\beta}^{BE}$) would use only the variance between individuals. As we will shortly see, RE is estimating something "in between". This sounds like an optimal compromise, but as we will also see, both sources of variance may produce very different answers. There are situations in which it is necessary not to mix them up. To understand this discussion, we have to know a bit more about between effects (BE) estimation.

Textbox 4.5 (BE estimation) We do not want to discuss between effects (BE) estimation in greater detail, because BE estimates are seldom published. But, as we will see, it is interesting to compare BE estimates with other panel estimators. The main idea of BE estimation is easily explained. It is simply a regression with data that include only the unit-specific means of all variables in the model. Starting again from the basic panel regression model, we average (4.5) across time points for each individual. The result is the following BE model:

$$\bar{y}_{i.} = \beta_0 + \beta_1 \bar{x}_{1i.} + \cdots + \beta_k \bar{x}_{ki.} + \gamma_1 z_{1i} + \cdots + \gamma_j z_{ji} + u_i$$

In deriving this result, we make use of the fact that idiosyncratic error is zero on average and that averaging time-constant variables across time, by definition, results in the original values $z_{1i}, \ldots, z_{ji}$ (the same applies to unobserved heterogeneity u_i). It should be noted, however, that the number of cases and, correspondingly, the degrees of freedom change. Compared to FE and RE, BE estimation is based on n and not on $n \cdot T$ cases.

For illustrative purposes, Table 4.11 (column "BE") also includes the BE estimates for our model of psychological distress. As you can see, all the RE estimates lie between the corresponding FE and BE estimates. Quite generally, pooled OLS and RE estimates will *always* lie between the corresponding FE and BE estimates, with the pooled OLS estimates being more close to the BE estimates. This is due to the fact that both the pooled OLS and the RE estimates are matrix-weighted averages of the BE and FE estimates, with OLS placing greater weight on the between effects (Baltagi, 2008, 17; Greene, 2008, 295). Let us switch to a simpler example in order to understand the intermediary position of RE estimation.

4.1 Modeling the Level of Y

Table 4.11 Determinants of psychological distress: FE, RE, pooled OLS, and BE (1980–1992)

Variable	FE Estimate	FE Std. Err.	RE Estimate	RE Std. Err.	Pooled OLS Estimate	Pooled OLS Std. Err.	BE Estimate	BE Std. Err.
Social selection (yes = 1)	(dropped)		0.2221	0.0455	0.1925	0.0329	0.1179	0.0740
Divorced (yes = 1)	0.1696	0.0492	0.2215	0.0472	0.3020	0.0575	0.5726	0.1885
Widowed (yes = 1)	0.1021	0.0877	0.1462	0.0790	0.1918	0.0837	0.2431	0.1813
Cohabiting (yes = 1)	−0.3475	0.1040	−0.3912	0.0998	−0.4935	0.1202	−0.7604	0.3597
Age (years)	0.0087	0.0017	0.0042	0.0012	0.0012	0.0010	−0.0007	0.0016
Female (yes = 1)	(dropped)		−0.0227	0.0302	−0.0300	0.0196	−0.0351	0.0303
Education (years)	−0.0234	0.0096	−0.0329	0.0049	−0.0365	0.0036	−0.0386	0.0058
Constant	−0.1195	0.1415	0.1825	0.0882	0.3613	0.0676	0.4702	0.1120
R^2	0.0162		0.0441		0.0466		0.0434	
$F(U)$ or X_1^2	4.43		1,464.19					
df_1, df_2	1,165	3,493	1					
$F(X, Z)$ or X_2^2	11.51		127.10		32.49		13.24	
df_1, df_2	5	3,493	7		7	4,656	7	1,158
r_u	−0.0623		0.0000					
$\hat{\sigma}_u$	0.5062		0.4325					
$\hat{\sigma}_e$	0.4678		0.4678					
$\hat{\rho}_{FE}, \hat{\rho}_{RE}$	0.5394		0.4609					
N	4,664		4,664		4,664			
n	1,166		1,166				1,166	
T	4		4					

Source: johnson-wu data (see Example 4.2)

Example 4.3 (efficiency data) The example has been borrowed from the former analysis of government spending, but it is a thought experiment that may not be true in reality. Nevertheless, it illustrates a situation in which different panel estimation methods yield different answers. The example has several features:

- We want to show that our panel estimation methods (FE, RE, BE, pooled OLS) make different use of the between and within variance of Y. Therefore, we need a manageable number of units that have been observed over time, such that there is variance both between units and over time. That is why we return to Example 4.1 and use a sample of only $n = 4$ fictitious countries that have been observed over a period of $T = 10$ years.
- We assume that the efficiency hypothesis is true, which asserts that globalization forces governments to downsize their expenditure. According to this hypothesis, we will construct two data sets, including a measure of government spending (Y) and imports from low wage countries, as an indicator of globalization (X):
  1. The first data set is named efficiency1 and is constructed in such a way that between and within country variance provide identical conclusions. In this situation, all panel estimators are more or less the same.
  2. The second data set is named efficiency2 and is constructed in such a way that between and within country variance provide different conclusions. As we will see, unobserved heterogeneity that correlates with X is an explanation for this difference. FE, RE, BE, and pooled OLS will provide different estimates. This raises the question as to which estimate is the correct one. As it turns out, the best estimate is the FE estimate that uses only the within country variance.

Figure 4.3 shows the first simulated data set efficiency1. Each measurement is indicated by a hollow circle. The four solid black points indicate average government spending at average imports from low wage countries for each of the four countries. For example, if we number the countries from the left to the right, country No. 1 in the upper-left corner—on average—has far fewer low wage imports and more public spending than country No. 4 in the lower-right corner. The data are perfectly in line with the efficiency hypothesis: (i) Countries importing more on average from low wage countries (like countries 3 and 4) have much less government spending on average than countries with far fewer imports from low wage countries (like countries 1 and 2). (ii) Also within each country, government spending decreases with increasing low wage imports. In this situation, all panel estimators are more or less the same. In decreasing order, they amount to $\hat{\beta}^{FE}_{lowwage} = -0.81$ (see the dashed lines running through the hollow circles), $\hat{\beta}^{RE}_{lowwage} = -0.81$ (solid line), $\hat{\beta}^{OLS}_{lowwage} = -1.00$ (line with long dashes), and $\hat{\beta}^{BE}_{lowwage} = -1.02$ (dashed line that is nearly identical to the OLS line). Note, in passing, that the order of the four

Fig. 4.3 Explaining between and within variance (uncorrelated u_i)

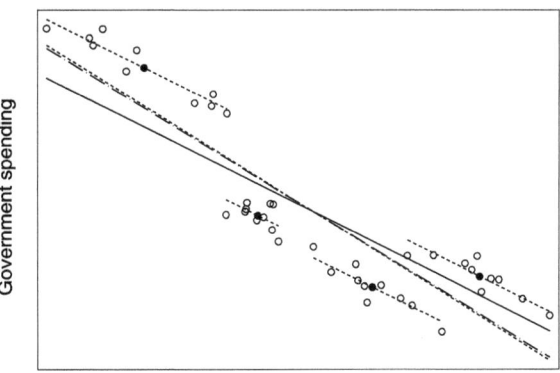

Imports from low wage countries
Source: efficiency1 data (see Example 4.3)

Fig. 4.4 Explaining between and within variance (correlated u_i)

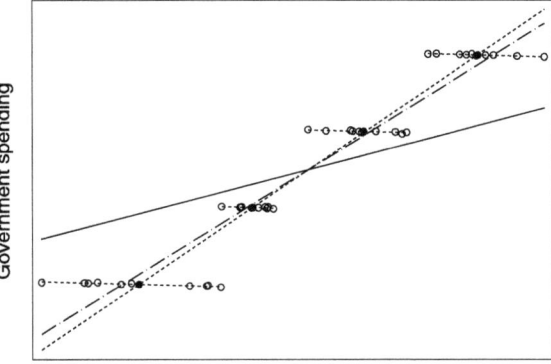

Imports from low wage countries
Source: efficiency2 data (see Example 4.3)

estimates is in line with our former statement about the "intermediary" position of RE estimates. The pooled OLS and the RE estimate lie between the FE and the BE estimate, with the pooled OLS estimate being closer to the BE estimate.

Now, let us turn to the second data set efficiency2, which is shown in Fig. 4.4. Things are pretty much the same as in efficiency1. Still, government spending decreases with increasing low wage imports *within* each country. However, at the country level, the situation is different. Contrary to the efficiency hypothesis, countries with—on average—a great deal of imports from low wage countries (like countries 3 and 4)—on average—also have a great deal of government spending. Obviously, there are some other factors operating at the country level that distort the relationship between government spending and low wage imports at the level of countries. This is a typical example of unobserved heterogeneity u_i at the unit level. Moreover, this unobserved heterogeneity correlates with the time-varying independent variable X (lowwage): $E(u_i|x_{1i1}, \ldots, x_{1iT}) \neq \beta_0$.

Very often, we will not know the source of this unobserved heterogeneity, or we have no measures of it. But let us speculate what these other factors could be in our example. Consider, for example, countries with capital-intensive and technology-oriented economies. Like other countries, they trade internationally, but may buy part of the necessary goods and services for their economy from low wage countries in particular, while earning their income by selling their high-tech products world-wide. For such a high-tech economy, these countries need a highly developed educational system and an infrastructure that supports technological change. If the governments of these countries pay large sums of money for their educational system and the high-tech sector of their economy, this would explain why government spending in these countries is above average for all countries. In other words: at the country level government spending increases with the technological development of the economy, which simultaneously implies more imports from low wage countries.

If we now apply different panel estimation methods to the efficiency2 data, we see that the technology effect is measured by the BE estimator which uses only the between-country variance, i.e., the country-specific averages (see the dashed line running close to the solid black points from the lower left to the upper right in Fig. 4.4: $\hat{\beta}_{lowwage}^{BE} = 4.53$). Within each country, however, government spending decreases with low wage imports, as expected by the efficiency hypothesis. This effect is measured by the FE estimator, which uses only the within-country variance, i.e., the time-demeaned data (see the dashed lines running through the data points for each country: $\hat{\beta}_{lowwage}^{FE} = -0.19$). The effect is rather small (the negative slope of the dashed lines is hardly visible), but it is the "true" effect of low wage imports on government spending that controls for the fact that certain countries trade with low wage countries more than others.

The data set efficiency2 has been constructed in such a way that unobserved country-specific heterogeneity u_i is correlated with the independent variable lowwage in the model. What happens to our pooled OLS and RE estimates now? Given our former discussion, pooled OLS estimates should be biased. This can be seen from the line with long dashes running from the lower-left to the upper-right corner in Fig. 4.4. Its slope equals the pooled OLS estimate ($\hat{\beta}_{lowwage}^{OLS} = 4.14$), which is nearly as large as the BE estimate. OLS is biased because it picks up most of the between-country variance and ignores a major part of the within-country variance. Concerning RE estimation, we must remember that this method also assumes independent unobserved heterogeneity (see Textbox 4.4). Therefore, its low wage estimate is biased as well (see the solid line in Fig. 4.4: $\hat{\beta}_{lowwage}^{RE} = 1.74$), but not as much as the OLS estimate. The RE estimate lies somewhere between the FE and the BE estimate: $\hat{\beta}_{lowwage}^{FE} < \hat{\beta}_{lowwage}^{RE} < \hat{\beta}_{lowwage}^{BE}$.[21] In other words: RE estimation is a mixture of between and within-unit variance of the dependent variable, and assumes that between (BE) and within effects (FE) are the same.

---

[21] The same is true for the OLS estimate, but it is far more similar to the BE estimate than the RE estimate.

4.1 Modeling the Level of Y

This conclusion also suggests a remedy to control for RE bias: If one uses the average amount of low wage imports in each country ($\overline{lowwage_{i.}}$) as an indicator for the technological development of its economy, and one includes this as an additional explanatory variable in the model, then one controls for the between country variance, and the RE estimate of lowwage should measure only the within effect. Indeed, such an extended model, estimated with RE, replicates the FE estimate ($\hat{\beta}^{FE}_{lowwage} = -0.19$):

$$spend_{it} = 24.66 + 4.72 \cdot \overline{lowwage_{i.}} - 0.19 \cdot lowwage_{it}$$

Another solution would be a model specification that allows between and within effects to be different. This is easily done by specifying a RE model with the averages as an indicator of between-unit variance and demeaned values of lowwage as an indicator of within-unit variance. This RE model replicates both the BE ($\hat{\beta}^{BE}_{lowwage} = 4.53$) and FE estimate ($\hat{\beta}^{FE}_{lowwage} = -0.19$):

$$spend_{it} = 24.66 + 4.53 \cdot \overline{lowwage_{i.}} - 0.19 \cdot (lowwage_{it} - \overline{lowwage_{i.}})$$

In the second section on p. 164, we will discuss these kinds of hybrid model that somehow combine the virtues of FE and RE estimation.

When to Apply Pooled OLS, FE, and RE? Although computationally the simplest technique, pooled OLS is also the most demanding estimation procedure among the three in terms of assumptions. It assumes that all unit-specific heterogeneity can be controlled for by the independent variables in the model, so that the remaining unexplained variance is simply "white noise". Social science theories and data hardly fulfill this assumption. Therefore, pooled OLS estimates are often biased because of correlated unobserved heterogeneity. Even if unobserved heterogeneity is independent of the variables in the model, it causes the error term to be serially correlated and, therefore, OLS standard errors will underestimate the true standard errors. Pooled OLS estimation can be a starting point, but it will seldom provide the final estimates.

Hence, in most panel applications, a choice has to be made between RE and FE estimation. It seemed as though from a substantive point of view, RE estimation would provide the most comprehensive conclusions from the data, because it allows us to estimate the effects of both time-constant and time-varying variables. But RE estimation—like pooled OLS—makes restrictive assumptions too. It admits unobserved unit-specific heterogeneity, but assumes it to be independent of the explanatory variables X and Z. From our last Example 4.3, we have learned that the failure of this assumption results likewise in biased estimates. This casts a negative light on our seemingly practical RE estimator.

But how can we decide whether RE estimates are biased? Our last example hints at an answer: If FE and RE estimates differ substantially, RE estimation might be biased. At this point, some people give up using RE estimation and stick with FE estimation only. Nevertheless, for two reasons, RE estimation is quite popular in panel research. Firstly, in some applications, the effects of time-constant independent variables are a central research question. The analysis of Johnson and Wu (2002) is an example. The effects of time-constant independent variables cannot be estimated with FE estimation. Secondly, we have to "demean" our data to arrive at FE estimates, and this results in a loss of degrees of freedom and produces correspondingly less precise estimates, especially if there is not much over time variation in the data.

Fortunately, according to Mundlak, the distinction between random and fixed effects models is "arbitrary and unnecessary" (Mundlak, 1978, 70). When the model is properly specified, he argues, the RE estimator is identical to the FE estimator. "Thus there is only one estimator" (Mundlak, 1978, 70). As we have seen in the previous section, it is easy to derive the FE estimate within a RE model, once we give up the restriction that between and within-unit effects should be the same. This is basically Mundlak's reconciliation of the FE and RE model. In the following section, we will apply this so-called hybrid model to Example 4.2. Furthermore, several overall test procedures have been proposed to test whether RE and FE estimates are significantly different. In the section on p. 166, we will explain one popular example, the so-called Hausman test. But it turns out that the hybrid model provides a much more comprehensive and flexible environment for these kinds of test.

A Hybrid Model At the end of the section on p. 157, we saw that it is possible to obtain an unbiased estimate of the effect of time-varying explanatory variables X in RE models, even if unobserved heterogeneity is correlated with X. The trick was to include the unit-specific means of X as additional explanatory variables. This can also be shown more formally. Let us assume that our dependent variable depends on two variables, one time-varying (X) and another time-constant (Z):

$$y_{it} = \beta_0 + \beta_1 x_{1it} + \gamma_1 z_{1i} + u_i + e_{it} \quad (4.26)$$

Erroneously, a researcher has neglected the time-constant variable Z and specified the following RE model:

$$y_{it} = \beta_0' + \beta_1' x_{1it} + u_i' + e_{it}' \quad (4.27)$$

Parameter estimates will now be different from (4.26) and thus, are symbolized as β_0' and β_1' instead of β_0 and β_1. If the unit-specific effect u_i' represents the effect of unobserved heterogeneity, it should have the expectation $E(u_i') = \gamma_1 z_{1i} + u_i$. As we know, omission of Z is problematic, if Z and X are correlated. This statistical dependence can be expressed with the following regression model:

$$z_{1i} = \delta_0 + \delta_1 \bar{x}_{1i.} + \varepsilon_i \quad (4.28)$$

in which $\bar{x}_{1i.}$ is the mean of all values of the time-varying variable X for unit i. ε_i is the usual OLS error term. In this model we use the unit-specific mean $\bar{x}_{1i.}$ instead of x_{1it} as the regressor variable, because $x_{1it} = (x_{1it} - \bar{x}_{1i.}) + \bar{x}_{1i.}$ and the regression coefficient in a regression of z_{1i} on $(x_{1it} - \bar{x}_{1i.})$ is zero. Substituting (4.28) into $E(u_i') = \gamma_1 z_{1i} + u_i$, computing the expectation and substituting the result

4.1 Modeling the Level of Y

into the misspecified equation (4.27) demonstrates that the unit-specific effect u'_i of the misspecified RE model depends on the mean $\bar{x}_{1i.}$ of the time-varying explanatory variable X:

$$y_{it} = \beta'_0 + \beta'_1 x_{1it} + (\gamma_1 \cdot (\delta_0 + \delta_1 \bar{x}_{1i.} + \varepsilon_i) + u_i) + e'_{it} \quad (4.29)$$

After rearranging terms and renaming some parameters we arrive at an extended regression model including—besides the original time-varying variable x_{1it}—its unit-specific mean $\bar{x}_{1i.}$:

$$\begin{aligned} y_{it} &= \beta'_0 + \beta'_1 x_{1it} + \gamma_1 \delta_0 + \gamma_1 \delta_1 \bar{x}_{1i.} + \gamma_1 \varepsilon_i + u_i + e'_{it} \\ &= (\beta'_0 + \gamma_1 \delta_0) + (\beta'_1) \cdot x_{1it} + (\gamma_1 \delta_1) \cdot \bar{x}_{1i.} + (u_i + \gamma_1 \varepsilon_i) + e'_{it} \\ &= \beta''_0 + \beta''_1 \cdot x_{1it} + \varphi''_1 \cdot \bar{x}_{1i.} + u''_i + e''_{it} \end{aligned} \quad (4.30)$$

The regression coefficient β''_1 will now equal the true regression coefficient β_1 in the population model (4.26). It is identical to the FE estimate that automatically controls for all unobserved heterogeneity that is correlated with X (including, for example, the neglected time-constant variable Z). Intuitively, this result can be explained as follows: although the extended regression model (4.30) does not include the "forgotten" variable Z, it controls for this unobserved heterogeneity by including another characteristic of each individual, i.e., the unit-specific mean of X.

We have done the same for all time-varying explanatory variables in our model of psychological distress (Example 4.2), and have reestimated the model with RE. The estimates are shown in column "Model 1" of Table 4.12. If we include the means of *all* time-varying explanatory variables, RE estimates of the effects of the time-varying explanatory variables in this hybrid model (HY1) are identical to the corresponding FE estimates (compare Table 4.11, column "FE").[22] Additionally, the model includes the effects of the time-constant variables that were of special interest to the researchers (e.g., socsel). The social selection effect ($\hat{\gamma}^{HY1}_{socsel} = 0.1179$) is still positive, but smaller and no more significant ($z = 1.59$, $p > 0.10$). Hence, Johnson and Wu's hypothesis that higher stress levels among the divorced partially result from selection of people with poor mental health into divorce is at stake in this more comprehensive model.

This raises the question of what the effects of time-constant explanatory variables Z actually measure in these kinds of hybrid model. It turns out that they replicate the BE estimates of the Z effects exactly (compare Table 4.11, column "BE"). From a substantive point of view, they are the effects of the Z on the dependent variable Y in a randomly chosen cross-section of the panel, controlling for the effects of the time-varying independent variables X. Instead of randomly choosing one cross-section, we can also use the mean of Y as the dependent variable, and the means of the X and

---

[22] It is essential to include the means of *all* X into the model. Otherwise, the FE estimates will *not* be replicated. Moreover, you *must* have a *balanced* panel. Otherwise, the effects of the time-varying explanatory variables in the hybrid model will only approximate the FE estimates. Depending on the amount of missing panel data, the differences may become quite large.

Table 4.12 Determinants of psychological distress: Hybrid models (1980–1992)

Variable	Model 1 RE		Model 2 RE	
	Estimate	Std. Err.	Estimate	Std. Err.
	Original variables		Demeaned variables	
Social selection (yes = 1)	0.1179	0.0740	0.1179	0.0740
Divorced (yes = 1)	0.1696	0.0492	0.1696	0.0492
Widowed (yes = 1)	0.1021	0.0877	0.1021	0.0877
Cohabiting (yes = 1)	−0.3475	0.1040	−0.3475	0.1040
Age (years)	0.0087	0.0017	0.0087	0.0017
Female (yes = 1)	−0.0351	0.0303	−0.0351	0.0303
Education (years)	−0.0234	0.0096	−0.0234	0.0096
	Mean variables		Mean variables	
Divorced (yes = 1)	0.4030	0.1948	0.5726	0.1885
Widowed (yes = 1)	0.1410	0.2014	0.2431	0.1813
Cohabiting (yes = 1)	−0.4129	0.3744	−0.7604	0.3597
Age (years)	−0.0094	0.0023	−0.0007	0.0016
Education (years)	−0.0152	0.0112	−0.0386	0.0058
Constant	0.4702	0.1120	0.4702	0.1120
R^2	0.0514		0.0514	
$X_2^2 (X, Z)$	150.21		150.21	
df_1, df_2	12		12	
X_2^2 (mean variables)	22.62		22.62	
df	5		5	
r_u	0.0000		0.0000	
$\hat{\sigma}_u$	0.4325		0.4325	
$\hat{\sigma}_e$	0.4678		0.4678	
$\hat{\rho}_{FE}, \hat{\rho}_{RE}$	0.4609		0.4609	
N	4,664		4,664	
n	1,166		1,166	
T	4		4	

Source: johnson-wu data (see Example 4.2)

the values of Z as the independent variables, in this cross-sectional OLS regression. That is basically what BE estimation is doing (see Textbox 4.5). But are these BE estimates of the effects of Z better estimates than the former RE estimates? We think they are, because effect estimates of time-constant Z, by definition, can only exploit the between variation. But in a panel context they should at least control for the time-varying independent variables, which they do by including the means of the X.

4.1 Modeling the Level of Y

Testing Differences Between RE and FE Estimates It is also interesting to take a look at the estimates and standard errors of the unit-specific means (mdivorce, mwidow1, etc.) in the hybrid model HY1. Significant effects indicate those effects that are possibly biased in the simple RE model from Table 4.11, because the effects of the unit-specific means measure how much the BE and FE estimates deviate from each other, while the simple RE model assumes that both effects are the same. For example, the effect of mdivorce is significant ($z = 2.07$, $p < 0.05$) and estimated as $\hat{\gamma}_{mdivorce}^{HY1} = 0.4030$ (see Table 4.12, column "Model 1"). The estimate equals the difference between the BE ($\hat{\beta}_{divorce}^{BE} = 0.5726$) and the FE estimate ($\hat{\beta}_{divorce}^{FE} = 0.1696$) of the divorce effect. Similarly, the effect of the variable mager is significant ($z = -4.04$, $p < 0.05$), indicating significant differences between BE and FE estimates also for the age effect. As an overall test of all BE–FE differences, and hence of biased RE estimates, one can also test the joint hypothesis that *all* unit-specific means have zero effects by using appropriate linear restrictions on the parameters. The corresponding Wald test results in a test statistic of $X_2^2 = 22.62$, which is highly significant ($p < 0.01$) with $df = q = 5$ degrees of freedom (q being the number of restrictions). We conclude that RE estimates differ significantly from FE estimates, because the RE assumption of equal between and within effects does not hold.

Some may also ask whether the estimated effects of the unit-specific means are of substantive interest. There is no general answer to this question, except to say that in some instances, they may provide important information. For example, in Fig. 4.4, the average level of low wage imports could be an indicator of each country's degree of economic development. A variable that includes the country-specific mean of low wage imports would then test the hypothesis regarding whether economically developed countries invest a higher proportion of their GDP on public spending. But in many other applications, the unit-specific means will simply function as a statistical tool to control for correlated unobserved heterogeneity.

Before we conclude this discussion, it should be noted that there are different ways in which one can specify a hybrid model. Some researchers prefer a slightly different parametrization than (4.30), which directly provides between- and within-unit effects. Let us call this alternative parametrization hybrid model 2 (HY2). It uses unit-specific means $\bar{x}_{1i.}$ and time-demeaned variables $(x_{1it} - \bar{x}_{1i.})$, because $\bar{x}_{1i.}$ and $(x_{1it} - \bar{x}_{1i.})$ are uncorrelated. This is easily achieved by adding and subtracting $\beta_1'' \bar{x}_{1i.}$ on the right hand side of (4.30):

$$y_{it} = \beta_0'' + \beta_1'' \cdot (x_{1it} - \bar{x}_{1i.}) + (\beta_1'' + \varphi_1'') \cdot \bar{x}_{1i.} + u_i'' + e_{it}'' \qquad (4.31)$$

With this parametrization, β_1'' measures the within-unit effect of X, while the regression coefficient of the mean variable, $(\beta_1'' + \varphi_1'')$, measures the between-unit effect. If we want to test whether FE and BE estimates differ within this parametrization, we have to test whether within-unit (β_1'') and between-unit ($\beta_1'' + \varphi_1''$) effects differ significantly, i.e., whether $\varphi_1'' = 0$. With Johnson and Wu's data, the corresponding linear restrictions yield exactly the same results as the former test for the unit-specific means ($X_2^2 = 22.62$, $df = 5$, $p < 0.01$; see Table 4.12 column "Model 2").

For example, in this second version of the hybrid model, the effect of mdivorce is estimated as $\hat{\gamma}^{HY2}_{mdivorce} = 0.5726$ and equals the BE estimate of the divorce effect, while the effect of the time-demeaned variable ddivorce is estimated as $\hat{\gamma}^{HY2}_{ddivorce} = 0.1696$ and equals the FE estimate of the divorce effect (compare Table 4.11 columns "BE" and "FE"). Testing the *difference* between these two estimates ($0.5726 - 0.1696 = 0.4030$) is identical to testing whether the mdivorce effect in the former hybrid model 1 ($\hat{\gamma}^{HY1}_{mdivorce} = 0.4030$) is significantly different from zero.

While the two versions of the hybrid model allow us to test whether BE and FE estimates are significantly different and thus, whether the RE assumption of identical between and within effects is contested, we can also directly test the differences between the RE and the FE model. This is essentially what Hausman's test does, which is a very general specification test that is applicable to various contexts, including our case of different panel estimators (Hausman, 1978). The general idea of the *Hausman test* is the following: If two estimators are consistent under a given set of assumptions, their estimates should not differ significantly. Let us call this set of assumptions A. Under a different set of assumptions, say B, this may not be true. If, in this case, only one of the two estimators provides consistent estimates, then the estimates from both estimators should differ significantly. Hausman showed that the standard error of these differences is a simple function of the variance-covariance matrices of each estimator. In our case, A equals a panel model, in which unobserved heterogeneity is uncorrelated with the independent variables in the model. In this situation, both RE and FE estimates are consistent, with RE estimates being more efficient than FE estimates. B pertains to a model with correlated unobserved heterogeneity, in which RE estimation provides biased results, while FE estimation is still consistent. For example, our analysis of psychological distress shows that more highly educated individuals show lower levels of distress, but the exact effect of education differs somewhat between FE ($\hat{\beta}^{FE}_{educr} = -0.0234$) and RE ($\hat{\beta}^{RE}_{educr} = -0.0329$) estimation (see Table 4.8). According to Hausman, the standard error of the difference between FE and RE estimates, $\hat{\sigma}_{(\hat{\beta}^{FE} - \hat{\beta}^{RE})}$, can be calculated from the standard errors of both estimates, $\hat{\sigma}_{\hat{\beta}^{FE}}$ and $\hat{\sigma}_{\hat{\beta}^{RE}}$, as follows:

$$\hat{\sigma}_{(\hat{\beta}^{FE} - \hat{\beta}^{RE})} = \sqrt{\hat{\sigma}^2_{\hat{\beta}^{FE}} - \hat{\sigma}^2_{\hat{\beta}^{RE}}}$$

This result could be used to compute a simple T test. However, since we are using results concerning the behavior of our estimates in large samples (consistency, etc.), it is better to use the Wald criterion. The square of the test statistic t,

$$X^2_3 = \left(\frac{(\hat{\beta}^{FE} - \hat{\beta}^{RE}) - 0}{\hat{\sigma}_{(\hat{\beta}^{FE} - \hat{\beta}^{RE})}} \right)^2$$

is distributed as χ^2 with $df = 1$ degree of freedom. Using the estimated standard errors of both education effects, $\hat{\sigma}_{\hat{\beta}^{FE}_{educ}} = 0.0096$ and $\hat{\sigma}_{\hat{\beta}^{RE}_{educ}} = 0.0049$, we arrive at a value of $X^2_3 = 1.1436^2 = 1.3078$, which is not significant when compared to an

4.1 Modeling the Level of Y

χ^2 distribution with $df = 1$ degree of freedom ($p > 0.10$). However, if we apply the same test to the estimated effects of the respondent's age, $\hat{\beta}_{age}^{FE} = 0.0087$ and $\hat{\beta}_{age}^{RE} = 0.0042$, a significant test statistic of $X_3^2 = 14.5197$ appears ($df = 1$, $p < 0.01$). This is in line with our former hybrid model 1, in which we found significant differences between the BE and FE estimates of the age effect (see the significance of the corresponding mean variable mager in Table 4.12, column "Model 1").

In the general case, when testing the joint significance of the differences between *all* RE and FE estimates, one uses the following Wald test:

$$X_4^2 = \left(\hat{\beta}^{FE} - \hat{\beta}^{RE}\right)' \cdot \left(\hat{\Psi}^{FE} - \hat{\Psi}^{RE}\right)^{-1} \cdot \left(\hat{\beta}^{FE} - \hat{\beta}^{RE}\right) \qquad (4.32)$$

which is a generalization of the former test procedure using matrix algebra. The vectors $\hat{\beta}^{FE}$ and $\hat{\beta}^{RE}$ include all the parameter estimates from FE and RE estimation, while the matrices $\hat{\Psi}^{FE}$ and $\hat{\Psi}^{RE}$ include the corresponding estimated variances and covariances of the estimates. X_4^2 is distributed as χ^2 with $df = l$ degrees of freedom (l being the number of coefficients tested). Applying this formula to the five comparable[23] estimates from our FE and RE model 1 (excluding the constant) results in a test statistic of $X_4^2 = 22.89$, which is highly significant, with $df = 5$ degrees of freedom ($p < 0.01$).

Although the Hausman test requires matrix algebra, it is easily applied, because many software packages include pre-programmed versions of (4.32). As already mentioned, the hybrid model tests the BE–FE differences, while the Hausman test focuses on the RE–FE differences. Hausman (1978, 1263) argued that a test of BE–FE differences has less power than a test of RE–FE differences. But Hausman and Taylor (1981) showed that both the former Wald tests (within the hybrid model) and the Hausman test yield asymptotically equivalent results. However, from a practical point of view, the hybrid model—by specifying linear restrictions—provides a much more flexible framework to test single parameters or groups of parameters.[24]

4.1.2.4 Wrapping Up: How to Choose Between the Different Models in Applied Panel Research?

Obviously, using our knowledge of the panel design of the data, and hence modeling the panel structure, is a far more convincing approach theoretically and statistically than simply treating the serial dependencies as a nuisance by using cluster-robust

---

[23] Naturally, this test focuses only on the effects of those variables that are estimated in both models, i.e., the time-varying explanatory variables X.

[24] It should also be noted that the inversion of $(\hat{\Psi}^{FE} - \hat{\Psi}^{RE})$ in (4.32) is sometimes problematic. One such situation is, when the model includes variables that show no between-unit variance, such as functions of time that are the same for all units. If the model would include only such variables, RE and FE estimates would be identical (see also the discussion surrounding the following Fig. 4.5). The other has to do with the two variance-covariance matrices $\hat{\Psi}^{FE}$ and $\hat{\Psi}^{RE}$. To estimate both of them, an estimate of σ_u is needed. As Table 4.11 shows, $\hat{\sigma}_u^{RE}$ and $\hat{\sigma}_u^{FE}$ can be different and this may cause the problem. Some software implementations of the Hausman test allow the user to specify which one of the two to use in the computation of (4.32). These practical problems do not exist, if one uses the hybrid model to test RE-FE differences.

standard errors. If the model of the data-generating process is correct, you will achieve less biased and more efficient parameter estimates. In addition, standard errors will take into account that the data do not include completely independent observations, and that they will, therefore, not provide you with overly optimistic test statistics that lead you to accept your (alternative) hypotheses too readily. That leads us to the first conclusion: that pooled OLS is not a useful estimation method in almost all cases. It supplies you with estimates that are a combination of differences that have been observed between units and over time. As we have learned, between- and within-unit variance may tell a different story about the effects of your variables of interest, especially if differences between units are affected by other factors for which you have not controlled in your model. If you think that your model controls for this unobserved heterogeneity, you should, at least, use cluster-robust standard errors to control for the remaining serial dependencies that are due to time-constant (unit-specific) random errors. If you would stick with the theoretical OLS standard errors, you would have to assume that your explanatory variables explain *all* of the between-unit variance (such that $\sigma_u^2 = 0$), which is, of course, a very unrealistic assumption.

Having said this, FE models seem to be the method of choice, because—by definition—they control for time-constant (unit-specific) unobserved heterogeneity, even if this heterogeneity is correlated with the variables in the model. Hence, if you are criticized for omitting an important determinant of your dependent variable, and if this omitted variable can safely be assumed to be time-constant (at least during the course of the panel), you can always argue that the effects of your time-varying explanatory variables X are unbiased with respect to any of those unobserved time-constant causes of Y. Of course, this advantage comes at a price:

1. Both your dependent and your independent variables have to vary over time, and if this within-variance is small, you may need a large sample to prove the significance of your explanatory variables.
2. Fixed effects are estimated either with unit-specific dummies or with time-demeaned data. Both procedures result in less degrees of freedom. In large household panels with several measurements over time this is usually no problem. But if you have only two measurements per unit, the number of observations $N = n \cdot 2$ is halved from the very beginning, because you have to estimate n coefficients for the dummies or, alternatively, you have to deduct n unit-specific means from the data. If on top of that the sample size n is small, your data may not have enough power to show significant effects.
3. Some scholars also think that FE estimation has the disadvantage of not providing any model forecasts. At first sight, this argument seems to be correct, because the estimated fixed effects $\hat{u}_i$ are specific to the units of the estimation sample. These scholars argue, if you would like to apply your FE estimates to units outside the estimation sample, you would not know their $\hat{u}_i$. This would only be feasible with RE estimation, in which one estimates parameters of the distribution of U and hence, can use random draws from this distribution for forecasting. These scholars, however, neglect the specific data FE estimation is using. It focuses on *demeaned* data and certainly, it can make forecasts for demeaned data. If

you would like to make forecasts for the original data, you would need a second (BE) model that would allow you to forecast the unit-specific means.
4. Finally, FE regression—by definition—is not the technique to estimate the effects of time-constant explanatory variables Z. It should be stressed, however, that FE regression controls for *all* (observed and unobserved) time-constant determinants of Y, even if it does not provide you with numerical estimates of their effects! Moreover, you can estimate how time-constant explanatory variables modify (interact with) the effects of your time-varying explanatory variables X (see Sect. 3.5.3.2; and in this case, you will get a numerical estimate of the interaction effect).

Nevertheless, sometimes, we lack the right time-varying explanatory variables X, and we have to use time-constant variables Z as indicators of their effects. Men, for example, show higher mortality rates, which are supposedly due to their riskier health behavior over the course of their lifetimes. Unfortunately, this (time-varying) information is not available in many surveys; hence (time-constant) gender must be used as its indicator (see the discussion in Sect. 3.5.3.1).

At this point, RE models may be a solution, because they allow us to estimate the effects of both time-constant and time-varying explanatory variables. Unfortunately, the simple RE model assumes that all unobserved heterogeneity at the unit level has been controlled with the time-constant variables Z in the model, such that all remaining errors at the unit level are simple random noise that is independent of the variables in the model. As we have seen, this assumption can be tested by comparing FE and RE estimates using a Hausman test.

In the former analysis of the johnson-wu data, this assumption did not hold. The Hausman test was highly significant (see the section on p. 166). This does not mean that the RE model completely fails to control for unobserved heterogeneity at the unit level. In fact, if one were to neglect the two time-constant variables in the model (social selection socsel and gender sexr), the Hausman test statistic becomes even larger: $X_4^2 = 44.16$, $df = 5$, $p < 0.01$. Of course, one could also debate whether the difference between the RE estimate ($\hat{\beta}^{RE}_{divorce} = 0.2215$) and the FE estimate ($\hat{\beta}^{FE}_{divorce} = 0.1696$) is of *practical* significance. But if you think that your data contain correlated unobserved heterogeneity, the RE model should be given up in favor of the FE model in your application.

When comparing pooled OLS, FE, and RE estimates, you should keep in mind that there are situations in applied research when the differences between the three estimators are hardly of any practical significance. Remember that RE and pooled OLS estimates are matrix-weighted averages of FE and BE estimates, with the pooled OLS estimates being closer to the BE estimates (see the section on p. 157). Hence, the RE estimates will always lie between the FE and the pooled OLS estimates: $\beta^{FE} < \beta^{RE} < \beta^{OLS}$ or $\beta^{FE} > \beta^{RE} > \beta^{OLS}$. This can also be seen by having a look at quasi-demeaning, $y_{it} - \theta \bar{y}_{i.}$, the computational basis of RE estimation. If the demeaning parameter θ would be one, this would result in FE estimation which uses demeaned data: $y_{it} - 1 \cdot \bar{y}_{i.}$. If the demeaning parameter θ would be zero, this would result in pooled OLS estimation which does not demean the data at all: $y_{it} - 0 \cdot \bar{y}_{i.}$.

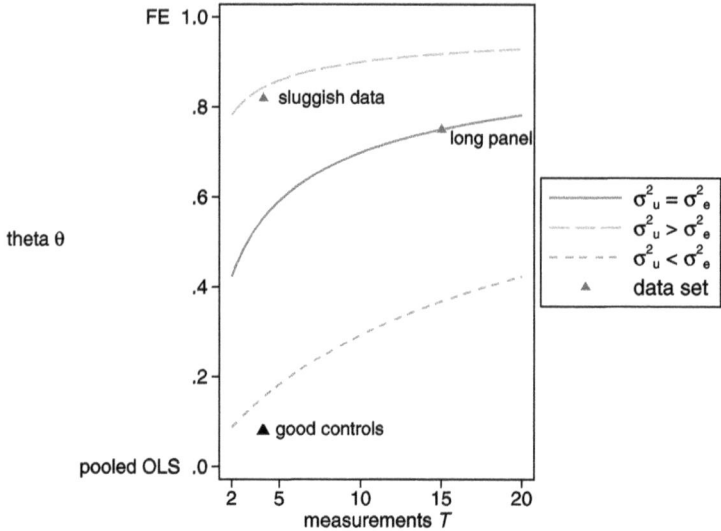

Fig. 4.5 Demeaning parameter θ with respect to number of measurements T

In practice, the demeaning parameter θ will never be exactly one or zero, but it may get very close to either value. Rearranging (4.23) into $\theta = 1 - \sqrt{1/(1 + T \cdot (\sigma_u^2/\sigma_e^2))}$ shows that the demeaning parameter θ depends on two characteristics of the data: (i) the number of measurements T (the "length" of the panel) and (ii) the ratio of between-unit to within-unit variance σ_u^2/σ_e^2. Since our models will always include some explanatory variables, these two error variances indicate the remaining unobserved time-constant and time-varying determinants of the dependent variable (including measurement errors) *after* controlling for the observed X and Z. In other words, σ_u^2 and σ_e^2 measure the *unexplained* variance at the unit and the measurement level. The effect of all three characteristics (T, σ_u^2, and σ_e^2) is illustrated in Fig. 4.5.

Obviously, with increasing number of measurements the RE estimator gets closer to the FE estimator. Take as an example the garmit data (Example 4.1) with $T = 33$: A RE model estimates $\hat{\theta} = 0.8481$ and RE and FE estimates are very similar (the Hausman test is weakly significant at the 5 % level: $X_4^2 = 14.69$, $df = 7$, $p = 0.04$). The longer the unit-specific time-series, the more the RE estimates are dominated by the time dimension of the data and the less dominant is the unit dimension; especially so, if between- and within-unit variance are of equal size (garmit data: $\hat{\sigma}_u^2 = 4.8307^2$, $\hat{\sigma}_e^2 = 4.2644^2$). Obviously, this happens independent of the correlation of unobserved heterogeneity with the variables in the model (i.e., independent of whether FE or RE should be applied from a statistical point of view).

Also, if the unexplained variance of our regression model is dominated by the between-unit variance (i.e., $\sigma_u^2 > \sigma_e^2$; see the long dashed curve in Fig. 4.5), the RE estimator gets closer to the FE estimator. To put it the other way round: If there is relatively little within-unit variance to explain, the FE estimator cannot perform much better. Finally, if the unexplained variance of our regression model is dominated by the within-unit variance (i.e., $\sigma_u^2 < \sigma_e^2$; see the short dashed curve in Fig. 4.5), then unobserved heterogeneity at the unit level hardly exists and one could even use the pooled OLS estimator. In the extreme case of $u_u^2 = 0$, pooled OLS, RE, and FE estimates will be identical. Pooled OLS, however, will provide wrong standard errors, because it ignores the serial correlations due to the panel design of the data.

Figure 4.5 also shows three prototypical data sets that you may encounter in your applied research. The first is a quite lengthy panel with $T = 20$ measurements. The second is a panel of moderate length ($T = 4$), which includes "sluggish" data in the sense that there is not much overtime (within-unit) variance to explain. Finally, the third data set, again with $T = 4$, illustrates a research in which you can luckily explain most of the between-unit variance and hence, unobserved heterogeneity is not of a problem. All remaining errors in the third data set are idiosyncratic and simply "white" noise. Even if you believe that FE estimation is always the preferred method for panel data, in these three situations alternative estimators obviously produce similar results and you may want to choose them, e.g., because they allow you to estimate the effects of time-constant explanatory variables.

Finally, there is an often neglected problem with respect to the statistical qualities of both the RE and FE estimator. In case of correlated unobserved heterogeneity, it is true that the FE estimator is unbiased, while the RE estimator is not. But if unobserved heterogeneity is independent of the variables in the model, the RE estimator is more efficient than the FE estimator. Now let us assume that in the first case the RE estimator is just a little bit biased, but since it is more efficient than the FE estimator, in a single sample, RE estimates may be much closer to the true population parameters, simply because they vary less across samples than the FE estimates (see the discussion in Sect. 7.2.1 and especially Fig. 7.3 on how to choose between different estimators). Inefficiency of the FE estimator is a particular problem, if within-unit variance is low and variables hardly change over time. Since you are never in the lucky situation of statistical theory, which assumes repeated sampling, your single sample may provide you with estimates quite different from the true population parameters. In that case, the fact that fixed effects are unbiased (i.e., correct on average) is no comfort for you. Hence, more research is needed that provides a more balanced view of both estimators that takes into account both unbiasedness *and* efficiency. This research should study the behavior of both estimators under different settings as they are typical in applied research.[25]

Fortunately, the RE model is much more general than traditional textbook introductions suggest. In fact, proper specification of the independent variables in a so-called hybrid model allows us to "combine" the virtues of FE and RE estimates. The basic idea of this hybrid model is to differentiate the effects of time-varying explanatory variables into their between- and within-unit components. In doing so, hybrid models replicate the FE estimates of the effects of the time-varying explanatory

---

[25] As an example, see the simulation study by Clark and Linzer (2012). The study also analyzes the power of the Hausman test. Unfortunately, it focuses on data sets with limited sample sizes as they are typical in political science research.

variables X. In addition, the effects of the time-constant explanatory variables Z are estimated in the same way as one would in a normal cross-sectional regression, while also controlling for the effects of the time-varying independent variables X (see the second section on p. 164). Moreover, the hybrid model provides a nice environment in which to test all kinds of difference between RE and FE estimates, in a much more flexible way than the Hausman test permits. All in all, the hybrid model should be the panel regression model to start with, and from which more restricted regression models can be derived.

4.1.3 Extensions

After having learned how to apply RE and FE, we already know the basic techniques for estimating panel regression models for continuous dependent variables. In this section, we briefly discuss two extensions to the former models that will easily come to your mind when you use RE and FE with respect to your data. For example, if the constant (intercept) can be a random variable that varies between units, as in RE, you may ask whether regression (slope) coefficients can also be random variables in order to model the variance of certain explanatory variables' effects between units (e.g., wages may increase quite differently for each individual with time t; see Fig. 3.2). Furthermore, you may be concerned about the previous treatment of the idiosyncratic error term, because you suspect that the unknown effects at the level of single measurements (summarized in e_{it}) somehow correlate over time due to the longer-lasting (although gradually diminishing) effects of time-point-specific shocks. Random slopes are discussed in Sect. 4.1.3.1, and autoregressive errors in Sect. 4.1.3.2. Since, up to now, all the examples have been based on balanced panel data, we also include a short discussion of unbalanced panel data (Sect. 4.1.3.3). Finally, we conclude this section by showing how panel regression models can be applied to panel data in wide format.

4.1.3.1 Models with Random Intercepts and Random Slopes

If u_i is conceptualized as a random variable in RE estimation, it can be interpreted as a randomly varying intercept in the regression model (4.20) that captures unmodeled unit-specific heterogeneity of Y's level (e.g., the heterogeneity of wage levels). A natural extension of this alternative conceptualization is the assumption that the effects of certain independent variables may as well vary randomly between units. In other words: a more general model assumes random intercepts as well as random slopes. In Sect. 3.5.3.3, we showed how regression coefficients (including the constant) can be specified as functions of random disturbances (to capture their unit-specific heterogeneity) and other independent variables (to capture their systematic behavior). We demonstrated that the assumption of random slopes automatically leads to regression models with heteroscedastic error terms, which challenges the OLS assumptions (see Textbox 4.1) and calls for more general estimation procedures. Such a general procedure (Maximum Likelihood) will be introduced in Sect. 4.2.3; therefore, we defer this discussion to a later example concerning value change (see Example 4.5).

4.1 Modeling the Level of Y

However, if we are not interested in numerical estimates of these random slopes and only want to control for them, there are some simple techniques available that we can apply using our present methodological expertise. As an example, think again of our analysis of hourly wages. We extend model (4.3) to allow for a linear time trend in wages ($\beta_0(t) = \beta_{0i} + \beta_{1i}t$), using the extended notation that we introduced in Textbox 3.2:

$$y_{it} = \beta_{0i} + \beta_{1i}t + \beta_2 x_{2it} + \cdots + \beta_k x_{kit} + \gamma_1 z_{1i} + \cdots + \gamma_j z_{ji} + e_{it} \quad (4.33)$$

Moreover, both the overall level (β_{0i}) as well as the growth rate (β_{1i}) are assumed to differ among individuals, and hence have an index i. In separate equations, we can specify how both coefficients depend on characteristics of the unit and on other (unknown) factors (see Sect. 3.5.3.3). This time we only specify a random factor:

$$\begin{aligned} \beta_{0i} &= \gamma_{00} + u_{0i} \\ \beta_{1i} &= \gamma_{10} + u_{1i} \end{aligned} \quad (4.34)$$

Inserting both equations into (4.33) and rearranging (4.33) results in the well-known model with a heteroscedastic error term:

$$y_{it} = \gamma_{00} + \gamma_{10}t + \beta_2 x_{2it} + \cdots + \beta_k x_{kit} + \gamma_1 z_{1i} + \cdots + \gamma_j z_{ji} + (u_{0i} + u_{1i}t + e_{it})$$

However, if we compute first differences,

$$\begin{aligned} y_{it} = &\ \gamma_{00} + \gamma_{10}t + \beta_2 x_{2it} + \cdots + \beta_k x_{kit} + \gamma_1 z_{1i} + \cdots + \gamma_j z_{ji} \\ &+ (u_{0i} + u_{1i}t + e_{it}) \\ y_{i,t-1} = &\ \gamma_{00} + \gamma_{10} \cdot (t-1) + \beta_2 x_{2i,t-1} + \cdots + \beta_k x_{ki,t-1} \\ &+ \gamma_1 z_{1i} + \cdots + \gamma_j z_{ji} + \left(u_{0i} + u_{1i} \cdot (t-1) + e_{i,t-1}\right) \end{aligned}$$

the model reduces to the familiar random effects model (without the time-constant explanatory variables Z):

$$\Delta y_{it} = \gamma_{10} + \beta_2 \Delta x_{2it} + \cdots + \beta_k \Delta x_{kit} + (u_{1i} + \Delta e_{it}) \quad (4.35)$$

In other words, if we estimate a RE model with differenced data, we are controlling for all time-constant characteristics of the units, plus possibly heterogeneous growth rates among the units. Of course, this technique rests on all the RE assumptions; in particular on the assumption of uncorrelated unobserved heterogeneity (see Textbox 4.4). Moreover, this technique is only feasible for slope coefficients of variables that can be eliminated by computing first differences (like, e.g., process time t).

4.1.3.2 More Complicated Error Processes

Up to now, we have assumed that time-varying unobserved variables and measurement errors, which were captured in the idiosyncratic error term e_{it}, are only time-point-specific "white noise" and do not have any longer lasting effects. Hence,

Corr(e_{it}, e_{is}) = 0 for any two time points t and s. This assumption may not be realistic in every case. For instance, a particular economic success in year t, which allows a company to pay its employees above-average wages in that year, may also have positive, though diminishing, effects in the following years, such that this company is able to pay a premium in the following years too. If we are unable to measure these success factors at the company level, these effects are part of the idiosyncratic error and the assumption Corr(e_{it}, e_{is}) = 0 no longer holds (see also the discussion in Sect. 3.6.2).

A simple model that captures these over time effects of time-varying unobserved variables and measurement errors is the following first-order autoregressive model for the idiosyncratic errors:

$$e_{it} = \rho e_{i,t-1} + v_{it}, \quad |\rho| < 1 \tag{4.36}$$

According to this model, part of the idiosyncratic error of the last time point $t - 1$ carries over to the present time point t. How much is carried over depends on the parameter ρ, which is a number between 0 and 1. In most cases, ρ is positive (see footnote 5 in Chap. 3). In addition to this carry-over effect, the idiosyncratic error is influenced by a random disturbance v_{it} that measures all the other unknown factors that are effective at time point t. v_{it} is assumed to behave like the idiosyncratic error in the former sections, i.e., having constant variance and being uncorrelated over time. Given these assumptions, it can be shown that idiosyncratic errors are correlated over time and that their correlation is a function of the autoregressive parameter ρ:

$$\text{Corr}(e_{it}, e_{i,t+s}) = \rho^s \tag{4.37}$$

As (4.37) shows, it is an exponential function of the time lag s between both time points. Since ρ is a number between 0 and 1, this correlation will decrease the longer the time lag. This conforms nicely to the former idea that time-point-specific shocks (like, e.g., economic success) have longer-lasting effects, which, however, diminish over time.

Such a model with autoregressive idiosyncratic errors e_{it} can be estimated assuming either random or fixed effects u_i for unobserved heterogeneity at the unit level. The methodology needed at this point borrows much from time-series analysis and is beyond the scope of this textbook. The general idea is to find an estimate of the autoregressive parameter ρ and then to "quasi-difference" the data, i.e., subtract not the full value of the previous Y, X, and Z values but only a fraction equal to ρ. This quasi-differencing gets rid of the autoregressive part of the idiosyncratic errors and leaves us with a traditional RE or FE model. Programmed versions of these autoregressive RE and FE models exist in many software packages so that they are easy to apply.

Earlier, we showed that the traditional FE and RE models imply an equal correlation structure (see (4.16) and (4.21)), which is sometimes also called an *exchangeable* correlation structure, because the expected amount of serial correlation of the error terms is assumed to be the same, irrespective of the time lag between

4.1 Modeling the Level of Y

the measurements. This rather restrictive implication of the traditional models is relaxed when assuming autoregressive idiosyncratic errors. If the composite error term consists of a unit-specific and an autoregressive idiosyncratic error component, $\varepsilon_{it} = u_i + \rho e_{i,t-1} + v_{it}$, then it can be shown that the correlation of the composite error amounts to

$$\text{Corr}(\varepsilon_{it}, \varepsilon_{i,t+s}) = \frac{\sigma_u^2 + \rho^s}{\sigma_u^2 + \sigma_e^2} \tag{4.38}$$

assuming, of course, that unobserved heterogeneity and idiosyncratic errors are independent of one another. Since ρ is a number between 0 and 1, this correlation is also a decreasing function of the time lag s between the measurements.

4.1.3.3 Unbalanced Panel Data

Since we have always used balanced panel data in our examples, you may ask what changes if the data are unbalanced. First of all, all panel regression models need at least two observations per unit. Units with only one observation will cancel out of the computations. Aside from that, all panel regression models can easily be adapted to unbalanced data (assuming, of course, that missing observations are missing completely at random). Concerning FE estimation, it makes no difference whether you demean the data with unit-specific means computed on (balanced) T observations per unit, or with unit-specific means computed on (unbalanced) T_i observations per unit. In the case of RE estimation, quasi-demeaning needs estimates of both error components, and formula (4.24) for estimating the variance of the unit-specific error term has to be adapted to the unbalanced data (see Greene, 2008, 203). First difference (FD) estimation, which will be discussed in the next section (see Textbox 4.6), is perhaps most affected by unbalanced data. FD estimation uses differences of all time-varying variables X and Y, which are computed by subtracting the values for observation $t-1$ from the values for observation t. If one of the two observations is missing, the difference scores ΔX and ΔY are missing too. Hence, in case of FD estimation, not only does one need at least two observations per unit, but they also have to come from two consecutive time points. Finally, as already mentioned in footnote 22 on p. 165, the hybrid model will only replicate the FE estimates, when it is estimated from balanced data.

4.1.3.4 Models for Data in Wide Format

Finally, we want to show how panel regression models look like and can be estimated when panel data are organized in wide format. As an example, we use again the johnson-wu data (see Example 4.2). Stored in wide format, the data set includes $N = 1,166$ records (one for each individual) but many more variables than in long format. This is so because the $t = 1, \ldots, 4$ measurements of the time-varying X and Y have to be stored in different variables: e.g., psydis1, psydis2, psydis3, and psydis4 for the dependent variable "psychological distress" and divorce1, divorce2, divorce3, and divorce4 for the explanatory variable "current family status: divorced". Time-constant Z, of course, have to be stored only once.

If each measurement of Y is a different variable in the data set, we cannot easily specify an (overall) regression model as we did with data in long format (see (4.5)). We rather have to specify the functional relationship between Y and X resp. Z for each measurement separately. In case of the former regression model for psychological distress, this results in four different equations:[26]

$$y_{i1} = \beta_0^1 + \beta_1^1 x_{1i1} + \cdots + \beta_k^1 x_{ki1} + \gamma_1^1 z_{1i} + \cdots + \gamma_j^1 z_{ji} + 1 \cdot u_i + e_{i1}$$
$$y_{i2} = \beta_0^2 + \beta_1^2 x_{1i2} + \cdots + \beta_k^2 x_{ki2} + \gamma_1^2 z_{1i} + \cdots + \gamma_j^2 z_{ji} + 1 \cdot u_i + e_{i2}$$
$$y_{i3} = \beta_0^3 + \beta_1^3 x_{1i3} + \cdots + \beta_k^3 x_{ki3} + \gamma_1^3 z_{1i} + \cdots + \gamma_j^3 z_{ji} + 1 \cdot u_i + e_{i3}$$
$$y_{i4} = \beta_0^4 + \beta_1^4 x_{1i4} + \cdots + \beta_k^4 x_{ki4} + \gamma_1^4 z_{1i} + \cdots + \gamma_j^4 z_{ji} + 1 \cdot u_i + e_{i4}$$
(4.39)

Using all time-varying variables Y and X measured at $t = 1$ (psydis1, divorce1, etc.) and the time-constant variables Z (socse1, female), one can estimate the parameters of the first equation. To estimate the parameters of the second equation, one needs the time-constant variables Z and all time-varying variables Y and X measured at $t = 2$; and so forth. However, the estimates of the regression coefficients will not be the same across equations, because each equation is estimated with different data that is specific to the corresponding time point. We have indicated this with an superscript attached to each regression coefficient: e.g., $\beta_1^1 \neq \beta_1^2 \neq \beta_1^3 \neq \beta_1^4$. Moreover, it is unclear how to estimate or control for time-constant unobserved heterogeneity U, if each estimation ($t = 1, \ldots, 4$) is based on only one cross-section. With panel data in long format, this was easy because each unit was represented T times in the data and we estimated an *overall* regression model. In sum, it is not only more cumbersome to specify a panel regression model in wide format, it is also not obvious how to replicate the estimates that have been found with the same data in long format.

We need a methodology to estimate the time-point-specific equations simultaneously. Moreover, this methodology must be able to place certain restrictions on the model parameters. For example, if we want to replicate the former RE model, regression coefficients should be the same across equations, unobserved heterogeneity U has to be uncorrelated with the explanatory variables in the model, both time-constant and time-varying explanatory variables should be allowed in the model, and the dependent variable should be a function of X, Z, U, and idiosyncratic errors E. These assumptions are shown graphically in the path diagram in the left panel of Fig. 4.6.

To keep the path diagram as simple as possible, it shows only one time-constant (Z_1) and one time-varying (X_1) explanatory variable. However, in accordance with our assumptions, the effects of X_1 and Z_1 on each measurement y_{it} of the dependent variable ($t = 1, \ldots, 4$) are identical and equal to β_1 resp. γ_1. Similarly, unobserved heterogeneity U has an identical effect on each measurement that—as specified in

---

[26] Why U is multiplied with a 1 will become clear later.

4.1 Modeling the Level of Y

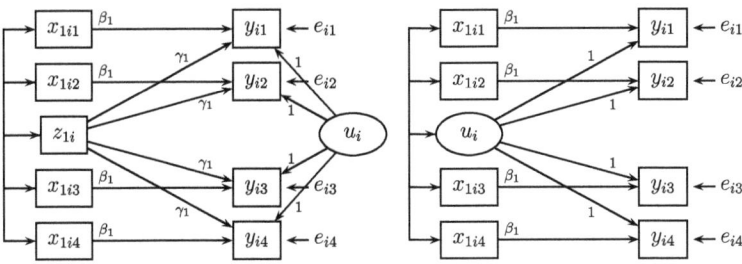

Fig. 4.6 Path diagram of a RE (*left*) and a FE model (*right panel*)

(4.39)—equals 1. Similar to an ordinary regression model that controls for the statistical associations among the explanatory variables, we also have to control for the covariances between the various measures of X_1 and Z_1. Following the practice of path diagrams, this is indicated by double-sided arrows between the various measurements of X_1 and between the measurements of X_1 and Z_1. Such double-sided arrows are absent between all measurements of E and between U and the explanatory variables. This is due to the RE assumptions, which posit (i) that unobserved heterogeneity is uncorrelated with the variables in the model and (ii) that idiosyncratic errors are not autocorrelated. The right path diagram in Fig. 4.6 shows the corresponding FE model. Now, unobserved heterogeneity U is correlated with the explanatory variables in the model (as indicated by the double-sided arrows) and time-constant explanatory variables are missing. More specifically, the effect of (known) time-constant explanatory variables are controlled together with unobserved heterogeneity U (which therefore and in line with our former notation should better be labelled V).

The two path diagrams provide a nice graphical summary of the RE and FE assumptions and as such, are instructive by themselves. However, the question remains how the regression coefficients can be estimated. In Sect. 3.6.3, we mentioned the methodology that is used to estimate such systems of regression equations: structural equation modeling (SEM). In this context, unobserved heterogeneity U is an (unobserved) latent variable, while X, Z, and Y are (observed) manifest variables. According to the conventions in SEM, the latent variable U is symbolized by a circle in the path diagram, while the manifest variables are represented with rectangles.

Within this textbook we do not have the space to explain how structural equation models are estimated. But we have indicated the underlying idea in Sect. 3.6.3.3: Given the assumptions specified in the path diagram, one can derive the variances and covariances of the manifest variables implied by the model. By minimizing the deviations between the *model-implied* and the *observed* variance-covariance matrix, one can estimate the regression coefficients that most likely have generated the statistical relationships between the observed variables.[27] As the term "likely" suggests, estimation is done using maximum likelihood (ML).

---

[27] Technically, it is easy to replicate the RE and FE estimates of Table 4.8 with SEM. See the web site (Sect. 7.3) for the corresponding syntax file.

The nice thing about SEM is the fact that all kinds of generalizations of the traditional RE and FE specifications can be estimated and tested against each other using the same kind of methodology. For example, in case of the johnson-wu data, one could go back to the five items that were used to generate the dependent variable "psychological distress" (see Example 4.2) and instead of using an additive index, one could specify a measurement model that accounts for the different reliabilities of the five items. One could also test whether the regression coefficients β and γ are indeed identical across measurements and in doing so, test the assumption whether the effects of particular explanatory variables change over time. Furthermore, models with lagged dependent variables or autocorrelated idiosyncratic errors are easily specified within this methodology (see Bollen and Brand (2010) for these and other possibilities within SEM). Annacker and Hildebrandt (2004) use this methodology to study the economic success of a sample of strategic business units over a period of six years.

4.2 Modeling the Change of Y

In this section, we will focus on linear models for the change of a continuous dependent variable Y that have been introduced in Sect. 3.5.2.1. We will start with an example that focuses on the change of Y between two consecutive panel waves, and apply a technique called *first difference (FD) estimation* (Sect. 4.2.1). This technique is very useful when analyzing the instantaneous change of Y. However, if change is gradual and follows a longer time path, FD estimation is not very practical. In that case, it is better to revert to a model at the levels of Y and analyze the change of the dependent variable with an impact function. This will be explained in Sect. 4.2.2. This leads to the more general question of how to analyze trends in the dependent variable over time. Section 4.2.3 will describe how to incorporate the different definitions of time (e.g., age, cohort, and period) into the regression model. Very often, these time trends differ between the units of analysis. Therefore, we will also show how to model random regression coefficients (intercepts and slopes) in a more general framework than in Sect. 4.1.3.1.

4.2.1 Analysis of Change Using Change Scores

In this section, we focus on models that explain change ΔY of the dependent continuous variable Y as a result of change ΔX of the explanatory variables X. ΔY and ΔX are also called *change scores*. In a model focusing on ΔY, it does not make sense to use any time-constant explanatory variables Z, because it is difficult to justify that change (ΔY) depends on something constant in time (Z). However, it is possible that the effects of the time-varying explanatory variables vary with respect to other characteristics of the units (including the time-constant variables Z). This would necessitate interaction effects between Z and ΔX (see the discussion in Sect. 3.5.3.2). As an example, we again use an analysis of the effects of partnership dissolution on subjective well-being, this time focusing on gender differences.

4.2 Modeling the Change of Y

Example 4.4 (genderdiff data)   It is well-known that divorce has negative consequences for the economic status of both marriage partners, especially so for women. Nevertheless, divorce is often instigated by the wife. Andreß and Bröckel (2007) argue that both observations are not necessarily a contradiction. Women may gain something that makes up for their economic loss, they say, and therefore, they analyze different aspects of subjective well-being. Given the higher economic losses of women, they hypothesize that measures of economic well-being decrease much more for women than for men after separation, while measures of overall life satisfaction should develop just the other way round, given that women gain in other aspects of life as indicated by their higher propensity to end an unhappy marriage. They use data from 16 waves of the German Socio-Economic Panel Study (SOEP) covering the period from 1984 to 1999. They focus on separation and not on divorce, because earlier analyses have shown that separation is connected with more economic changes than legal divorce, which follows separation, sometimes several years later, when the economic situation has already stabilized.

More specifically, their analysis includes all married couples separating between 1984 and 1999. The analysis focuses on the first separation within that period and uses information from all available panel waves before, after, and including the first separation. All in all, the sample consists of 418 separated men and 450 separated women contributing information to the analysis from between 1 and 16 panel waves. Obviously, panel attrition was higher for men than for women. Since the research question refers to men and women in general, it was not necessary to restrict the sample to complete couple data, which would have implied an even smaller sample size, if only those individuals were used for which the partner was also represented in the data. All analyses were restricted to individuals younger than 55 years of age, making sure that all events being studied were experienced during the (economically) active life of the sample members. This age restriction reduced the sample size to 837 separating individuals.

The SOEP collects information on subjective well-being by asking every panel member about satisfaction in different life domains (housing, job, income, health, life, etc.). Answers can be provided on an 11-point scale ranging from 0 (completely dissatisfied) to 10 (completely satisfied). The authors used data on income satisfaction as a measure of economic well-being, and satisfaction with life as an overall measure of subjective well-being. The main explanatory variable is the event of separation, and a variable T measures the time point of each measurement relative to the individual's year of separation (with $t = 0$ indicating the year of separation). Control variables include the individual's gender, age, education (in years of education), employment status (a dummy for being gainfully employed), parental status (a dummy for having a child below 18 years of age in the household), income (equivalized household income), residential mobility (a dummy for having changed the address), and a dummy for whether a new partner is present.

> Compared to the other examples, this data set represents an unbalanced panel, because each individual contributes a different number of measurements depending on how long he or she has been observed before and after the separation. Moreover, some individuals may have missing data in between, because they did not participate in the survey in all years. Hence, some individual time series may include gaps.

In the following, we will not try to replicate all of Andreß and Bröckel's findings completely. We will focus only on the change in overall life satisfaction. Similar to the authors, we want to show that overall life satisfaction does not change as negatively for women as it does for men. For didactic purposes, we restrict the sample to those observations that have valid values for all explanatory variables and for both satisfaction measures. We also exclude the (few) individuals that have gaps in their time series and whose time series consist of only one measurement. Both restrictions are necessary for computing first differences. The second one is obvious: In order to compute first differences one needs at least two measurements over time. The first one is due to the fact that FD estimation cannot easily deal with unbalanced data (see Sect. 4.1.3.3). After applying all these restrictions, our sample consists of 7,619 observations altogether for 705 individuals. Hence, we will not replicate Andreß and Bröckel's estimates exactly, which were based on 9,066 observations for 837 individuals. But the main substantive conclusions remain the same.

In order to estimate the effect of separation, we generate a dummy variable named separated that indicates whether the individual is separated at the given time point. Within each individual's time series, it is 0 up to the year before separation and 1 from the year of separation onwards (i.e., separated = 1 if $t \geq 0$). In addition, we generate a variable named sepsex that measures the interaction between gender and separation. It has the value 1 from the year of separation onwards, if the individual is female. Hence, it measures the differential separation effect for women. If we difference both variables, they will have values of 1 in the year of separation (0 otherwise) and thus indicate the change in marital status for both genders and for women only. If we regress the change of life satisfaction (the differenced satisfaction score Δy_{it}) on both differenced explanatory variables (excluding a regression constant):

$$\Delta y_{it} = \beta_1 \Delta separated_{it} + \beta_2 \Delta sepsex_{it} + \Delta e_{it} \qquad (4.40)$$

β_1 estimates the change in life satisfaction due to separation for men, while β_2 estimates how much more (or less) change is expected for women. According to the estimates in Table 4.13, life satisfaction decreases significantly by about 0.4 scale points for men ($t = -3.55$, $p < 0.01$), while it hardly changes for women ($\hat{\beta}_1 + \hat{\beta}_2 = -0.4033 + 0.3626 = -0.0407$). The differential effect for women ($\hat{\beta}_2 = 0.3626$) is highly significant ($t = 2.32$, $p = 0.02$; but the sum of both effects (-0.0407) is not significantly different from zero). These estimates are based on

4.2 Modeling the Change of Y

Table 4.13 Change in overall life satisfaction due to separation (FD estimation)

Variable	Estimate	Std. Err.	Estimate	Std. Err.
Separated	−0.4033	0.1138	−0.5240	0.1259
Separated*Female	0.3626	0.1562	0.4527	0.1615
Age (years)			−0.0574	0.0253
Education (years)			−0.1065	0.0697
Employed			0.4166	0.0697
Parenting			−0.1339	0.0912
Income			0.0024	0.0015
Changed home			0.1501	0.0470
New partner			0.6693	0.1075
R^2		0.0018		0.0161
N		6,914		6,914
n		705		705
T_{min}, T_{max}	2	16	2	16

Source: genderdiff data (see Example 4.4)

6,914 (and not 7,619) observations, because differencing results in missing values for the first measurement in each individual's time series. There is no observation before the first measurement; hence it is impossible to compute a change score. In other words: When using FD estimation, one loses as many degrees of freedom as there are units in the data set (in our case: 705 and $7{,}619 - 705 = 6{,}914$). And: if we had allowed gaps in the individual time-series, the number of missing values would have increased even further.

As shown in Sect. 3.5.2.1, there is a close connection between a model in levels and a model in first differences. This is also the case for the model we have just estimated. Model (4.40) can be derived by differencing the following model in levels

$$y_{it} = \beta_0 + \beta_1 separated_{it} + \beta_2 sepsex_{it} + u_i + e_{it} \quad (4.41)$$

This connection shows us that focusing on change (both in the dependent and the explanatory variables) also controls for unobserved heterogeneity at the unit level. First differencing not only eliminates the constant β_0, but also the error term u_i. Hence, the former estimates $\hat{\beta}_1$ and $\hat{\beta}_2$ measure separation effects controlling for all (known and unknown) time-constant characteristics of the individuals.

However, they do not control for time-varying characteristics, whose change over time may also affect the change in the dependent variable. For example, in our case, one could argue that a new job that provides additional income and social contacts may also increase the individual's life satisfaction, and that women may seek out a new job more often than men. Therefore, before differencing equation (4.41), we should augment the model with the time-varying control variables that have been supplied with the data (see Example 4.4).

Before discussing the results of this augmented FD model, let us think about a potential problem of model identification. Besides the dummy variable separated, the augmented model includes the variable age. Both variables are functions of time and whenever one includes several of such "time variables" into a panel regression model that only exploits the overtime (within-unit) variance of the data, as FD does, one should check whether they are linearly dependent. Fortunately, this is not the case, because separated is a step function of time having identical values at several points in time (separated = 1 if $t \geq 0$ and separated = 0 if $t < 0$).

Table 4.13 shows the estimates of this augmented model. The effects of the control variables are not of interest here, but their effects are in line with Andreß and Bröckel's hypotheses. We are interested in the separation effect on life satisfaction for men and women, and the corresponding estimates are even slightly larger than in the simple model, which does not include the control variables. The hypothesis that life satisfaction develops less negatively for women than for men thus finds even greater support when controlling for all the other life changes that accompany a separation. The following Textbox 4.6 summarizes the main assumptions of FD estimation.

Textbox 4.6 (FD assumptions) FD estimation controls for unit-specific heterogeneity by computing change scores for all variables (Y, X, Z, U, E) in the model (e.g., $\Delta y_{it} = y_{it} - y_{i,t-1}$). As a side effect, time-constant explanatory variables cancel out of the model, because $\Delta Z = 0$ (similar to $\Delta U = 0$). In other words, FD estimation is based on the following linear model for panel data: $\Delta y_{it} = \beta_1 \Delta x_{1it} + \cdots + \beta_k \Delta x_{kit} + \Delta e_{it}$. It includes no regression constant, if the original (undifferenced) model included no time trend and assumed $\beta_0(t) = \beta_0$. It should be noted that computing change scores (first differences) eliminates the first observation for each unit. Hence, the data set includes $i = 1, \ldots, n$ units with $t = 2, \ldots, T$ observations each.

The statistical properties of FD estimation rest on assumptions that are very similar to those of OLS estimation:
1. The units $i = 1, \ldots, n$ in the panel data set are a simple random sample from a cross-section of a well-defined population.
2. The model is linear in its parameters $\beta_0, \beta_1, \ldots, \beta_k$.
3. Each independent variable $x_{1it}, \ldots, x_{kit}$ changes over time and is not a linear function of the other independent variables.
4. Idiosyncratic error is independent of the variables in the model and independent of unit-specific unobserved heterogeneity: $E(e_{it}|x_{1i2}, \ldots, x_{1iT}, \ldots, x_{ki2}, \ldots, x_{kit}, u_i) = 0$. It should be noted that the assumption expects independence from *all* measurements of each variable over time, a characteristic that has become known as *strict exogeneity*.
5. The differenced idiosyncratic error has constant variance, given any value of the independent variables: $\text{Var}(\Delta e_{it}|x_{1i2}, \ldots, x_{1iT}, \ldots, x_{ki2}, \ldots, x_{kit}) = \sigma^2_{\Delta e}$.
6. The differenced idiosyncratic error is uncorrelated between any two observations t and s ($t \neq s$) of unit i, given any value of the independent variables: $\text{Corr}(\Delta e_{it}, \Delta e_{is}|x_{1i2}, \ldots, x_{1iT}, \ldots, x_{ki2}, \ldots, x_{kit}, u_i) = 0$.

7. The differenced idiosyncratic error is normally distributed with mean 0 and variance $\sigma^2_{\Delta e}$: $\Delta e_{it} \sim Normal(0, \sigma^2_{\Delta e})$, given any value of the independent variables.

The first assumption is necessary for making statistical inferences about the population using the sampled data. The second assumption ensures that we can use least squares estimation to estimate the parameters of the model. If the model were not linear in its parameters, we would have to use other estimation techniques. The third assumption guarantees that a numerical value exists for each regression coefficient. The fourth assumption is the most important one, because it ensures that FD estimates $\hat{\beta}_0, \hat{\beta}_1, \ldots, \hat{\beta}_k$ are *unbiased*: $E(\hat{\beta}_0) = \beta_0$, $E(\hat{\beta}_1) = \beta_1, \ldots, E(\hat{\beta}_k) = \beta_k$. Assumption 4, however, is far more demanding than assumption 4 for OLS with cross-section data (see Textbox 4.1). Strict exogeneity (as opposed to contemporaneous exogeneity) assumes that there are no feedback mechanisms caused by unobserved effects over time. Furthermore, if assumptions 5 and 6 are true, FD estimates are the best linear unbiased estimates (BLUE) of $\beta_0, \beta_1, \ldots, \beta_k$. Finally, assumption 7 guarantees that we can use standard test procedures and confidence intervals based on the normal distribution. Otherwise, we have to rely on asymptotic approximations, which are only true in large samples. Discussions of these FD assumptions and proofs of the statistical properties of FD can be found in Wooldridge (2010, 279).

4.2.2 Analysis of Change Using Impact Functions

Although the last section provided convincing proof that women *in the year of separation* suffer less than men with respect to their overall life satisfaction, the analysis is contested by the argument that life satisfaction may develop differently for both genders. If, for instance, women as initiators of the separation process cope with the concomitants earlier than men, we have simply measured the gender difference at the wrong point in time. Figure 4.7 shows that this is actually the case for the genderdiff data. Life satisfaction decreases in parallel for both genders before separation, but only until one year before the separation date. After that time point, life satisfaction continues to decrease for men, while it remains constant for women. After the separation date, life satisfaction increases again for both genders, but does not reach its original level during marriage, at least for men.[28] Given this observation, it is not surprising that the former FD model, which focuses on change between

---

[28]The figure shows a slightly different trend for men than the corresponding figure in Andreß and Bröckel (2007), in which men—from the third year after separation—have about the same degree of life satisfaction as women. This is due to the larger sample they used, and to the fact that they weighted the data for this descriptive plot.

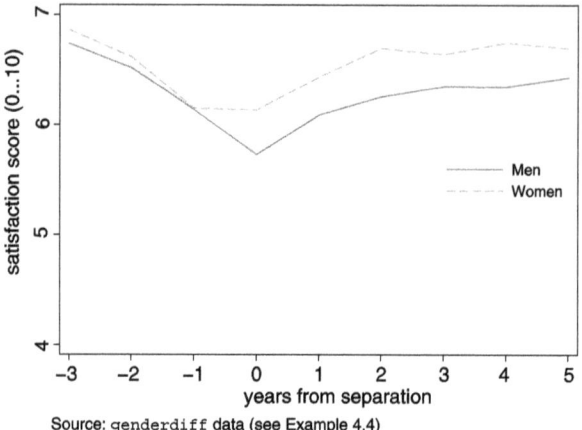

Fig. 4.7 Mean life satisfaction before and after separation

$t = -1$ and $t = 0$ (when $\Delta separated_{it} = 1$), does not show a significant change for women. Therefore, it is a much more convincing strategy to model the development of life satisfaction over the whole separation process and not only in the year of separation.

Obviously, FD models are not very useful to perform these kinds of analyses, because they focus on yearly change and not the long-term trend. Therefore, we follow Andreß and Bröckel who use a model at levels like (4.5) and who, instead of specifying a regression constant β_0, analyze the change in satisfaction by a suitable impact function $\beta_0(t)$. As defined in Sect. 3.5.2.1, an impact function $\beta_0(t)$ is a function of t that measures the trend of the dependent variable Y before and after an event of interest (in our case: separation). Hence, t is defined relative to the event of interest with $t = 0$ when the event occurs. The most general way to do this, is to use dummies for all the measurements before, after, and including the event. This is the approach of Andreß and Bröckel (2007), who use seven dummies altogether (D_{it}, with $t = -3, \ldots, +3$) and their interactions with gender (I_{it}, again with $t = -3, \ldots, +3$) in the following regression model:

$$y_{it} = \beta_0 + \alpha_{-2} D_{i,-2} + \alpha_{-1} D_{i,-1} + \alpha_0 D_{i,0} + \alpha_1 D_{i,1} + \alpha_2 D_{i,2} + \alpha_3 D_{i,3}$$
$$+ \delta_{-3} I_{i,-3} + \delta_{-2} I_{i,-2} + \delta_{-1} I_{i,-1} + \delta_0 I_{i,0} + \delta_1 I_{i,1} + \delta_2 I_{i,2} + \delta_3 I_{i,3}$$
$$+ \sum_{k=1}^{7} \beta_k x_{kit} + u_i + e_{it} \qquad (4.42)$$

D_{it} equals 1 if the measurement pertains to the year t, with two exceptions: $D_{i,-3} = 1$ if $t \leq -3$ and $D_{i3} = 1$ if $t \geq 3$. Correspondingly, I_{it} equals 1 if $D_{it} = 1$ *and* the individual is female. To ease the notation, we have used α's and δ's to symbolize the effects of the dummies and their interactions with gender. $\sum_{k=1}^{7} \beta_k x_{kit}$ captures the influence of the seven control variables (age, employment, parenting, etc.). By implication, the regression constant β_0 measures the level of life satisfaction for men three or more years before separation, when all control variables are zero. Hence, the effects $\alpha_{-2}, \alpha_{-1}, \ldots$ of the time dummies measure the extent to which life satisfaction is lower for men in the years from $t = -2$ onwards, and the interaction effects $\delta_{-3}, \delta_{-2}, \ldots$ measure how much women deviate from this trend. All unknown time-constant characteristics (unobserved heterogeneity) are controlled by the error term u_i. And finally, the model includes an idiosyncratic error term for measurement errors and all unknown time-varying characteristics. Depending on our assumptions regarding unobserved heterogeneity (correlated or uncorrelated with the X), this model can be estimated with FE or RE.

Again, because the set of control variables includes the age of the respondent, we should first discuss whether the model is identified. Compared to the simple FD model in the previous section, we now model the effect of separation with a nonparametric function of time that is different for each point in time during the observation period.[29] Therefore, in a model that only exploits the overtime (within-unit) variance of Y, as FE and FD do, we cannot estimate the effect of a second variable like age that is also a function of time. It is important that the user of statistical software does these identification checks before specifying the corresponding FD or FE estimation command. Otherwise, the software will arbitrarily exclude one variable from the model.

The identification problem does not exist for RE estimation, which—as we know—exploits both the between- and the within-unit variance. In other words, it does not only recognize that all respondents get older each year, it also takes into consideration that respondents have different ages. To make a fair comparison between FE and FD estimates on the one side and RE estimates on the other, one should differentiate the variable age into her between- and within-unit components by measuring the respondent's age at a certain point in time (e.g., the year of separation) and how he or she ages over time. The effect of the first (time-constant) component—by definition—cannot be estimated with FE or FD, while the effect of the respondent's aging is included in the time dummies, whose effects can be estimated both with RE, FE, and FD estimation.

Table 4.14 shows the estimates of the correctly specified RE (with age at separation) and FE model (without age). Before going into the methodological details, we will briefly comment on the overall substantive results, which are not very different between the various estimation methods and are also in line with the conclusions drawn by Andreß and Bröckel (2007), despite the slightly more restricted sample. The effects of the control variables show the expected signs with which

---

[29] This is not exactly true for how we defined the time dummies. Since we measured all observations three and more years before separation with the same dummy ($D_{i,-3} = 1$ if $t \leq -3$), the impact function does not change for the more distant observations before separation. The same is true for the more distant observations after separation ($t \geq 3$) for which $D_{i3} = 1$. Hence, the following linear dependence is not perfect and in principle, we could estimate a FE or FD model that also includes the variable age. We have not done this, because this age effect would be estimated from the observations that are at least three years away from the event of separation.

Table 4.14 Life satisfaction before and after separation (RE, FE, and FD estimation)

Variable	RE		FE		FD	
	Estimate	Std. Err.	Estimate	Std. Err.	Estimate	Std. Err.
$D_{i,-2}$	−0.4035	0.1057	−0.3707	0.1067	−0.1881	0.1282
$D_{i,-1}$	−0.7789	0.1025	−0.7501	0.1040	−0.5801	0.1736
$D_{i,0}$	−1.2665	0.1078	−1.2524	0.1099	−1.1617	0.2123
$D_{i,1}$	−0.9747	0.1166	−0.9442	0.1190	−0.8638	0.2459
$D_{i,2}$	−0.8626	0.1270	−0.8227	0.1297	−0.7260	0.2806
$D_{i,3}$	−0.8060	0.0985	−0.7529	0.1036	−0.6463	0.3177
$I_{i,-3}$	0.1381	0.1140				
$I_{i,-2}$	0.1359	0.1550	0.0302	0.1453	−0.0042	0.1746
$I_{i,-1}$	0.0704	0.1498	−0.0214	0.1418	−0.0680	0.2370
$I_{i,0}$	0.5352	0.1530	0.4329	0.1465	0.3883	0.2872
$I_{i,1}$	0.4525	0.1619	0.3456	0.1569	0.3955	0.3313
$I_{i,2}$	0.5260	0.1727	0.4163	0.1694	0.4424	0.3779
$I_{i,3}$	0.3347	0.1256	0.2118	0.1253	0.3258	0.4281
Age at separation (years)	−0.0192	0.0058				
Education (years)	0.0075	0.0179	−0.0301	0.0459	−0.1131	0.0694
Employed	0.4515	0.0581	0.4007	0.0612	0.3995	0.0655
Parenting	0.0214	0.0537	0.0064	0.0579	−0.1381	0.0911
Income	0.0072	0.0013	0.0057	0.0014	0.0024	0.0015
Changed home	0.1410	0.0532	0.1464	0.0539	0.1801	0.0475
New partner	0.5415	0.0734	0.5065	0.0771	0.5415	0.1094
Constant	6.8554	0.2991	6.7972	0.5171		
R^2					0.0223	
$R^2_{overall}$	0.0844		0.0722			
R^2_{within}	0.0513		0.0516			
$R^2_{between}$	0.1454		0.123			
$\hat{\sigma}_u$	1.143		1.306			
$\hat{\sigma}_e$	1.555		1.555			
$\hat{\rho}_{RE}, \hat{\rho}_{FE}$	0.3508		0.4137			
N	7,619		7,619		6,914	
n	705		705		705	
T_{min}, T_{max}	2	16	2	16	2	16

Source: genderdiff data (see Example 4.4)

we are already familiar from the former FD analysis (see Table 4.13). Gainful employment, income, residential mobility, and a new partner significantly increase life satisfaction, while education and parenting do not have a significant effect. Age has a significant negative effect. The time dummies nicely show how life satisfaction decreases for men up to the year of separation. After that year, life satisfaction

4.2 Modeling the Change of Y

slightly improves for men, but never returns to the starting level it reached during the partnership (all estimates are negative and significant). The interaction effects are all positive and significant starting from the year of separation, which indicates that women are not as dissatisfied as men, especially after the separation from their partner.

The differences between the RE and FE estimates are rather small, so we do not discuss them in greater detail (the significance of the differences could be tested with the methods discussed in the section on p. 166). However, you should note that FE does not estimate the effect of the interaction $I_{i,-3}$ between gender and the first time dummy. It has to do with the fact that FE estimates a regression constant indirectly by demeaning the data in a very specific way (see the discussion of (4.10)). In doing so, it controls for the overall mean of all variables; hence, it cannot distinguish whether that overall mean is different for men and women. Again, it is important that you do not specify this interaction in your FE program. Otherwise, the program will arbitrarily omit some other variable from the model.

Since FD estimation also controls for unobserved heterogeneity at the level of units, as we have just seen in Sect. 4.2.1, we have estimated model (4.42) also with FD to learn a little bit more about the differences between the various panel regression models. Similar to FE estimation, FD does not assume that unobserved heterogeneity is independent of the variables in the model (see Textbox 4.6). For similar reasons, which we have just discussed, it also does not estimate an effect of age and the interaction variable $I_{i,-3}$. As Table 4.14 shows, the FD estimates point in the same direction as the FE (and RE) estimates, but the most striking difference is that most of them are not significant. Most T statistics are smaller in the FD than in the FE model.[30] This is a quite common experience, and is why FD estimation is less often used, at least when analyzing micro panels where the time dimension is rather short. Textbox 4.7 compares the characteristics of FE and FD estimation.

> *Textbox 4.7* (FE versus FD estimation)   Before discussing the pros and cons of FE and FD estimation from a statistical point of view, you should note that both estimation procedures yield identical results when applied to a two-wave panel (i.e., when $T = 2$). However, when $T \geq 3$, FE and FD estimators are not the same. Nevertheless, in large samples, both estimators should be the same, because both estimation methods provide consistent estimates if the

---

[30] As already noted in footnote 9 on p. 138, estimating regression models without a constant results in not very useful R^2 and F statistics. Hence, the R^2 of the FD model (0.0223) should not be compared with the overall R^2 values of the FE and RE model. If one is only interested in the effects of the time-dependent explanatory variables X, Wooldridge (2009, 466) suggests to stick with the undifferenced time dummies and to specify a model that includes a constant and $(T-2)$ (original) time dummies. The estimated effects of the explanatory variables X will be the same as in the "completely" differenced model.

first four (identical) assumptions in Textboxes 4.2 and 4.6 are true. Hence, the choice between both methods depends on the efficiency of each estimator, and efficiency in both cases depends (aside from heteroscedasticity) on the serial correlation of the idiosyncratic errors (e_{it} in case of FE, Δe_{it} in case of FD). If e_{it} is uncorrelated, FE is more efficient than FD; if Δe_{it} is uncorrelated, FD is more efficient than FE. While it is easy within the FD framework to test for the latter assumption, a simple test for uncorrelated e_{it} within FE estimation is not available. Other potential problems result from the strict exogeneity assumption and measurement errors. If the strict exogeneity assumption is violated, the FE estimator is less biased than the FD estimator. On the other hand, the FD estimator performs better when T is large and n is small (e.g., in macro panels). But both estimators are equally negatively affected by measurement error (see Sect. 3.6.3). In sum, it is difficult to choose between both estimators on statistical grounds. Even more so, it is important to recognize the substantive differences between both estimation procedures. FD estimation focuses on the change between only two (adjacent) panel measurements (t and $t-1$), while FE estimation takes all the measurements ($t = 1, \ldots, T$) into account. For research questions related to (instantaneous) change of Y, FD seems to be the most adequate method, while for research questions related to the level of Y, FE seems to be more useful.

Although a (discontinuous) impact function with dummies is the least restrictive with respect to the trend of Y (it assumes any kind of trend), it is also the most demanding in terms of the number of parameters to be estimated. Furthermore, in many applications, a more parsimonious impact function may also fit the data. By using appropriate linear restrictions for the effects $\alpha_{-2}, \alpha_{-1}, \ldots$, we can test whether life satisfaction follows a linear downward trend before separation and a linear upward trend after separation (the latter being different from the trend before separation). In our case, the four linear restrictions look like this:

$$\alpha_{-2} = \alpha_{-1} - \alpha_{-2}$$

$$\alpha_{-1} - \alpha_{-2} = \alpha_0 - \alpha_{-1}$$

$$\alpha_1 - \alpha_0 = \alpha_2 - \alpha_1$$

$$\alpha_2 - \alpha_1 = \alpha_3 - \alpha_2$$

and the corresponding F statistic for the FE model is not significant ($f = 0.51$, $df_1 = 4$, $df_2 = 6896$, $p = 0.730$), indicating that a linear model with a trend break in the year of separation would also fit the data (the same applies to the corresponding Wald statistic in the RE model: $X_2^2 = 1.84$, $df = 4$, $p = 0.765$). Similarly, we could test whether men and women are equally (dis)satisfied before separation and whether they differ by a constant degree after separation, using the following six

4.2 Modeling the Change of Y

restrictions for the interaction effects (in case of the FE model, we drop the first one because the effect of the first interaction variable is not estimated):

$$(\delta_{-3} = 0)$$
$$\delta_{-2} = 0$$
$$\delta_{-1} = 0$$
$$\delta_0 = \delta_1$$
$$\delta_0 = \delta_2$$
$$\delta_0 = \delta_3$$

Again, the test statistics indicate that this simplification fits the data (FE: $f = 0.60$, $df_1 = 5$, $df_2 = 6896$, $p = 0.703$; RE: $X_2^2 = 4.00$, $df = 6$, $p = 0.677$). However, if we want to generate estimates of those linear trends and the gender difference after separation, we have to respecify the model.

Johnson and Wu (2002) specify such a linear impact function in their analysis of psychological distress and divorce. Starting from model (4.20) with all the variables that are shown in Table 4.8, they specify an impact function using the two variables that measure the time from the current interview to the marital disruption (todiv), and the time from the current interview since the marital disruption (frmdiv). Hence, instead of $\beta_0(t) = \beta_0$, model (4.20) now includes

$$\beta_0(t) = \beta_0 + \beta_1 \cdot todiv_{it} + \beta_2 \cdot frmdiv_{it}$$

β_1 measures the linear trend in psychological distress before marital disruption and β_2 a similar trend after marital disruption. Table 4.15 shows the results. Similar to the analyses in Sect. 4.1.2.2, the differences between FE and RE estimates are rather small.[31] All previously discussed effects have similar estimates. The two trend variables show the expected sign: Before divorce (todiv), there is an upward trend in psychological distress, and after divorce (frmdiv), there is a downward trend, which is, however, not significant from zero. Note also that by using these two time trends, which measure the exact temporal distance of each interview (panel wave) from the date of the respondent's divorce, the analysis controls for the different spacing of the panel waves (see Example 4.2). Given our former discussion of model identification with different time variables, you may also wonder how it is possible to estimate the effect of age and time (relative to the date of marital disruption) in the same model. The explanation is very simple: While age is measured in years, time to and from divorce are measured in months and hence, are no perfect linear function of age.

---

[31] Again, the largest differences are observed for the variable age (ager) and being divorced (divorce).

Table 4.15 Psychological distress before and after divorce

Variable	FE		RE	
	Estimate	Std. Err.	Estimate	Std. Err.
Social selection (yes = 1)	(dropped)		0.1469	0.0482
Divorced (yes = 1)	0.2583	0.0661	0.3013	0.0623
Widowed (yes = 1)	0.1110	0.0874	0.1528	0.0788
Cohabiting (yes = 1)	−0.3207	0.1038	−0.3683	0.0997
Age (years)	0.0089	0.0017	0.0043	0.0012
Female (yes = 1)	(dropped)		−0.0230	0.0301
Education (years)	−0.0239	0.0096	−0.0331	0.0049
Time from disruption	−0.0064	0.0182	−0.0065	0.0178
Time to disruption	0.0844	0.0162	0.0770	0.0161
Constant	−0.1312	0.1410	0.1822	0.0881
$R^2_{overall}$	0.0253		0.0454	
R^2_{within}	0.0240		0.0211	
$R^2_{between}$	0.0268		0.0610	
$\hat{\sigma}_u$	0.5033		0.4325	
$\hat{\sigma}_e$	0.4661		0.4678	
$\hat{\rho}_{FE}, \hat{\rho}_{RE}$	0.5383		0.4609	
N	4,664		4,664	
n	1,166		1,166	
T	4		4	

Source: johnson-wu data (see Example 4.2)

4.2.3 Analysis of Trends

Instead of analyzing the effect of events with change scores and impact functions, scholars are often interested in the long-term change in their dependent variables and how this relates to different definitions of time. A typical example is the analysis of value change and how values and norms of individuals change among different generations, age groups, and points in time. More generally speaking, the interest is in showing how the level of Y relates to the (starting) date when the unit came into existence (the *cohort* effect), to the time elapsed between the time point of the current measurement and the former starting date (the *age* effect), and finally to the time point of the measurement itself (the *period* effect). When analyzing value change in a sample of individuals, the three effects would be measured by the year of birth, the age, and the time point of the current panel wave. If one were interested in the economic success of a sample of start-up firms, one would analyze with a panel of such firms how economic success Y relates to calendar time (the time point of the panel wave), the founding year, and the age of each firm. This type of analysis is called *cohort analysis*. It can be applied both to pooled cross-sectional

4.2 Modeling the Change of Y

and panel data. In this section, we want to illustrate its application to panel data with an example related to value change.

Example 4.5 (`postmat` data) Klein and Pötschke (2004) analyze the change of post-materialistic value orientations using data from the German Socio-Economic Panel Study (SOEP) spanning a period of 12 years. According to Inglehart (1971), Western industrialized societies, depending on their degree of modernization, experience a transformation of individual values, switching from materialist values, emphasizing economic and physical security, to a new set of post-materialist values, which instead emphasize autonomy and self-expression. These kinds of value orientations are measured with a survey question, in which the respondent is asked to rank the following four political aims: (i) maintaining order in the nation, (ii) giving people more say in important political decisions, (iii) fighting rising prices, and (iv) protecting freedom of speech. Items (i) and (iii) are used as indicators of materialism, items (ii) and (iv) as indicators of post-materialism. From these items Klein and Pötschke generate an index of post-materialism ranging from 1 (materialism) to 4 (post-materialism), with the values 2 and 3 indicating intermediate levels of materialism and post-materialism. Although the number of values that this index can take on is (very) limited, they analyze it as a continuous variable, and we follow their practice. Basically, there are three different explanations for why value orientations should change:

- The first one assumes that values and norms are formed during socialization in one's early years of life, and then remain rather stable over one's life course. Hence, if there are different value orientations among individuals, this occurs because individuals belong to different generations with different socialization experiences.
- The second one assumes that value change is also at work once the process of (primary) socialization has ended. More specifically, this position assumes that changes during the life course are the major driving forces behind value change. For example, the fact that individuals—after finishing their education—start a career and establish a family is supposed to make them more materialistic.
- Finally, the third explanation attributes value change to the socio-economic context in which individuals currently live. Hence, all individuals—irrespective of their year of birth or their age—are equally affected by the current options and constraints of their wider social context, and, therefore, will adapt their values and norms accordingly.

In reality, Klein and Pötschke argue, all three explanations may be at work at the same time. At least it is useful to start with this more general assumption and then test whether simpler (monocausal) explanations fit the data.

Obviously, the three explanations refer to three variables as possible explanations for the degree of post-materialistic value orientations Y: year of birth, age, and survey year. Hence, their analysis tries to disentangle the effects of generation, maturation, and time period. Moreover, they use education and gender as control variables, because cohorts and age groups differ in their composition with respect to their education and gender. They also assume that (intra-individual) value change (hence, the maturation effect) varies with respect to gender and education. Highly educated or male individuals are assumed to show more pronounced maturation effects, because they experience important life course changes within a shorter period of time, while starting from a higher level of post-materialism. Finally, they control for the national inflation rate, because the post-materialism index includes an item on rising prices, and large price increases in certain years may show up as period effects in the model.

Klein and Pötschke analyze SOEP data for the years 1984, 1985, 1986, and 1996. Their sample for analysis includes 5,418 survey respondents that have participated in all four panel waves. Hence, in principle, it is a balanced panel, with $4 \cdot 5,418 = 21,672$ observations; however, due to item non-response, several observations and a few individuals could not be used in the analysis. Therefore, the following models will be based on an unbalanced panel of $n = 5,415$ individuals and $N = 21,204$ observations. The variables of post-materialism, age, survey year, and inflation are time-varying (with inflation showing no between-unit variation), while the variables year of birth (cohort), education, and gender are time-constant.

In the following, we first want to show the identification problems that result if we test several definitions of time within the same model. These identification problems exist independently of the kind of panel regression model (pooled OLS, FE, RE, FD); hence, you can check them within the computationally simplest statistical model (e.g., pooled OLS). In a second step, we want to replicate Klein and Pötschke's findings, and since they assume that maturation effects differ among individuals, we make heavy use of variance components models that allow for these kinds of random slopes. More specifically, we use *maximum likelihood* (ML) to estimate these variance components models, and, in order to apply this estimation method, we have to assume that all unobserved heterogeneity is independent of the variables in the model (hence, the classical RE assumption; see Textbox 4.4).

Starting with the identification problems, you can easily verify with any kind of regression program that the following model testing for cohort (generation), age (maturation), *and* period effects is not identified:

$$y_{it} = \beta_0 + \beta_1 t + \beta_2 \cdot age_{it} + \gamma_1 \cdot year\_birth_i + u_i + e_{it} \qquad (4.43)$$

because age is a linear function of the survey date and the year of birth: $age = t - year\_birth$. Hence, the regression program will omit one of the three variables from the model. In other words: it is easy to estimate two of the three effects, but

4.2 Modeling the Change of Y

without additional countermeasures it is impossible to estimate all three of them. This identification problems exists irrespective of whether we treat age, cohort, and period as continuous explanatory variables (as is done in (4.43)) or as categorical ones (i.e., using dummies for different years of age, birth, and survey measurement).

There are different solutions to this identification problem. From a technical point of view, their common feature is that they all destroy the exact linear dependency between age, year of birth, and t. From a substantive point of view, all of them argue that year of birth, age, or survey measurement *as such* are not of interest. What is of interest is the fact that a person belongs to a certain generation (or cohort), is in a certain period of his or her life course, or experiences certain challenges in the wider socio-economic context.

- Hence, one solution is to distinguish among different cohorts of individuals born in certain historical periods and to replace year of birth with dummies for the different cohorts. Alternatively, one could distinguish among different age groups. In both cases, it is impossible to compute the exact age of a person from the survey year t and cohort membership (or alternatively, to compute the year of birth from t and the age group), because now cohort, as well as age group, include each *several* years of birth (age).
- Another solution would be to compute the age variable in months by using the exact (monthly) survey date. If year of birth and t are measured in years, the linear relationship $age = t - year\_birth$ also no longer holds.
- Finally, instead of age, year of birth, or t, one can use other variables that represent the assumed generation, maturation, or period effects much more directly. For instance, if one assumes that period effects are basically a result of inflation (due to the price item), then one should include the inflation rate instead of t into the model.

Nevertheless, in many cases, the "new" explanatory variables will be highly collinear with the other variables in the model, because even when exact linear dependencies are avoided, variables may still be almost dependent. Consider, for example, a cohort defined as all individuals born between 1920 and 1922. It is true that you do not know their exact age in the survey year 1996, but the range of possible values (74–76) is very limited; hence the degree of independent variation is very low. Based on German history, Klein and Pötschke (2004, 449) define six different cohorts, which comprise between 7 and 12 years of birth (e.g., the World War I and post-war generation is defined by all individuals born between 1922 and 1934). Using dummies $d_{2i}, d_{3i}, \ldots$ for the different cohorts (except the first one) instead of year of birth

$$y_{it} = \beta_0 + \beta_1 t + \beta_2 \cdot age_{it} + \gamma_2 d_{2i} + \gamma_3 d_{3i} + \gamma_4 d_{4i} + \gamma_5 d_{5i} + \gamma_6 d_{6i} + u_i + e_{it} \quad (4.44)$$

results in an identified model, but the cohort dummies do not have any significant effects. The reason for this result is the fact that the set of dummies and the variable age are highly collinear, which means that the estimation procedure has difficulties in allocating the effects to one of them and is thus unable to provide precise estimates (with small standard errors). Note, in passing, that similar to year of birth, the cohort dummies are time-constant variables and only have an index i as a result.

Because of these multicollinearity problems, the authors argue that the individual's (absolute) age is already controlled for by cohort membership $(d_{2i}, d_{3i}, \ldots)$ and year of measurement (t), which are already in the model. To measure the maturation effect, they propose to generate a new variable, which they call process time $(p = t - 1984)$ that measures how much older each individual gets between the four survey years. However, with the values $p_{i,1984} = 0$, $p_{i,1985} = 1$, $p_{i,1986} = 2$, and $p_{i,1996} = 12$ the model is not identified, because now t and p are linearly dependent, as you can see when you insert the definition of p into the equation:

$$y_{it} = \beta_0 + \beta_1 t + \beta_2 \cdot p_{it} + \sum_{j=2}^{6} \gamma_j d_{ji} + u_i + e_{it}$$

$$= \beta_0 + \beta_1 t + \beta_2 \cdot (t - 1984) + \sum_{j=2}^{6} \gamma_j d_{ji} + u_i + e_{it} \quad (4.45)$$

You cannot estimate two times the effect of t. Therefore, the authors replace the period effect t with the corresponding national inflation rate (i_rate):

$$y_{it} = \beta_0 + \beta_1 \cdot i\_rate_t + \beta_2 \cdot (t - 1984) + \sum_{j=2}^{6} \gamma_j d_{ji} + u_i + e_{it} \quad (4.46)$$

Note that, similar to the period variable t, the variable i_rate does not need an index i, because it varies over time, but not between individuals.

Basically, Klein and Pötschke's analysis rests on the specification of generation, maturation, and period effects that are shown in (4.46). As controls, they add the time-constant variables of education (a dummy for the German university-entrance diploma "Abitur") and gender (a dummy for men), and estimate the model with a non-linear (quadratic) maturation effect:

$$y_{it} = \beta_0 + \beta_1 \cdot i\_rate_t + \beta_2 \cdot (t - 1984) + \beta_3 \cdot (t - 1984)^2 + \sum_{j=2}^{6} \gamma_j d_{ji}$$

$$+ \gamma_7 \cdot abitur_i + \gamma_8 \cdot male_i + u_i + e_{it} \quad (4.47)$$

In principle, this model can be estimated with FE, which would have the advantage that all unknown variables at the individual level could be correlated with the variables in the model. However, FE would not provide us with any numerical effects of the cohorts, education, and gender, it only controls for these observed and all the unobserved (u_i) characteristics of the individuals. Moreover, if we assume that some of the slope coefficients vary between the units, as the authors do when they hypothesize that maturation is different for men and highly educated individuals, then we have to use a more general estimation procedure, such as maximum likelihood (ML).

4.2 Modeling the Change of Y

Unfortunately, ML estimation of models with random intercepts and random slopes for continuous Y is much too involved to be demonstrated here in its mathematical details. But its main idea is easily understood. ML uses those parameter values as estimates of the true population parameters that maximize the probability (more precisely, the likelihood) of observing the given sample with its values of Y, X, and Z. In order to do that, it needs an assumption about the probability distribution of the dependent variable Y. In the case of a continuous Y, as in our example, it uses the density function instead of the probability function, and that is why the general term for the method is maximum *likelihood* (and not maximum *probability*). Only in the case of categorical Y, a probability function is used (see Sect. 5.1.1.2 for an application). For the kinds of model that are relevant here, the following Textbox 4.8 summarizes the main assumptions.

Textbox 4.8 (ML estimation of random effects models)   Similar to the assumptions of the other estimation procedures, the following ML assumptions specify the conditions under which estimates can be computed, are unbiased, and can be generalized to the population. ML is a very general estimation procedure that can be applied to all kinds of statistical models. As mentioned before, it finds estimates by maximizing the density of Y, X, and Z. Therefore, an essential ingredient of this estimation technique is the correct specification of the density function. In case of random effects models for continuous Y, one usually assumes that the error terms U and E are normally distributed and independent of the variables X and Z in the model, whose values are assumed to be given. Given this assumption, Y is also a normally distributed random variable, since Y is a function of the variables X and Z assumed to be fixed and the variables U and E assumed to be random (hence, the distribution of U and E determines the distribution of Y). More specifically, the statistical properties of ML estimation rest on the following assumptions:
1. The data are a simple random sample of a well-defined population.
2. Each explanatory variable is neither a constant nor a linear function of the other explanatory variables.
3. The observations in the sample are independent of one another and are identically distributed.
4. The parameters of the density function are correctly specified.

Similar to the other estimation procedures using least squares, the first assumption is necessary for making statistical inferences about the population using the sampled data. The second assumption guarantees that a numerical value exists for each regression coefficient (technically, the model is identified). The third assumption ensures that we can use the same density function for each observation (because they are identically distributed) and obtain the overall likelihood of the whole sample by using formulas that combine the densities of independent observations. Obviously, panel data do not include independent observations. But the independence assumption is not at stake, if we are willing to assume that all serial correlations are controlled by X, Z and U. Once it is clear which density function can be applied to all observations and how to combine the single densities, we can deal with the question of how to include the explanatory variables. We are not able to show this here, but it turns out that the parameters of the normal distribution can be modeled as a function of X and Z as specified in the regression model (e.g., (4.47)). Here, the aforementioned fourth assumption comes in. If this function is misspecified, e.g., if an important explanatory variable is left out, then the estimates will be biased. Hence, the fourth assumption is the most important one, because it makes sure that our ML estimates are *unbiased*. As in the case of the other estimation methods, unbiasedness means that our estimates are correct; not necessarily in a single sample, but on average, i.e., across a multitude of samples, they are identical to the "true" population parameters. Whether these estimates are then *efficient* and how to compute their standard errors are even more difficult questions that cannot be dealt with in this textbook (but see Sect. 7.2.2). Statistical theory shows, however, that ML estimates are normally distributed in large samples and similarly efficient as estimates from other estimation procedures. Therefore, tests of significance can use the normal and related distributions as the testing distributions (see Sect. 7.2.2).

A difficult question is how the ML estimates can be computed, because there is no analytical solution to the maximization problem (see also Sect. 5.1.1.2). One needs methods of numerical optimization, and there are two different ways of specifying the likelihood to be maximized. One is called full maximum likelihood (FML), the other restricted maximum likelihood (REML). In the first case, both the regression coefficients and the variance components ($\sigma_{0i}^2, \sigma_{1i}^2$, etc.) are included in the likelihood function. In the second case, the likelihood includes only the (co)variance components, while the regression coefficients are estimated in a second step. According to statistical theory, REML should provide better estimates (FML is supposed to provide biased estimates of the (co)variance components), but from a practical point of view, the differences between REML and FML estimates are often negligible. In the following, we use FML, because it allows to compare models with the likelihood ratio test.

Model (4.47) includes only a random intercept (due to u_i) and hence, could be estimated with the methodology we already know (FGLS, see Sect. 4.1.2.2). However, we use ML from the beginning, because we also want to include random slopes, and here, ML is the more general estimation technique with better statistical qualities. Table 4.16 shows our estimation results; as you can see, the ML estimates (model 2) do not differ to the forth digit from the FGLS estimates (model 1). Statistical theory tells us that both estimates are asymptotically equivalent. The estimates show that (in line with the hypotheses) post-materialism increases significantly in

4.2 Modeling the Change of Y

the younger cohorts. Men and individuals having an "Abitur" are significantly more post-materialistic than women and individuals with lower educational degrees than "Abitur". Post-materialism is non-linearly related to maturation, but the effect is not significant. Finally, as expected, post-materialism decreases with rising inflation rates.

In the next step of our analysis, we test whether both maturation effects vary significantly among individuals and hence necessitate the inclusion of random slopes. In order to test for this possibility, we assume that both regression coefficients are unit-specific (have an index i) and consist of fixed and random parts. Using the extended notation that we introduced in Textbox 3.2, these assumptions can be specified as follows:

$$\begin{aligned} \beta_{2i} &= \gamma_{20} + u_{2i} \\ \beta_{3i} &= \gamma_{30} + u_{3i} \end{aligned} \qquad (4.48)$$

γ_{20} and γ_{30} represent the fixed part, while u_{2i} and u_{3i} represent the random part. Similar to u_i in (4.47), u_{2i} and u_{3i} are assumed to be normally distributed random variables with constant variance. Reinserting (4.48) into (4.47) and rearranging shows that the model is now heteroscedastic (in line with the extended notation we have renamed u_i into u_{0i} and γ_j into γ_{0j}, $j = 2, \ldots, 8$):

$$y_{it} = \beta_0 + \beta_1 \cdot i\_rate_t + \gamma_{20} \cdot (t - 1984) + \gamma_{30} \cdot (t - 1984)^2 + \sum_{j=2}^{6} \gamma_{0j} d_{ji}$$
$$+ \gamma_{07} \cdot abitur_i + \gamma_{08} \cdot male_i + u_{0i} + u_{2i} \cdot (t - 1984) + u_{3i} \cdot (t - 1984)^2 + e_{it}$$
$$(4.49)$$

and calls for a general estimation procedure such as ML. Model 3 in Table 4.16 shows that the variance of both maturation effects (estimated as $\hat{\sigma}_2^2 = 0.0444$ and $\hat{\sigma}_3^2 = 0.0003$) is significantly different from zero[32] and therefore, both slope coefficients should be modeled by characteristics of the individuals. Klein and Pötschke use the following models:

$$\begin{aligned} \beta_{2i} &= \gamma_{20} + \gamma_{21} d_{5i} + \gamma_{22} d_{6i} + \gamma_{23} \cdot abitur_i + u_{2i} \\ \beta_{3i} &= \gamma_{30} + \gamma_{31} \cdot male_i + u_{3i} \end{aligned} \qquad (4.50)$$

which assume that the linear part of the maturation effect is different for younger (belonging to cohort 5 and 6) and highly educated individuals, while the quadratic part of the maturation effect is different for males. Model 4 in Table 4.16 shows the regression estimates of this extended model. Since most effects hardly change,

---

[32] We do not interpret the substantive effects of model 3, because they are similar to those of the former model 2.

Table 4.16 Determinants of post-materialism (1984–1996) (FGLS and ML estimates)

Variable	FGLS Model 1 Estimate	FGLS Model 1 Std. Err.	ML Model 2 Estimate	ML Model 2 Std. Err.	ML Model 3 Estimate	ML Model 3 Std. Err.	ML Model 4 Estimate	ML Model 4 Std. Err.
Cohort 2	0.1721	0.0411	0.1721	0.0412	0.1687	0.0408	0.1681	0.0408
Cohort 3	0.3385	0.0397	0.3385	0.0397	0.3440	0.0394	0.3433	0.0394
Cohort 4	0.5012	0.0411	0.5012	0.0412	0.4989	0.0409	0.4984	0.0409
Cohort 5	0.6089	0.0399	0.6089	0.0399	0.5948	0.0396	0.6600	0.0418
Cohort 6	0.6352	0.0493	0.6352	0.0494	0.6206	0.0490	0.6919	0.0537
Male	0.1427	0.0186	0.1427	0.0187	0.1353	0.0185	0.1608	0.0207
Abitur	0.8062	0.0306	0.8062	0.0306	0.7834	0.0303	0.9006	0.0356
Inflation	−0.0557	0.0104	−0.0557	0.0104	−0.0553	0.0096	−0.0553	0.0096
$(t - 1984)$	−0.0088	0.0174	−0.0088	0.0174	−0.0086	0.0164	−0.0015	0.0164
$(t - 1984)^2$	0.0006	0.0013	0.0006	0.0013	0.0005	0.0013	0.0008	0.0013
Cohort 5 · $(t - 1984)$							−0.0140	0.0028
Cohort 6 · $(t - 1984)$							−0.0152	0.0047
Abitur · $(t - 1984)$							−0.0249	0.0039
Male · $(t - 1984)^2$							−0.0005	0.0002
Constant	1.7501	0.0459	1.7501	0.0460	1.7594	0.0447	1.7142	0.0450
$\hat{\sigma}_0^2$	0.3057		0.3071	0.0090	0.4153	0.0190	0.4122	0.0190
$\hat{\sigma}_2^2$					0.0444	0.0107	0.0441	0.0107
$\hat{\sigma}_3^2$					0.0003	0.0001	0.0003	0.0001
$\hat{\sigma}_e^2$	0.6073		0.6067	0.0068	0.5226	0.0102	0.5227	0.0102
ln L			−27,743.181		−27,632.592		−27,585.929	
N	21,204		21,204		21,204		21,204	
n	5,415		5,415		5,415		5,415	
T_{min}, T_{max}	1	4	1	4	1	4	1	4

Source: `postmat` data (see Example 4.5)

we focus on the interpretation of the interaction effects. The linear and the quadratic maturation effect are still not significant. However, for the two youngest cohorts and individuals with "Abitur", post-materialism decreases significantly with maturation; and for men, the quadratic term is negative and significant, indicating that men grow increasingly materialistic as they mature. Some of the estimates are rather small. You should consult Klein and Pötschke (2004) for a graphical illustration of the practical relevance of the estimated parameters.

At the end of this discussion of variance components models, we would like to mention a specification problem that is always prevalent, when one assumes more than one variance component (besides the variance of the idiosyncratic error term). Model 4, for example, estimated three different variance components: σ_0^2, σ_2^2, and σ_3^2. In that case we also have to think about the covariance of the random coefficients. Remember our `wagepan` data and Fig. 3.2, where we said that every individual has its own income trajectory with a unit-specific intercept and a unit-specific slope. These individual regression coefficients are not independent of each other: If the slope is very flat, the corresponding intercept is rather large (and vice versa). Hence, we also have to estimate the covariances between the three random variables U_0, U_2, and U_3. The most complex assumption is that each covariance is different from the other, and this assumption (an *unstructured* covariance matrix) was used when estimating models 3 and 4. Furthermore, with panel data, we may have to deal with serially correlated idiosyncratic errors (see Sect. 3.6.2) and hence, may also want to assume a certain covariance structure for the e_{it} (e.g., an autoregressive one).

4.3 Conclusion and Further Reading

In this chapter we have discussed how to analyze the level and the change of a continuous dependent variable with panel data. We showed that traditional OLS regression does not account for the fact that panel data include serially dependent information due to the repeated measurements for each unit of analysis. A simple alternative was to use empirical standard errors that are robust against the clustering of observations within units. A theoretically more convincing alternative is, of course, to model the panel design of the data. This is feasible by introducing a unit-specific error term U that captures all (unknown) time-constant characteristics of the units that influence each measurement and hence, is one source of the serial correlations in the data. Different techniques—FE, FD, and RE estimation—are available to deal with this unit-specific error term, depending on our assumptions about U (e.g., whether it is assumed to be independent or correlated with the variables in the model). The nice thing about this unit-specific error term is also that it allows us to control for unobserved heterogeneity at the unit level. Therefore, compared to cross-sectional models, which are plagued by omitted variable bias, panel regression models are able to deal with unobserved determinants of Y that are specific to the unit and hence, time-constant. The basics of FE, FD, and RE models can be found in any introductory econometrics textbook (e.g., Wooldridge, 2009). Baltagi

(2008), Cameron and Trivedi (2005), Hsiao (2003), Greene (2008) and Wooldridge (2010) include more advanced treatments. FE models and especially the combination between FE and RE models, which we called hybrid models, are discussed at length in Allison (2009).

We also mentioned various extensions of these panel regression models, including autocorrelated error terms, random coefficients, impact functions, and panel regression models for data in wide format or with missing data (unbalanced panels). Models with random coefficients are discussed extensively in the literature on hierarchical linear or multi level models. For example, Hox (2010) and Snijders and Bosker (2011) both include a chapter on longitudinal data. Bollen and Brand (2010) show how RE and FE models can be applied to data in wide format using structural equation models (SEM). SEM also provide a nice environment, in which it is easy to implement all kinds of extensions of the simple panel regression models (see Bollen and Brand, 2010) and to study the change of Y using growth curves (see Bollen and Curran, 2006; Duncan et al., 2006).

Panel Analysis of Categorical Variables

In this chapter, the focus shifts to panel models for categorical dependent variables. Categorical variables—as opposed to continuous variables—have only a few discrete values. They can be measured on nominal, ordinal, or metric scale. To illustrate the rationale behind panel data models for categorical variables, we focus first on categorical variables having only two distinct values (i.e., binary or dichotomous variables). Models for categorical variables having more than two values will be treated later on in specific sections of this chapter. As examples for binary dependent variables, think of an analysis of women's labor force participation, party membership, or a company's pension plan. In the first example, the dependent variable indicates whether or not women take a position in the primary labor market. In the example of party membership, the dependent variable measures whether or not individuals are aligned with any party or independent, and the final example exemplifies a model in which the dependent variable indicates whether or not a company offers an occupational pension scheme.

The first part of the chapter is concerned with models focusing on the level of Y. As discussed in Chap. 3, models for the level of categorical variables focus on the probability of observing a certain value (category) of Y. More specifically, in Sect. 3.5.1.2, we introduced a model in which the probability, $\Pr(y_{it} = q)$, of observing category q for unit i at time point t is a function of a set of independent variables, which may be either time-constant (Z) or time-varying (X):

$$\Pr(y_{it} = q) = G\big(\beta_0(t) + \beta_1 x_{1it} + \cdots + \beta_k x_{kit} + \gamma_1 z_{1i} + \cdots + \gamma_j z_{ji}\big) \quad (5.1)$$

$G(\cdot)$ is a suitable transformation function that ensures that the right-hand side of the equation provides values that are within the proper limits of probabilities (i.e., $0 \le \Pr(y_{it} = q) \le 1$). Like in the previous chapters, the right-hand side of the model includes k time-varying independent variables $x_{1it}, \ldots, x_{kit}$ and j time-constant variables $z_{1i}, \ldots, z_{ji}$. $\beta_1, \ldots, \beta_k$ and $\gamma_1, \ldots, \gamma_j$ denote the corresponding regression coefficients. The term $\beta_0(t)$ determines the overall level of the probability. Since its level may change over time, $\beta_0(t)$ can be any function of time to control for possible time trends. If there is no time trend, it reduces to the familiar regression constant

$\beta_0(t) = \alpha_0$. In the more general case, $\beta_0(t)$ can be a linear, quadratic, exponential or non-parametric function of time t (see Sect. 3.5.1.1). This model is called a *discrete response model* and is extensively discussed in Sect. 5.1. Going back to the examples from above, typical research questions would be: Who is more likely to participate in the labor market? Who is more likely to be a party member? What are the characteristics of companies that offer pension plans?

In the second part of the chapter, we will discuss models for the change of Y. In Chap. 3, we proposed to model the change of categorical variables by focusing on the conditional transition probability of making a change from category p to category q of the dependent variable, given the unit has been observed in category p at all former measurements. As a shortcut, we introduced $h_p(t) = \Pr(y_{it} = q | y_{i1} = \cdots = y_{i,t-1} = p)$, which is also termed the (discrete-time) hazard rate (of leaving state p). A model for the conditional transition probability resp. hazard rate would look like this (see Sect. 3.5.2.2):

$$h_{ip}(t) = G\big(\beta_0(t) + \beta_1 x_{1it} + \cdots + \beta_k x_{kit} + \gamma_1 z_{1i} + \cdots + \gamma_j z_{ji}\big) \quad (5.2)$$

Again, $G(\cdot)$ is a suitable transformation function to make sure that the right-hand side of the equation provides values that are within the proper limits of probabilities (i.e., $0 \leq h_{ip}(t) \leq 1$). Note also the index i for the hazard rate, which is necessary because model (5.2) assumes that the hazard varies depending on the characteristics X and Z of each unit i. This model is called a *discrete-time event history model* and is extensively discussed in Sect. 5.2. Typical research questions would be: Who is more likely to re-enter the labor market after maternity leave? Under which conditions are party members more prone to resign? Do people employed in companies with private pension plans retire earlier than those in companies without such plans? Here, we study decisions about whether and, if so, when to do something.

Obviously, categorical variables include much more than dichotomous variables. Categorical variables can have more than two categories and the categories can be measured on a nominal, ordinal, or metric scale (or a mixture of them). Discussing all the specialized models that have been developed for these different types of categorical variable would certainly go beyond the scope of this introductory textbook. But both parts of this chapter include a section on extensions (Sects. 5.1.3 and 5.2.4) that will mention some of these models and make references to the corresponding literature. The chapter concludes with some suggestions for further reading (Sect. 5.3).

5.1 Modeling the Level of Y: Discrete Response Models for Panel Data

This section illustrates how to estimate discrete response models with panel data (i.e., models that focus on the probability of observing a certain value [category] of the dependent variable). Like in the previous chapter on continuous variables, we start in Sect. 5.1.1 with traditional cross-sectional models for the analysis of categorical data, including the linear probability (Sect. 5.1.1.1), the logistic (Sect. 5.1.1.2),

5.1 Modeling the Level of Y: Discrete Response Models for Panel Data

and the probit regression model (Sect. 5.1.1.3). As already mentioned in Chap. 3, the efficiency of traditional models is at stake if the data include dependent observations. As a remedy, we will introduce robust standard errors that control for the hierarchical nature of the data (observations within units). In doing so, we arrive at more conservative test results of our parameter estimates reflecting the fact that we do not have as many independent data points as the mass of panel data suggests.

However, robust standard errors only try to improve the estimates of the standard errors and leave the parameter estimates unchanged. Moreover, they control for any kind of serial dependence in the data, while you may have a specific idea about what causes this serial correlation. Hence, a more appropriate method adopts the traditional models to the panel structure of the data. This is the task of Sect. 5.1.2 and it is based on the assumption that the stochastic part of the model consists of two independent components: one operating at the level of units and the other operating at the level of measurements. This assumption has two advantages:

1. It allows us to compute more efficient estimates of the parameters and their standard errors than in traditional models even with robust standard errors. This method is called *random effects (RE) estimation* (see Sect. 5.1.2.2). However, it still hinges on the exogeneity assumption (A.1) (see Sect. 3.6.4), which is at stake if we have omitted important explanatory variables of $\Pr(y_{it} = q)$ that correlate with the variables in the model.
2. It allows us to control for omitted variables bias at the unit level, when the unobserved variables at the unit level correlate with the variables in the model, which is not possible with cross-section data. This method is called *fixed effects (FE) estimation* (see Sect. 5.1.2.1) and it is to be preferred over RE if the exogeneity assumption does not hold. However, if the exogeneity assumption is true (the omitted variable bias does not exist), it is less efficient than RE estimation.

Having explained these basic estimation techniques for binary categorical variables, we then turn to some extensions to other types of (polytomous) categorical variable (Sect. 5.1.3). To illustrate how to estimate a discrete choice model with panel data, we refer to a study by Heineck and Schwarze (2004) on the determinants of secondary job holding, also called moonlighting.

Example 5.1 (heineck-schwarze data) Secondary employment is a rather common phenomenon in most of the Western industrialized countries and the driving forces behind it are particularly relevant in the analysis of atypical forms of employment. The focus of the analysis by Heineck and Schwarze (2004) is on the propensity of men and women in Germany and Great Britain to have a second job.

Moonlighting is an inherently complex issue in labor supply. Heineck and Schwarze (2004) develop two motives: the hours-constraints motive and the heterogeneous-jobs motive. Key indicators for the hours-constraints motive are working time preferences and actual working hours. The idea is that some workers face labor supply constraints in their first job and hence, have an incentive to hold a second job, in particular to overcome economic hardship.

At the same time, there are also instances in which workers earn a sufficient income but nevertheless opt for moonlighting since they are not satisfied with the quality of their first job. To capture this heterogeneous-jobs motive, several indicators are used: satisfaction with job security, total pay, and work itself as well as employment prospects in the first job.

To simplify matters in this textbook, we do not use the full sample analyzed by Heineck and Schwarze (2004), but concentrate on British men only. Our subset focuses only on those original sample members with full interviews and a permanent full-time job. This decision can also be justified on theoretical grounds, since temporary and part-time employment does not play a prominent role for men in Great Britain. The data set includes 2,338 men with altogether 13,328 observations describing their job histories in the years 1991–2000. Contrary to the examples in Chap. 4, this data set is an unbalanced panel since the sampled men contribute differing numbers of panel waves to our data.

We do not replicate the full model presented by Heineck and Schwarze (2004). In our illustrative example, one of the variables of interest is the question of whether the respondent would like to change the number of working hours in the first job. From the answers, the authors generate two dummy variables that capture both the desire to work more hours and less hours (with no desire for any change being the reference category). Hence, both dummies—including a continuous variable measuring the actual weekly working hours in the first job—are indicators for the hours-constraints motive. Furthermore, the model includes two dummies measuring whether a person is not satisfied with the pay of the first job and the work itself. These latter two dummies are meant to reflect the heterogeneous-jobs motive. Besides that, the model includes various control measures (for details, see Table 5.1). All of the explanatory variables vary over time and the coded data reflect their current values at the time of the interview. Thus, contrary to the data sets used in the previous chapter, there are no time-constant explanatory variables in our example data set. However, if we would use the full data set of Heineck and Schwarze, we would include information on gender and nationality for each sample member—two typical time-constant variables. Note also that due to missing values for some of the variables, in the multivariate analyses, the analyzed data reduce to $n = 2,127$ men with altogether 10,687 observations in the observation period.

5.1.1 Ignoring the Panel Structure

The data set is arranged in such a way that we observe the job histories of British men between 1991 and 2000. In each of the panel waves, we know whether these individuals were double jobholders or not. The dependent variable in this illustrative example is equal to 1 if a man holds a second job and 0 otherwise. Our intention

5.1 Modeling the Level of Y: Discrete Response Models for Panel Data

Table 5.1 Measures used in the analysis of the heineck-schwarze data

Variable	Description
Variables of interest	
Hours-constraints motive	two dummies for the desire to work more hours and the desire to work fewer hours
Working hours	weekly working hours in the first job
Heterogeneous-jobs motive	two dummies for being not satisfied with pay and being not satisfied with work itself
Control variables	
Labor income	logarithm of first job income
Non-labor income	logarithm of non-labor income
Sector of employment	dummy for having a public employer
Job tenure	logarithm of job tenure (in years)
Age and age squared	age in years
Children	dummy for having a child under 15 in the household
Occupational class	dummy for being a service class worker

is now to analyze the probability to observe the response $y_{it} = 1$ at a given point in time: Which conditions captured by our explanatory variables make it more likely to be in secondary employment at a given point in time?

Like in the previous chapter on continuous dependent variables, a first step in the analysis could be to pool all available information from the $t = 1, \ldots, 10$ panel waves for all $i = 1, \ldots, 2{,}127$ individuals without missing information and treat them as if they represent independent information for $n = 10{,}687$ individuals.[1] We could then apply all the techniques that are available for the analysis of cross-sectional data. In that case, we have to deal with two problems. One of them is the well-known fact that repeated measurements of the same units over time are not independent of each other. The other pertains to the specific nature of the dependent variable that now has only two distinct values. The latter problem is easily illustrated by way of comparison with our former analysis of continuous dependent variables; for example, the analysis of psychological distress in the johnson-wu data (Example 4.2). Examination of the scatter plots in Fig. 5.1 for psychological distress against age and moonlighting against first job wage reveals the specific nature of our categorical dependent variable. While it is possible to observe different levels of psychological distress in the johnson-wu data, the heineck-schwarze data do not show the probability of moonlighting; they rather show who is moonlighting ($Y = 1$) and who is not ($Y = 0$). Thus, the first problem is how to estimate a model for (unobserved) probabilities from an observed binary variable.

---

[1] Note that it is an unbalanced panel and hence, the number of observations does not equal $n = 10 \cdot 2{,}127 = 21{,}270$.

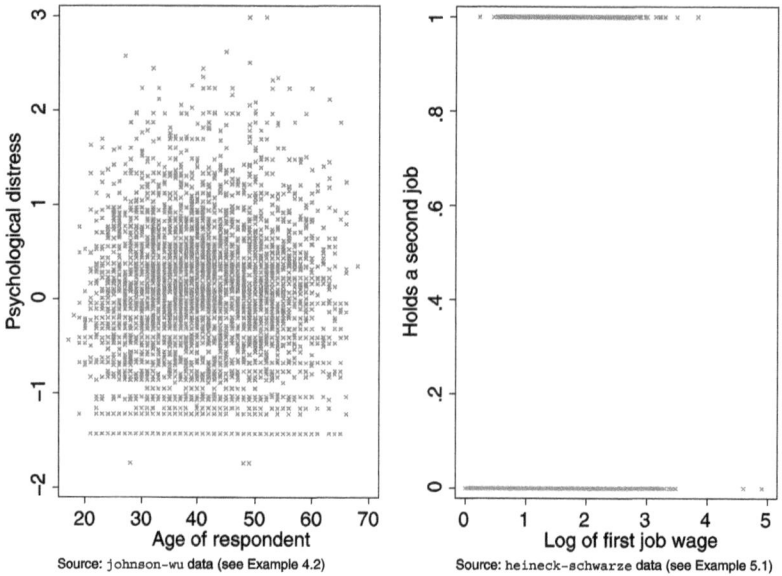

Fig. 5.1 Scatter plots of a continuous and dichotomous Y

The other problem is related to the fact that our analysis is based on pooled data. In other words, there are multiple observations for each individual. In contrast to a single cross-section, the observations in this panel data set are not independent of each other. In our case, a plausible assumption is that individuals holding a second job this year, will do the same in the following year. Table 5.2 reports the observed transitions between each pair of panel waves.[2] We can quite clearly see statistical dependencies between the observations in our example data. Of those men who moonlight at time $t - 1$, about 62 % also hold a secondary job at time t. On the other hand, about 97 % of the men without a second job at time $t - 1$ do not moonlight at time t. In other words, moonlighters and non-moonlighters are a rather constant group over time, at least over a two-year period. Part of this stability may be explained by the explanatory variables in Table 5.1. But possibly, some unobserved heterogeneity remains that renders the residuals of our traditional cross-sectional models to be serially correlated. This raises the question of whether the model still provides unbiased and efficient estimates. Before dealing with this panel data problem, let us first understand how to solve the first problem of analyzing binary dependent variables. We do that by ignoring the panel structure of the data and treating the data as statistically independent observations.

---

[2]Note that Table 5.2 shows year-to-year transitions of the moonlighting variable. At the end of each unit's observation period T_i, we lose one observation, because we do not observe $T_i + 1$, and hence, cannot compute the transition from T_i to $T_i + 1$. Therefore, the table includes $13,328 - 2,338 = 10,990$ observations. Note also that this table focusses only on the dependent variable and does not suffer from missing values for the explanatory variables. It also ignores gaps in the sequences of the dependent variable. In other words: The table uses all available information for Y.

5.1 Modeling the Level of Y: Discrete Response Models for Panel Data

Table 5.2 Transitions between panel waves

	No sideline job at time t	Moonlighting at time t	Total
No sideline job at time $t-1$	97.19 (9,884)	2.81 (286)	100.00 (10,170)
Moonlighting at time $t-1$	37.68 (309)	62.32 (511)	100.00 (820)
Total	92.75 (10,193)	7.25 (797)	100.00 (10,990)

Source: heineck-schwarze data (see Example 5.1)

5.1.1.1 The Pooled Linear Probability Model

As a starting point, let us assume that the transformation function $G(\cdot)$ in our probability model (5.1) is the identity function $G(a) = a$. In that case, model (5.1) is very similar to the traditional linear regression model, except the fact that the left-hand side includes a probability (since we have a binary dependent variable, we use the probability of observing $q = 1$):

$$\Pr(y_{it} = 1) = \beta_0(t) + \beta_1 x_{1it} + \cdots + \beta_k x_{kit} + \gamma_1 z_{1i} + \cdots + \gamma_j z_{ji} \quad (5.3)$$

Why is it possible to interpret this linear regression model as a linear probability model (LPM)? The answer is quite easy. Without an error term, the right-hand side of the equation predicts the expected value, $E(y)$, of the dependent variable (see (3.11) in Chap. 3). Furthermore, the expected value of a binary variable with two categories coded 0 and 1 equals the probability of observing its value 1. For example, the mean of the moonlighting dummy in the heineck-schwarze data equals 0.0759, which means that the overall probability of observing a person in a secondary job is about 7.6 % (1,012 out of the 13,328 observations are moonlighters). Thus, we can use the familiar linear regression model to make statements about the effects of our explanatory variables on the probability of moonlighting. The regression coefficients β and γ measure the change of that probability for a given change in our explanatory variables X and Z. In case these are dummy variables, the corresponding regression coefficient shows us how much that probability differs in the group of individuals identified by that dummy (e.g., those who are unsatisfied with their first job's pay) from the probability of moonlighting in the reference group (those who are satisfied with their pay).

Table 5.3 shows the OLS estimates of the full model based on the $N = 10,867$ observations with non-missing information on the explanatory variables. For ease of exposition, the model assumes $\beta_0(t) = \beta_0$ (but see footnote 2 on p. 121). In line with the expectation of Heineck and Schwarze (2004), we find evidence for the hours-constraints motive: Compared to those who are satisfied with their working hours, those who would like to work more hours are 3.2 percentage points more likely to moonlight and this effect is highly significant ($p < 0.01$). Turning to the actual working hours, we get, as one would expect, a significant negative effect ($p < 0.01$). However, from a practical point of view, this effect does not seem to be very large. An increase of weekly working time by ten hours would decrease the probability to hold a secondary job by one percentage point.

Table 5.3 Determinants of secondary job holding (pooled LPM)

Variable	Estimate	Std. Err.
Desire to work more hours	0.0323	0.0113
Desire to work fewer hours	−0.0013	0.0055
Weekly working hours in first job	−0.0010	0.0003
Is not satisfied with pay	0.0138	0.0056
Is not satisfied with work itself	0.0030	0.0066
Logarithm of first job wage	−0.0236	0.0078
Logarithm of non-labor income	0.0023	0.0019
Has public employer	0.0271	0.0073
Logarithm of job tenure	0.0111	0.0029
Age	0.0022	0.0020
Age squared	−0.0000	0.0000
Has children under 15	0.0159	0.0058
Being a service class worker	0.0192	0.0060
Constant	0.1005	0.0395
F	9.30	
df_1, df_2	13	10,673
R^2	0.011	
N	10,687	

Source: heineck-schwarze data (see Example 5.1)

To visualize this result, it is possible to draw a plot of the observed responses and the predicted probabilities. Predicted probabilities are calculated using (5.3) for different hours of weekly working time in the first job and for two different groups (those who desire more hours to work and those who desire less hours to work), while the other independent variables are fixed at their respective means in the estimation sample or set to zero in the case of dummy variables. The solid line in Fig. 5.2 represents men who would like to work more hours, while the dashed line stands for the comparison group (i.e., men who would like to work less hours).

The question that arises from a statistical point of view is now: Is the linear probability model a good model to predict such response probabilities? The estimators of the linear probability model are consistent and unbiased, if the usual OLS assumptions are met (see Textbox 4.1 in Chap. 4). Yet, it is important that, when using OLS regression for a categorical dependent variable, $\Pr(y_{it} = 1)$ is not constrained to lie in the unit interval. For example, for men of age 60 who work quite a lot, say 50 hours during the week, and, hence, desire to work less hours, we predict a probability of moonlighting that is below 0 %, if they earn a sufficient income in the upper quartile of the income distribution (all other variables fixed at their mean resp. dummies set to zero). Obviously, this is not a meaningful value for a probability.

Another problem is the linearity assumption. Mean weekly working time amounts to 45.9 hours and is located at the upper end of the abscissa in Fig. 5.2. At that point, the probability of moonlighting is already quite low (below 6 %). Now increase working hours unit by unit. What happens to the predicted response

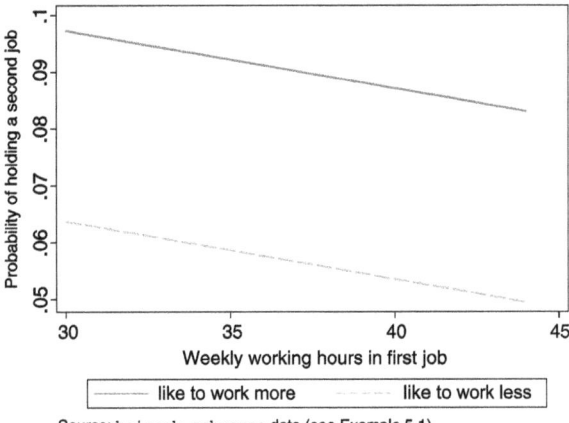

Fig. 5.2 Observed responses and predicted probability of secondary job holding (pooled LPM)

probability $\Pr(y_{it} = 1)$? Regardless of the initial working hours, $\Pr(y_{it} = 1)$ always decreases by the same amount. This is hardly a realistic assumption, especially when the probability of moonlighting is already quite low. In that case, you always have to anticipate bottom effects, because a probability cannot exceed the unit interval (and ceiling effects if the probability is already quite high and you move in the opposite direction). In other words, the change in the probability of moonlighting should become increasingly smaller with each unit change of the independent variables.

Furthermore, the linear probability model violates the homoscedasticity assumption of OLS (see Textbox 4.1 in Chap. 4). This is due to the fact that our dependent variable is a dummy variable and as such can take on only the values zero and one. It is easy to show that the OLS error term ε_{it} does not have constant variance but rather varies systematically with the values of the independent variables. More specifically, it can be shown that $\text{Var}(\varepsilon_{it}) = \Pr(y_{it} = 1)[1 - \Pr(y_{it} = 1)]$ and since $\Pr(y_{it} = 1)$ is a function of the independent variables (see (5.3)), $\text{Var}(\varepsilon_{it})$ is also a function of X and Z and thus not homoscedastic. Hence, even when the linear probability model provides consistent and unbiased estimators, it comes at the expense of biased standard errors.

Finally, we can use Example 5.1 to illustrate the problems of applying the linear probability model to panel data. It is not only that the OLS error term ε_{it} is not homoscedastic. The observations in our data set are also not independent. Thus, if we compute the residuals of our former OLS regression model (which are estimates of the OLS error term) and correlate these residuals for any two panel waves for all men in our estimation sample, we still find a strong positive correlation of $r_{LPM} = 0.6762$, although we control for quite a lot of explanatory variables that characterize moonlighters and non-moonlighters. The same is true for all pairs of time points, which can be seen from the yearly serial correlations shown in Table 5.4. We therefore conclude that our linear probability model also suffers from serially correlated errors, since it does not recognize the panel structure of the data. This violates the no autocorrelation assumption of OLS estimation (see Textbox 4.1

Table 5.4 Correlation of residuals from the pooled LPM

Year	1991	1992	1993	1994	1995	1996	1997	1998	1999	2000
1991	1.0000									
1992	0.6138	1.0000								
1993	0.5472	0.6514	1.0000							
1994	0.5571	0.6157	0.6852	1.0000						
1995	0.4631	0.5572	0.5776	0.7482	1.0000					
1996	0.3594	0.3877	0.3833	0.5836	0.7206	1.0000				
1997	0.3771	0.3966	0.3423	0.5089	0.6374	0.6539	1.0000			
1998	0.3042	0.3506	0.3276	0.4663	0.6220	0.5843	0.7060	1.0000		
1999	0.2345	0.3529	0.3226	0.4857	0.5212	0.5541	0.6047	0.6769	1.0000	
2000	0.2670	0.2801	0.3717	0.4116	0.4354	0.5733	0.5633	0.6550	0.6316	1.0000

Source: heineck-schwarze data (see Example 5.1)

in Chap. 4) and therefore, it is no longer ensured that the estimated standard errors are unbiased.

All in all, there are strong reasons to believe that firstly the linear probability model is not very well suited to binary dependent variables where the probability of observing one or the other category is either very low or very high, because in those regions of the probability space, statistical relationships are often non-linear. Secondly, when the linearity assumption is not at stake, OLS parameter estimates are unbiased, but standard errors are definitely biased because of heteroscedastic and serially correlated error terms.

Like in the case of continuous dependent variables (see Sect. 4.1.1.2), there is an overall remedy against the latter inefficiencies: robust standard errors. Estimates of OLS standard errors can be made robust both against heteroscedasticity and serial correlation (see Sect. 7.2.1). Table 5.5 compares the point estimates and standard errors of the linear probability model and two regression models using robust standard errors. One model uses heteroscedasticity robust standard errors and the other takes into account (in addition to heteroscedasticity) the statistical dependencies among observations in the heineck-schwarze data (observations clustered within units). The regression coefficients are exactly the same in all models; however, the standard errors differ. The robust standard errors are higher and become even larger once adjusted for serial correlation.[3] As a consequence, the number of working hours loses its statistical significance.[4] It is also easy to see that our indicators for the hours-constraints and heterogeneous-jobs motive are no longer statistically significant since standard errors are more than twice as large as the

---

[3]When we are using cluster-robust standard errors, the degrees of freedom, $df_2 = n - 1$, of the overall F test depend on the number of clusters (i.e., $n = 2,127$ units) and not on the number of observations ($N = 10,687$) in the data set.

[4]Using the exact parameter estimate (-0.0010171) and its estimated standard error (0.0005391), we arrive at a t value of $t = -1.89$, which is not significant ($p = 0.059$).

5.1 Modeling the Level of Y: Discrete Response Models for Panel Data

Table 5.5 Determinants of secondary job holding (pooled LPM with robust standard errors)

Variable	Theoretical standard errors		Robust standard errors		Cluster robust standard errors	
	Estimate	Std. Err.	Estimate	Std. Err.	Estimate	Std. Err.
Desire to work more hours	0.0323	0.0113	0.0323	0.0134	0.0323	0.0167
Desire to work fewer hours	−0.0013	0.0055	−0.0013	0.0055	−0.0013	0.0077
Weekly working hours in first job	−0.0010	0.0003	−0.0010	0.0003	−0.0010	0.0005
Is not satisfied with pay	0.0138	0.0056	0.0138	0.0059	0.0138	0.0076
Is not satisfied with work itself	0.0030	0.0066	0.0030	0.0069	0.0030	0.0094
Logarithm of first job wage	−0.0236	0.0078	−0.0236	0.0083	−0.0236	0.0149
Logarithm of non-labor income	0.0023	0.0019	0.0023	0.0019	0.0023	0.0028
Has public employer	0.0271	0.0073	0.0271	0.0082	0.0271	0.0149
Logarithm of job tenure	0.0111	0.0029	0.0111	0.0030	0.0111	0.0049
Age	0.0022	0.0020	0.0022	0.0021	0.0022	0.0034
Age squared	0.0000	0.0000	0.0000	0.0000	0.0000	0.0000
Has children under 15	0.0159	0.0058	0.0159	0.0059	0.0159	0.0092
Being a service class worker	0.0192	0.0060	0.0192	0.0062	0.0192	0.0108
Constant	0.1005	0.0395	0.1005	0.0412	0.1005	0.0643
F	9.30		8.80		3.36	
df_1, df_2	13	10,673	13	10,673	13	2,127
R^2	0.011		0.011		0.011	
N	10,687		10,687		0,687	
n					2,127	
T_{min}, T_{max}					1	10

Source: `heineck-schwarze` data (see Example 5.1)

corresponding regression coefficients. Heteroscedasticity (which is a necessary implication of using a linear probability model) is less of a problem than serial correlation, which is somewhat expected, since we are using panel data. This simple example already gives us an indication of how statistical dependencies in panel data can yield invalid statistical inferences.

5.1.1.2 Pooled Logistic Regression

Before we go on with models that explicitly take into account the panel structure, we first try to cope with the principal limitations of the linear model when it comes to the analysis of probabilities. We still assume that we have a pooled data set of $n = 10{,}687$ independent observations, but we need a non-linear function $G(\cdot)$, which ensures that the predicted response probabilities $\Pr(y_{it} = 1)$ are between zero and one. In principle, any statistical distribution function does this job, because— by definition—distribution functions return a probability using as input an often continuous random variable ranging between minus and plus infinity. Because of its mathematical simplicity, one popular option is the logistic distribution function

$G(a) = \exp(a)/(1+\exp(a))$. This results in the following logistic regression model for the pooled heineck-schwarze data:

$$\Pr(y_{it} = 1) = \frac{\exp(\beta_0(t) + \beta_1 x_{1it} + \cdots + \beta_k x_{kit} + \gamma_1 z_{1i} + \cdots + \gamma_j z_{ji})}{1 + \exp(\beta_0(t) + \beta_1 x_{1it} + \cdots + \beta_k x_{kit} + \gamma_1 z_{1i} + \cdots + \gamma_j z_{ji})} \quad (5.4)$$

This model is essentially non-linear, which makes it difficult to interpret the effects of the independent variables. Thus, to obtain an understanding of how $\Pr(y_{it} = 1)$ changes with different values of X and Z, let us assume a very simple model including only one time-constant explanatory variable and no time trend: $\Pr(y_{it} = 1) = \exp(\beta_0 + \gamma_1 z_{1i})/(1 + \exp(\beta_0 + \gamma_1 z_{1i}))$. To illustrate how $\Pr(y_{it} = 1)$ varies, we draw a plot with Z_1 varying between -5 and $+5$, assuming that $\beta_0 = 0$ and $\gamma_1 = 1$. In the left panel of Fig. 5.3, it is easy to see that a logistic curve with these parameters is s-shaped (see the solid curve). Two aspects of the logistic model are important to note here. First, with rising Z_1 we can see that $\Pr(y_{it} = 1)$ gets close to 1 but will never approach 1. If Z_1 declines, $\Pr(y_{it} = 1)$ gets close to 0 without approaching zero. Hence, predictions of this logistic regression model lie between the proper limits of the unit interval. Second, from the solid curve, we also see that the change in $\Pr(y_{it} = 1)$ for a given increase in Z_1 depends on the level of Z_1. In the lower and upper range of Z_1, a unit increase of Z_1 results in rather small changes of $\Pr(y_{it} = 1)$, while in the middle range, a unit increase of Z_1 results in rather large changes of $\Pr(y_{it} = 1)$. This can be seen more formally by computing the first derivative with respect to Z_1, which is not only a function of γ_1 but also of Z_1:

$$\frac{\partial \Pr(y_{it} = 1)}{\partial z_{1i}} = \gamma_1 \cdot \frac{\exp(\beta_0 + \gamma_1 z_{1i})}{1 + \exp(\beta_0 + \gamma_1 z_{1i})} \cdot \left(1 - \frac{\exp(\beta_0 + \gamma_1 z_{1i})}{1 + \exp(\beta_0 + \gamma_1 z_{1i})}\right) \quad (5.5)$$

In other words, the size of the effect of Z_1 depends on the level of Z_1 itself. This is essentially what *non-linearity* means, while in the linear probability model, the derivative would simply equal γ_1, indicating that in the linear case, the effect of Z_1 is always the same, irrespective of the level of Z_1. The other curves in Fig. 5.3 show how the relationship varies with different values of β_0 and γ_1. A negative regression coefficient γ_1 implies a negative relationship between $\Pr(y_{it} = 1)$ and Z_1 (see left panel of Fig. 5.3), while the regression constant β_0 affects the overall level of $\Pr(y_{it} = 1)$ (see right panel). Larger regression coefficients (e.g., $\gamma_1 = 2$ instead of $\gamma_1 = 1$) lead to steeper curves, indicating that the explanatory variable Z_1 has a stronger effect on the response probability (see left panel).

The first conclusion from this discussion is that it is easy to make statements about the direction of the relationship between an explanatory variable and the response probability (by simply using the sign of the regression coefficient), but when you want to quantify the effect, things become complicated. In general, in multivariate logistic regression models the effect of a unit change of one explanatory variable on the response probability is contingent upon the level of the variable itself and on the level of *all* other variables in the model.[5] Hence, the model is not only

---

[5] In that case, the two multiplicands in (5.5) would include the full logistic regression model with all independent variables X and Z.

5.1 Modeling the Level of Y: Discrete Response Models for Panel Data

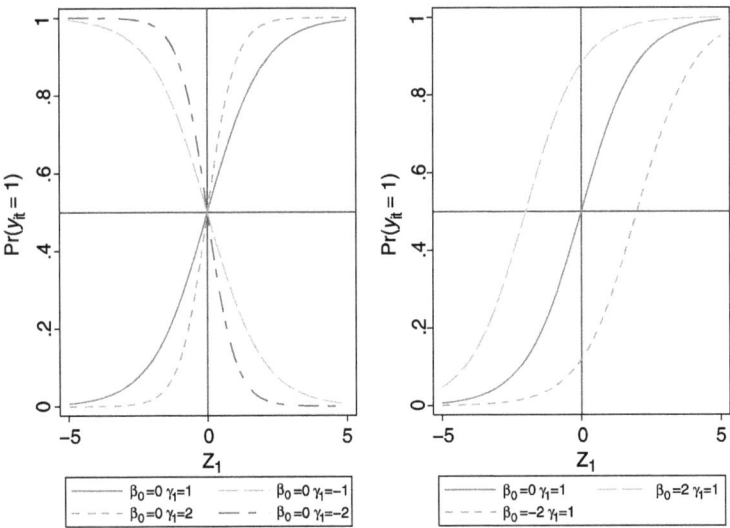

Fig. 5.3 The logistic regression function

non-linear, it is also *multiplicative*, meaning that the effect of a certain explanatory variable depends on the values (and the effects) of all other variables in the model. If you want to know how the predicted response probability varies with a particular explanatory variable, you can use graphical techniques like *conditional effect plots* to illustrate this non-linear and multiplicative relationship. A conditional effect plot simply draws predicted response probabilities for different values of this variable while holding all other independent variables in the model at fixed values (e.g., at their mean values). An example of such a plot is given below (Fig. 5.4 on p. 223).

There are also two summary measures that are useful for understanding whether and to what extent we expect a change in the response probability as variables change. One is the so-called *marginal effect*, which is identical to the former first derivative (5.5). In the multivariate case, the marginal effect of a specific explanatory variable (say, X_j) looks like this:

$$\text{Marginal effect} = \frac{\partial \Pr(y_{it} = 1)}{\partial x_{jit}} = \beta_j \cdot \Pr(y_{it} = 1) \cdot \left(1 - \Pr(y_{it} = 1)\right) \quad (5.6)$$

with $\Pr(y_{it} = 1)$ defined as in (5.4). Since $\Pr(y_{it} = 1)$ is a function of all X and Z, the marginal effect also depends on all variables in the model. To compute the marginal effect, all explanatory variables are often held at their mean values. A second summary measure is the *discrete change*. Assume that we are particularly interested in the effect of a variable X_1. We can then compute two probabilities $\Pr(y_{it} = 1 | x_{1it}, \ldots, x_{kit}, z_{1i}, \ldots, z_{ji})$ and, after increasing X_1 by some quantity Δx_{1it}, $\Pr(y_{it} = 1 | x_{1it} + \Delta x_{1it}, \ldots, x_{kit}, z_{1i}, \ldots, z_{ji})$. The discrete change is then given by the difference of those two probabilities. The concept of marginal effects captures the instantaneous rate of change; the concept of discrete change indicates the expected amount of change in the response probability for a given change in an explanatory variable.

While the logistic regression model is non-linear in the response probability, some simple algebraic transformations of (5.4) show that it is linear in a quantity called the logit:

$$\ln\left(\frac{\Pr(y_{it}=1)}{1-\Pr(y_{it}=1)}\right)=\beta_0(t)+\beta_1 x_{1it}+\cdots+\beta_k x_{kit}+\gamma_1 z_{1i}+\cdots+\gamma_j z_{ji} \quad (5.7)$$

The ratio $\Pr(y_{it}=1)/(1-\Pr(y_{it}=1))$ is termed the *odds* of observing category $y_{it}=1$ rather than category $y_{it}=0$. It relates the two response probabilities of observing either the one or the other category of the dependent binary variable. The log-odds, or for short the *logit*, is defined as the natural logarithm of the odds $\Pr(y_{it}=1)/(1-\Pr(y_{it}=1))$.[6]

Logits and odds open up new possibilities for motivating and interpreting the logistic regression model. For example, the linear probability model (5.3) had the disadvantage that the right-hand side of the equation could provide values outside the unit interval (in principle, the LPM provides values between minus and plus infinity). With logits rather than probabilities now on the left-hand side of the equation, this is no longer a restriction, since logits are continuous and at least in principle vary between minus and plus infinity. Thus, the logistic regression model is sometimes also called the logit model, but it should be stressed that both terms refer to the same model and that (5.4) is easily transformed into (5.7) and vice versa. In principle, we can use the logistic regression coefficients to make statements about how change in a particular explanatory variable affects the logit of moonlighting. Since the model is linear in the logits, interpretation would be as simple as in the linear probability model. But the problem is that the concept of logits is complicated and therefore, hard to communicate in the general public.

This is different for the concept of odds, which are well known from betting and biometric research. In the `heineck-schwarze` data, we find 1,012 observations in which someone is moonlighting, while there are 12,316 observations in which no one holds a second job. Thus, the overall odds of observing a moonlighter rather than an employee with only one job amount to 0.082 ($=1,012/12,316$). More generally, odds can obtain values between zero and plus infinity. When the odds are lower than 1, the probability to hold a second job is below 50 %. Quite the opposite is true when the odds are greater than 1. In that case, the probability to hold a second job would be greater than 50 %. If the odds would equal 1, there would be as many moonlighters as non-moonlighters. Applying the exponential function to both sides of (5.7) shows how the logistic regression model looks like in terms of odds:

---

[6]Equation (5.7) shows that the logistic regression model is linear-additive with respect to the logits. In the context of generalized linear models, the right-hand side of (5.7) is also called the *linear predictor* of the model. Obviously, the logit transformation links the response probabilities to the linear predictor. This is why the logit transformation is also called a *link function*.

5.1 Modeling the Level of Y: Discrete Response Models for Panel Data

$$\frac{\Pr(y_{it}=1)}{1-\Pr(y_{it}=1)}$$
$$= \exp\big(\beta_0(t) + \beta_1 x_{1it} + \cdots + \beta_k x_{kit} + \gamma_1 z_{1i} + \cdots + \gamma_j z_{ji}\big)$$
$$= \exp\big(\beta_0(t)\big) \cdot \exp(\beta_1)^{x_{1it}} \cdot \ldots \cdot \exp(\beta_k)^{x_{kit}} \cdot \exp(\gamma_1)^{z_{1i}} \cdot \ldots \cdot \exp(\gamma_j)^{z_{ji}} \quad (5.8)$$

While the model has been linear-additive in logits, it now becomes non-linear and multiplicative. Nevertheless, interpretation is quite straightforward. We use the antilogarithms of the logistic regression coefficients, $\exp(\beta_1), \ldots, \exp(\beta_k), \exp(\gamma_1),$ $\ldots, \exp(\gamma_j)$, to make statements on how the odds change given a unit change of a particular explanatory variable. For example, the odds of observing a moonlighter change by a (multiplicative) factor of $\exp(\beta_1)$, if X_1 is increased by one unit:

$$\frac{\Pr(y_{it}=1|x_{1it}+1)}{1-\Pr(y_{it}=1|x_{1it}+1)}$$
$$= \exp\big(\beta_0(t)\big) \cdot \exp(\beta_1)^{(x_{1it}+1)} \cdot \ldots \cdot \exp(\beta_k)^{x_{kit}} \cdot \exp(\gamma_1)^{z_{1i}} \cdot \ldots \cdot \exp(\gamma_j)^{z_{ji}}$$
$$= \exp(\beta_1) \cdot \big(\exp(\beta_0(t)) \cdot \exp(\beta_1)^{x_{1it}} \cdot \ldots \cdot \exp(\gamma_1)^{z_{1i}} \cdot \ldots \cdot \exp(\gamma_j)^{z_{ji}}\big)$$
$$= \exp(\beta_1) \cdot \left(\frac{\Pr(y_{it}=1|x_{1it})}{1-\Pr(y_{it}=1|x_{1it})}\right) \quad (5.9)$$

Instead of arguing with multiplicative factors, we can also compute the percentage change of the odds, which amounts to $(\exp(\beta_1)-1) \cdot 100$. The antilogarithm $\exp(\beta_1)$ is also called an *odds ratio* because it measures the ratio of two odds: one when X_1 is increased by one unit and the other when X_1 has its original value. In the following, we will mostly use odds ratios to interpret the results of logistic regression models.

But how can we estimate logistic regression coefficients? Using OLS to estimate the linear-additive model (5.7) is not feasible, since the logit (like the probability) is not directly observable and it is not possible to transform the binary dependent variable in a way that makes it similar to a logit.[7] Hence, estimates of a logistic regression model have to be found using the method of *maximum likelihood* (ML).

In the case of a categorical dependent variable, the underlying idea of ML is easy to understand, but the mathematical computations are quite complicated and require numerical optimization. ML estimates of the regression coefficients are defined as those estimates that—given the independent variables X—maximize the joint probability of observing the values of the dependent variable Y. Section 7.2.2 explains this approach in greater detail. It builds on a mathematical function, called the *likelihood function*, that measures the overall probability of the observed data given certain values for the regression coefficients (see (7.17)). For our panel data set with

---

[7]OLS (or more precisely, weighted least squares) can only be applied if *all* (independent and dependent) variables in the model are categorical. These kinds of model have been proposed by Grizzle et al. (1969) and extensively treated in the textbook by Forthofer and Lehnen (1981).

time-constant and time-varying independent variables the likelihood function looks like this:[8]

$$L(\beta, \gamma) = \prod_{i=1}^{n} \prod_{t=1}^{T} [\Pr(y_{it} = 1)]^{y_{it}} \cdot [1 - \Pr(y_{it} = 1)]^{(1-y_{it})}$$

$$= \prod_{i=1}^{n} \prod_{t=1}^{T} \left[\frac{\exp(\beta_0(t) + \beta_1 x_{1it} + \cdots + \gamma_1 z_{1i} + \cdots)}{1 + \exp(\beta_0(t) + \beta_1 x_{1it} + \cdots + \gamma_1 z_{1i} + \cdots)} \right]^{y_{it}}$$

$$\cdot \left[1 - \frac{\exp(\beta_0(t) + \beta_1 x_{1it} + \cdots + \gamma_1 z_{1i} + \cdots)}{1 + \exp(\beta_0(t) + \beta_1 x_{1it} + \cdots + \gamma_1 z_{1i} + \cdots)} \right]^{(1-y_{it})} \quad (5.10)$$

Practically, one maximizes the natural logarithm of the likelihood function (see (7.18) in Sect. 7.2.2):

$$\ln L(\beta, \gamma)$$

$$= \sum_{i=1}^{n} \sum_{t=1}^{T} y_{it} \cdot \ln \left(\frac{\exp(\beta_0(t) + \beta_1 x_{1it} + \cdots + \gamma_1 z_{1i} + \cdots)}{1 + \exp(\beta_0(t) + \beta_1 x_{1it} + \cdots + \gamma_1 z_{1i} + \cdots)} \right)$$

$$+ \sum_{i=1}^{n} \sum_{t=1}^{T} (1 - y_{it}) \cdot \ln \left(1 - \frac{\exp(\beta_0(t) + \beta_1 x_{1it} + \cdots + \gamma_1 z_{1i} + \cdots)}{1 + \exp(\beta_0(t) + \beta_1 x_{1it} + \cdots + \gamma_1 z_{1i} + \cdots)} \right)$$

$$(5.11)$$

If we use plausible starting values for the regression coefficients, the values of the independent variables X and Z[9] and the values of the dependent binary variable Y for each observation, (5.11) provides the natural logarithm of the overall probability of the observed sample given the starting values β^0 and γ^0. For example, a good starting point is the assumption that none of the independent variables has an effect (hence, all regression coefficients except the constant are zero). Consequently, the starting value for the regression constant has to be chosen in such a way that the model replicates the overall proportion of moonlighters (the proportion of observations with $y_{it} = 1$). For the heineck-schwarze data, with these starting values, we arrive at a value of $L(\beta^0, \gamma^0) = -2,868.0943$. Obviously, the result depends on the chosen starting values, which rely on the unrealistic assumption of no explanatory variable having an effect. You may try other values and in doing so, try to maximize $L(\beta, \gamma)$. Logistic regression programs solve this task easily by using numerical optimization algorithms.

---

[8]In case of an unbalanced panel data set the upper limit T of the second multiplication index t varies between units and therefore should be written as T_i. This, however, does not alter the general principle.

[9]We provide the formula with the complete panel regression model, although the heineck-schwarze data do not include any time-constant Z (see Example 5.1).

5.1 Modeling the Level of Y: Discrete Response Models for Panel Data

The parameters β and γ that maximize the log likelihood are called *ML estimates*. ML assumes that all observations are independent of each other and identically distributed. The probability distribution for each observation is specified in the likelihood function. Given that this probability distribution is correctly specified, ML estimates are consistent and asymptotically normally distributed. Variances and covariances of the estimates (and hence, their standard errors) are derived from the second derivatives of the maximized log likelihood function. The behavior of ML estimates in small samples, however, is not always known, and hence, testing rests on large sample theory. Section 7.2.2 summarizes the main testing procedures and fit measures that we will use in the case of ML estimation. The following Textbox 5.1 discusses the ML assumptions in greater detail.

Textbox 5.1 (ML assumptions) Although the assumptions for ML estimation may look different from the OLS assumptions in Textbox 4.1, they have similar functions. They specify the conditions under which estimates can be computed, are unbiased and can be generalized to the population. ML is a very general estimation procedure that can be applied to all kinds of statistical model. It finds estimates by maximizing the probability of the observed sample values of Y, X, and Z. In the general case of continuous Y, it maximizes the density and not the probability of the observations. Therefore, an essential ingredient of this estimation technique is the correct specification of the probability distribution function (density function). More specifically, the statistical properties of ML estimation rest on the following assumptions:

1. The data are a simple random sample of a well-defined population.
2. Each explanatory variable is neither a constant nor a linear function of the other explanatory variables. Furthermore, in the case of categorical dependent variables, Y must vary for each linear combination of the independent variables in the model.
3. The observations in the sample are independent of each other and identically distributed.
4. The parameters of the probability distribution resp. density function and the function itself are correctly specified.

Similar to OLS, the first assumption is necessary for making statistical inferences about the population using the sampled data. The second assumption guarantees that a numerical value exists for each regression coefficient (technically, the model is identified). In the case of a dependent variable Y that has only few distinct values (a categorical variable), we also have to make sure that Y has enough independent variation. If there is a linear combination of the independent variables that predicts one value of the categorical Y perfectly, then there is no unique ML solution (a situation that has become known as the problem of *separability*).

The third assumption makes sure that we can use the same probability model for each observation (because they are identically distributed) and obtain the overall likelihood of the whole sample by using formulas combining the probabilities of independent observations. The product rule for the probability of two independent observations A and B states that you can simply multiply the single probabilities of observing A and B: $\Pr(A \cap B) = \Pr(A) \cdot \Pr(B)$. To obtain the overall probability of *all* observations, you simply multiply *all* single probabilities. However, the independence assumption is not a serious limitation. Alternatively, if observations would be dependent, as they often are with panel data, one can use the product rule for dependent observations: $\Pr(A \cap B) = \Pr(A) \cdot \Pr(B|A)$. One only needs a specification of the conditional probability of observing B, given that A has occurred: $\Pr(B|A)$. Later, when we model the panel structure of the data, we will show how that can be done (see Sect. 5.1.2).

Once it is clear which probability model can be applied to all observations and how to combine the single probabilities, we can deal with the question of how to include the explanatory variables. To be a little bit more concrete, the former likelihood function (5.10) for the logistic model applied the Bernoulli distribution[10] to each observation and included the explanatory variables by specifying the main parameter of the Bernoulli distribution (i.e., the probability of observing the first category of the dichotomous variable Y, as a function of X and Z). Here, the aforementioned fourth assumption comes in. If this function is misspecified (e.g., if an important explanatory variable is left out), then the estimates will be biased. Hence, the fourth assumption is the most important, because it makes sure that our ML estimates are *unbiased*. Similar to OLS estimation, unbiasedness means that our estimates are correct; not necessarily in a single sample, but on average (i.e., across a multitude of samples), they are identical with the "true" population parameters. Whether these estimates are then *efficient* and how to test their significance are more difficult questions that cannot be dealt with in this textbook. Statistical theory shows, however, that ML estimates are normally distributed in large samples and are similarly efficient as estimates from other estimation procedures. Therefore, tests of significance can use the normal and related distributions as the testing distributions.

Table 5.6 shows the ML estimates of a logistic regression model applied to the `heineck-schwarze` data. Again, the desire to work more hours has a significant positive effect on moonlighting ($p < 0.01$), while the number of weekly working hours has a significant negative effect ($p < 0.01$). But this model erroneously as-

---

[10]The Bernoulli distribution is used to model the probabilities of observing each of the two categories of a dichotomous random variable.

5.1 Modeling the Level of Y: Discrete Response Models for Panel Data

Table 5.6 Determinants of secondary job holding (pooled logistic model)

Variable	Theoretical standard errors		Cluster robust standard errors	
	Estimate	Std. Err.	Estimate	Std. Err.
Desire to work more hours	0.3678	0.1375	0.3678	0.1738
Desire to work fewer hours	−0.0156	0.0819	−0.0156	0.1160
Weekly working hours in first job	−0.0158	0.0047	−0.0158	0.0091
Is not satisfied with pay	0.1865	0.0799	0.1865	0.1049
Is not satisfied with work itself	0.0395	0.0924	0.0395	0.1277
Logarithm of first job wage	−0.3553	0.1165	−0.3553	0.2307
Logarithm of non-labor income	0.0334	0.0277	0.0334	0.0427
Has public employer	0.3513	0.0960	0.3513	0.1751
Logarithm of job tenure	0.1625	0.0427	0.1625	0.0710
Age	0.0474	0.0305	0.0474	0.0516
Age squared	−0.0009	0.0004	−0.0009	0.0007
Has children under 15	0.2250	0.0828	0.2250	0.1258
Being a service class worker	0.2786	0.0853	0.2786	0.1522
Constant	−2.3621	0.5770	−2.3621	0.9472
Pseudo R^2	0.0211		0.0211	
LR or X_1^2	121.23		49.44	
df	13		13	
ln L	−2,807.4795		−2,807.4795	
N	10,687		10,687	
n			2,127	
T_{min}, T_{max}			1	10

Source: heineck-schwarze data (see Example 5.1)

sumes that we have 10,687 independent observations. In fact, we analyze 2,127 men who are observed several times during the observation period rendering the independence assumption to be false. Hence, the consistency of the estimates is at stake and standard errors are underestimated. Fortunately, like with OLS estimation, it is possible to compute robust standard errors that are adjusted to the serial correlations in the panel data. Table 5.6 also shows these estimation results and compares them with those obtained from pooled logistic regression. Not surprisingly, both methods produce identical coefficients, but standard errors differ. More precisely, the satisfaction with pay and the number of weekly working hours lose their statistical significance once the panel data structure is taken into account ($p > 0.05$). Only the desire to work more hours retains its significant positive effect on moonlighting ($p < 0.05$).

As already mentioned, the difficulty with the coefficients obtained from logistic regression is that it is hard to grasp what a change in the logit means. However, we discussed some alternatives that make things much easier. For instance, rather than

Table 5.7 Determinants of secondary job holding (odds ratios for pooled logistic model)

Variable	$\hat{\beta}$	$\exp(\hat{\beta})$	%	$\exp(\hat{\beta} \cdot \hat{\sigma}_x)$	%
Desire to work more hours	0.3678*	1.4446	44.5	1.0893	8.9
Desire to work fewer hours	−0.0156	0.9845	−1.5	0.9924	−0.8
Weekly working hours in first job	−0.0158	0.9843	−1.6	0.8638	−13.6
Is not satisfied with pay	0.1865	1.2051	20.5	1.095	9.5
Is not satisfied with work itself	0.0395	1.0403	4.0	1.0161	1.6
Logarithm of first job wage	−0.3553	0.7009	−29.9	0.8618	−13.8
Logarithm of non-labor income	0.0334	1.0339	3.4	1.049	4.9
Has public employer	0.3513*	1.4209	42.1	1.1319	13.2
Logarithm of job tenure	0.1625*	1.1764	17.6	1.1659	16.6
Age	0.0474	1.0486	4.9	1.6237	62.4
Age squared	−0.0009	0.9991	−0.1	0.4822	−51.8
Has children under 15	0.2250	1.2524	25.2	1.1185	11.9
Being a service class worker	0.2786	1.3213	32.1	1.1482	14.8
Constant	−2.3621*				

Note: $* \ p < 0.05; ** \ p < 0.01; *** \ p < 0.001$
Source: `heineck-schwarze` data (see Example 5.1)

looking at the coefficients, you may want to interpret the predicted change in odds, given a unit change in your explanatory variable. Since the explanatory variables in our model are measured differently, the coefficients are not directly comparable. To facilitate a within-model comparison, you can calculate the expected change in odds for a standard deviation increase in your explanatory variable. The output in Table 5.7 presents these transformations.

To illustrate how to interpret the results of the pooled logistic regression model, we restrict ourselves to two indicators for the hours-constraints motive, namely the dummy for "would like to work more hours" and weekly working hours in the first job. As already mentioned, the logistic regression coefficient of the dummy indicates that compared to those who are completely satisfied with their working hours, men who want to work more hours are significantly more prone to hold a second job. Technically speaking, the logit changes by 0.368, holding all other explanatory variables in our estimation model constant. Alternatively, we can use the odds ratio interpretation and state that—keeping all other variables constant—the odds of moonlighting increase by a factor of $\exp(0.368) = 1.445$. Yet another way of interpreting the finding is to look at the percentage change in odds: Compared to those who are completely satisfied with their working hours, the odds of secondary job holding are $(1.445 - 1) \cdot 100 = 44.5\ \%$ higher for men who want to work more hours.

This interpretation is easily conveyed to continuous explanatory variables such as working hours in the first job. Here, we could simply infer that with each additional hour of work in the first job, the odds of being a double job holder decrease by a factor of 0.984 or $(0.984 - 1) \cdot 100 = -1.6\ \%$. Among the continuous explanatory vari-

5.1 Modeling the Level of Y: Discrete Response Models for Panel Data

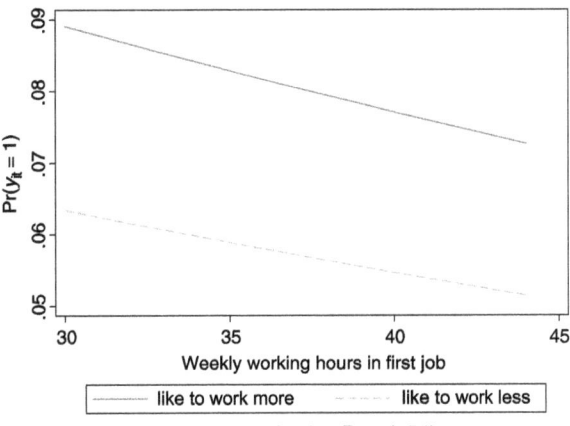

Fig. 5.4 Conditional effects plot using the estimates of the pooled logistic model

ables, age seems to have the strongest effect, when looking at the (comparable) standard deviation increases. However, the effect of age is not statistically significant.

We can also draw a plot to illustrate whether and to what extent the predicted probabilities for moonlighting vary over working hours in the first job for people with and without the hours-constraints motive. To this end, we fix all continuous variables in our model at their mean and set all dummies equal to zero except for "would like to work more hours". The conditional probability is then derived from (5.4) using all the parameter estimates and the selected values of the explanatory variables. Figure 5.4 shows the resulting conditional effect plot. The figure gives support to the assumption of Heineck and Schwarze (2004) that workers who face labor supply constraints in their first job are more likely to moonlight. At the same time, we see that the more working hours in the first job, the less likely secondary job holding.

5.1.1.3 Pooled Probit Regression

An alternative way to specify the transformation function $G(\cdot)$ in our probability model (5.1) is to use the standard normal distribution function, which is expressed as an integral and abbreviated with the Greek symbol $\Phi(\cdot)$ (capital "Phi"):

$$G(a) = \int_{-\infty}^{a} \frac{1}{\sqrt{2\pi}} \exp\left(\frac{-u^2}{2}\right) du \equiv \Phi(a) \qquad (5.12)$$

This results in the following probit regression model for the pooled heineck-schwarze data:

$$\Pr(y_{it} = 1) = \Phi\left(\beta_0(t) + \beta_1 x_{1it} + \cdots + \beta_k x_{kit} + \gamma_1 z_{1i} + \cdots + \gamma_j z_{ji}\right) \qquad (5.13)$$

Again the model is non-linear, and we need the first partial derivatives to see how $\Pr(y_{it} = 1)$ changes with different values of X and Z. For a very simple model including only one time-constant explanatory variable and no time trend, the first derivative with respect to Z_1 looks like this:

$$\frac{\partial \Pr(y_{it} = 1)}{\partial z_{1i}} = \gamma_1 \cdot \phi(\beta_0 + \gamma_1 z_{1i}) = \gamma_1 \cdot \left(\frac{1}{\sqrt{2\pi}} \exp\left(-\frac{(\beta_0 + \gamma_1 z_{1i})^2}{2} \right) \right) \quad (5.14)$$

in which the Greek symbol $\phi(\cdot)$ ("phi") stands for the standard normal density function. Again, the change of the response probability is not only a function of the effect of the explanatory variable, γ_1, but also of the level of the explanatory variable Z_1, which enters through the normal density. Like in the case of logistic regression, the effect of a unit change in an explanatory variable on the response probability is contingent upon the level of the variable itself and—in the case of multivariate models—on the level of *all* other variables in the model. In the multivariate case, the *marginal effect* of a specific explanatory variable (say, X_j) equals

$$\frac{\partial \Pr(y_{it} = 1)}{\partial x_{jit}} = \beta_j \cdot \phi\bigl(\beta_0(t) + \beta_1 x_{1it} + \cdots + \beta_k x_{kit} + \gamma_1 z_{1i} + \cdots + \gamma_j z_{ji}\bigr) \quad (5.15)$$

Hence, all the basic rules of interpretation that we have learned about logistic regression also apply to probit regression.

A distribution function like the logistic or the standard normal return a probability and use as input a continuous variable that ranges from minus to plus infinity. The inverse of a distribution function does the same just the other way round: It returns a continuous variable given a probability. The inverse of the standard normal distribution is called the *probit* function, which gave probit regression its name:

$$\text{probit}\bigl(\Pr(y_{it} = 1)\bigr) = \beta_0(t) + \beta_1 x_{1it} + \cdots + \beta_k x_{kit} + \gamma_1 z_{1i} + \cdots + \gamma_j z_{ji} \quad (5.16)$$

Hence, a probit regression model is linear-additive in the probits.[11] Unfortunately, there is no transformation of probits that is easily communicated to the general public (as it is possible for odds in the case of logits). But you may think of an underlying latent variable that determines whether you observe $y_{it} = 1$ or $y_{it} = 0$.[12] In the heineck-schwarze data, this could be the utility of moonlighting and the higher its value, the higher the probability that a person has a second job. If this utility is normally distributed, the predicted values of (5.16) measure these utilities. More specifically, the predicted values tell you, for a given combination of X and Z, how many standard deviations the predicted utility is above or below the average utility. This concept of a latent variable is very popular in econometrics (see Textbox 5.2 below for more details).

We have estimated a probit regression model for the heineck-schwarze data using robust standard errors that control for the serial correlation in the panel data. Table 5.8 shows the results and compares them with the former results of the linear probability and logistic regression model (again with cluster robust standard errors).

---

[11] In other words, the probit transformation links the response probabilities to the linear predictor (the right-hand side of (5.16)). In the language of generalized linear models, the probit transformation is another *link function* for discrete response models.

[12] A similar argument can be made for the logistic regression model. In that case, the underlying latent variable would be logistically distributed.

5.1 Modeling the Level of Y: Discrete Response Models for Panel Data

Table 5.8 Determinants of secondary job holding (pooled linear probability, logistic, and probit model with cluster robust standard errors)

Variable	LPM		Logistic model		Probit model	
	Estimate	Std. Err.	Estimate	Std. Err.	Estimate	Std. Err.
Desire to work more hours	0.0323	0.0167	0.3678	0.1738	0.1903	0.0903
Desire to work fewer hours	−0.0013	0.0077	−0.0156	0.1160	−0.0134	0.0560
Weekly working hours in first job	−0.0010	0.0005	−0.0158	0.0091	−0.0074	0.0043
Is not satisfied with pay	0.0138	0.0076	0.1865	0.1049	0.0915	0.0514
Is not satisfied with work itself	0.0030	0.0094	0.0395	0.1277	0.0210	0.0626
Logarithm of first job wage	−0.0236	0.0149	−0.3553	0.2307	−0.1627	0.1101
Logarithm of non-labor income	0.0023	0.0028	0.0334	0.0427	0.0158	0.0204
Has public employer	0.0271	0.0149	0.3513	0.1751	0.1782	0.0889
Logarithm of job tenure	0.0111	0.0049	0.1625	0.0710	0.0804	0.0341
Age	0.0022	0.0034	0.0474	0.0516	0.0204	0.0249
Age squared	0.0000	0.0000	−0.0009	0.0007	−0.0004	0.0003
Has children under 15	0.0159	0.0092	0.2250	0.1258	0.1111	0.0624
Being a service class worker	0.0192	0.0108	0.2786	0.1522	0.1359	0.0742
Constant	0.1005	0.0643	−2.3621	0.9472	−1.3437	0.4550
R^2 or Pseudo R^2	0.0112		0.0211		0.0210	
F or X_1^2	3.36		49.44		48.98	
df_1, df_2	13	2,127	13		13	
$\ln L$			−2,807.4795		−2,807.9019	
N	10,687		10,687		10,687	
n	2,127		2,127		2,127	
T_{min}, T_{max}	1	10	1	10	1	10

Source: heineck-schwarze data (see Example 5.1)

In terms of direction and significance of the effects, the conclusions from all three models are virtually the same. The only exception is the desire to work more hours, which is only significant in the logistic and the probit regression model.

However, the size of the coefficients differs quite substantially. Before drawing any conclusions from these differences, you should remember that both the logistic and the probit regression model are multiplicative non-linear models and consequently, the estimated effect of an explanatory variable depends both on the level of that specific variable *and* the level and effects of all the other explanatory variables in the model (see (5.6) and (5.15)). The linear probability model, on the other hand, is linear and additive and hence, the effect of an explanatory variable equals the estimated regression coefficient irrespective of the variable's level and the level of all the other variables in the model. Therefore, effects of the three types of model should be compared using marginal effects and not using simple regression coefficients. Moreover, since marginal effects depend on the level and the estimated regression coefficients of all the explanatory variables in the model, they should be evaluated at

Fig. 5.5 Linear probability, logistic and probit regression model

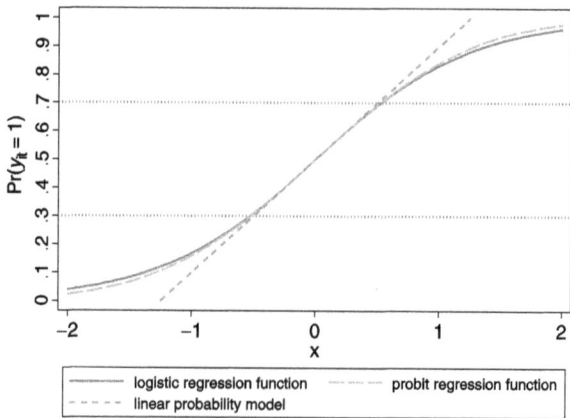

a common point of comparison (e.g., when all explanatory variables equal zero and hence, the linear predictor $\eta = \beta_0(t) + \beta_1 x_{1it} + \cdots + \beta_k x_{kit} + \gamma_1 z_{1i} + \cdots + \gamma_j z_{ji}$ is zero too). For $\eta = 0$, (5.6) equals $\beta_j \cdot 0.5 \cdot 0.5$ and (5.15) equals $\beta_j \cdot 0.4$ (the marginal effect of a linear probability model will always equal $\beta_j = \beta_j \cdot 1$). Hence, if the marginal effects of the logistic and the probit model should be equal, the estimated regression coefficient of a logistic regression model should be roughly 1.6 times larger than the corresponding estimate of the probit model. Furthermore, logistic and probit estimates should be multiplied with 0.25 and 0.4, respectively, to make them comparable with the linear probability model:

$$\hat{\beta}_{\text{Logit}} \approx 1.6 \cdot \hat{\beta}_{\text{Probit}} = \frac{0.4}{0.25} \cdot \hat{\beta}_{\text{Probit}}$$

$$\hat{\beta}_{\text{LPM}} \approx 0.25 \cdot \hat{\beta}_{\text{Logit}} = \frac{0.25}{1} \cdot \hat{\beta}_{\text{Logit}} \quad (5.17)$$

$$\hat{\beta}_{\text{LPM}} \approx 0.4 \cdot \hat{\beta}_{\text{Probit}} = \frac{0.4}{1} \cdot \hat{\beta}_{\text{Probit}}$$

Take the estimate for the desire to work more hours as an example. From probit regression, we obtain 0.1903. Multiplied by 1.6, we get 0.3045, which comes much closer to the estimate 0.3678 in the logistic model. If the distribution of the dependent variable would be less skewed (remember, the overall probability of having a second job was only 7.6 %), the difference between the logistic and the probit regression coefficients would be even much smaller.

To sum up, from a theoretical point of view, logistic and probit regression should be preferred to the linear probability model. Whether you use a logistic or a probit regression model is often merely a matter of taste. If the distribution of the dependent variable is not too skewed, the practical differences between all three approaches are usually quite small. To illustrate this, suppose we estimate a probit model in which $\Pr(y_{it})$ varies with one time-constant explanatory variable Z_1 and assume that $\beta_0 = 0$ and $\gamma_1 = 1$. Figure 5.5 compares this probit regression function with the regression functions resulting from the corresponding linear and logistic regression model.[13] As visualized in Fig. 5.5, both the logistic and the probit function are almost linear in the middle range of probabilities ($0.3 < \Pr(y_{it}) < 0.7$). Therefore, within this mid-range of probabilities, logistic and probit regression models return very similar results and, moreover, in this range the linear probability model also gives a good approximation of both non-linear models. Therefore, in case of symmetrically distributed dichotomous dependent variables, the linear probability model is an attractive alternative, because it is much easier to interpret than logistic or probit regression.

5.1.2 Modeling the Panel Structure

Obviously, it is a wrong assumption to treat the 10,687 observations in the `heineck-schwarze` data as representing 10,687 independent units. We called this mode of analysis pooled linear, pooled logistic, or pooled probit regression, depending on the specific model used. Treating the serial correlations of the repeated observations as a nuisance factor by using robust standard errors is not very convincing because it only treats the symptoms and not the causes of these statistical dependencies (see the discussion for continuous variables in Sect. 4.1.1.2). A much more valid approach would be to extend the model in a way that it recognizes the panel structure of the data. In Chap. 3, we mentioned three reasons for the serial correlations: (i) characteristics of the units that are constant over time (Z), (ii) characteristics of the units that change over time (X), which are, however, serially correlated, and (iii) lagged effects of the dependent variable Y (see Sect. 3.4.2). Hence, in principle, serial correlation can be controlled by including these variables X, Z, and lagged Y into the model. But quite probably, not all of these variables are known to the researcher and there will always be some unobserved heterogeneity both at the unit level and at the level of measurements. As shown in Chap. 3, pooled models for continuous variables can easily be extended to account for this unobserved heterogeneity (see Sect. 3.6.1.2):

$$y_{it} = \beta_0(t) + \beta_1 x_{1it} + \cdots + \beta_k x_{kit} + \gamma_1 z_{1i} + \cdots + \gamma_j z_{ji} + u_i + e_{it} \quad (5.18)$$

The stochastic part of this continuous model, $\varepsilon_{it} = u_i + e_{it}$, distinguishes between two components: (i) u_i: unobserved predictors of Y that are specific to the unit and therefore time-constant, (ii) e_{it}: unobserved predictors of Y (including measurement error) that are specific to the time point *and* the unit. Since we are dealing with categorical dependent variables in this chapter, we have to discuss whether a similar extension is meaningful for our discrete response model (5.1). Of course, there should be a unit-specific error term u_i. But one may ask whether a time-varying (idiosyncratic) error term e_{it} is necessary if the model focuses on probabilities and hence, is inherently stochastic (see Textbox 5.2). As it turns out, discrete response

---

[13] To make the regression functions comparable, we have applied the simple rule of thumb shown in (5.17) to the corresponding coefficients β_0 and γ_1 of the linear and the logistic regression model.

models for panel data seldom include time-varying error terms. Therefore, the basic panel data model for categorical variables looks like this:

$$\Pr(y_{it} = q) = G\big(\beta_0(t) + \beta_1 x_{1it} + \cdots + \beta_k x_{kit} + \gamma_1 z_{1i} + \cdots + \gamma_j z_{ji} + u_i\big) \quad (5.19)$$

and includes only the unit-specific error term u_i.

Textbox 5.2 (Latent variable specification for discrete response model) The fact that discrete response models do not need an idiosyncratic error term can also be motivated with a slightly different specification. Assume that there is a latent variable Y^* underlying the observed categorical dependent variable Y. As already mentioned, in the `heineck-schwarze` data, this could be the utility of moonlighting and the larger its value, the larger the probability that a person has a second job. Let this latent variable be a linear-additive function of our explanatory variables X and Z, unobserved heterogeneity u_i at the unit level and a time-dependent (idiosyncratic) error term e_{it}:

$$y_{it}^* = \beta_0(t) + \beta_1 x_{1it} + \cdots + \beta_k x_{kit} + \gamma_1 z_{1i} + \cdots + \gamma_j z_{ji} + u_i + e_{it} \quad (5.20)$$

This is exactly the same specification that we used for continuous dependent variables (see (5.18)). If this latent variable exceeds a certain value, say $y_{it}^* > 0$, we observe $y_{it} = 1$ (the person holds a second job); otherwise, we observe $y_{it} = 0$ (the person has no second job). This definition is often written as follows:

$$y_{it} = 1\big(\beta_0(t) + \beta_1 x_{1it} + \cdots + \beta_k x_{kit} + \gamma_1 z_{1i} + \cdots + \gamma_j z_{ji} + u_i + e_{it} > 0\big) \quad (5.21)$$

$1(\cdot)$ is an indicator function returning a value of 1, if the logical expression in brackets is true; otherwise, it returns a value of 0. Now let us assume that the idiosyncratic error is independent of the variables in the model and has a certain distribution that allows us to compute the probability that e_{it} exceeds a certain value. For example, we could choose a value that makes the logical expression in (5.21) true and thus, returns the probability of observing $y_{it} = 1$:

$$\Pr(y_{it} = 1) = \Pr\big(e_{it} > -(\beta_0(t) + \beta_1 x_{1it} + \cdots + \beta_k x_{kit} + \gamma_1 z_{1i} + \cdots + \gamma_j z_{ji} + u_i)\big) \quad (5.22)$$

If this distribution is symmetric, (5.22) can also be written as

$$\Pr(y_{it} = 1) = \Pr\big(e_{it} < \beta_0(t) + \beta_1 x_{1it} + \cdots + \beta_k x_{kit} + \gamma_1 z_{1i} + \cdots + \gamma_j z_{ji} + u_i\big)$$
$$= F\big(\beta_0(t) + \beta_1 x_{1it} + \cdots + \beta_k x_{kit} + \gamma_1 z_{1i} + \cdots + \gamma_j z_{ji} + u_i\big)$$

which equals the cumulative distribution function $F(\cdot)$ of the idiosyncratic error term. The standard normal or the logistic distribution function are two possible options for such symmetric distribution functions. This shows us that

5.1 Modeling the Level of Y: Discrete Response Models for Panel Data

the discrete response model (5.19) can be traced back to a linear-additive model for an underlying latent variable that, among other things, is a function of an error term that includes unit-specific u_i and idiosyncratic errors e_{it}. If this idiosyncratic error is normally distributed with mean 0 and variance 1, the probit regression model results; if it is logistically distributed, the logistic regression model results. The logistic distribution also has a mean of 0, but a variance of $\pi^2/3$.

As already mentioned, u_i is a measure of time-constant unobserved heterogeneity at the unit level and different estimation strategies exist depending on our assumptions about this heterogeneity. A simple starting point is the assumption that unobserved heterogeneity is uncorrelated with the variables in the model. In other words, u_i is a sort of random disturbance at the individual level. This is called a *random effect* in the literature, and random effects (RE) estimation is used to assess the effects of the explanatory variables in the model. For many applications, assuming uncorrelated heterogeneity, however, is not a very realistic assumption. In our example of moonlighting, a possible unobserved variable is health status, which is also positively correlated with the time-varying explanatory variables in the model (e.g., with the desire to work fewer hours). If one assumes health status to be time-constant, this unobserved variable is a possible candidate for u_i. With such correlated heterogeneity, we have to use other techniques that have become known as fixed effects (FE) estimation. In that context, the unit-specific error, u_i, is termed a *fixed effect*, stressing the fact that it is typical for unit i and fixed over time.

The distinction between random and fixed effects also has to do with the fact that in the first case, u_i is assumed to be a realization of a random variable with a given distribution, while in the second case, u_i is assumed to be a parameter that is specific to the sampled unit i (and hence, may be different in another sample). Therefore, some scholars argue that statistical inference for the population is only possible with random effects, while fixed effects would always be sample-specific and thus unsuitable for statistical inferences. As we will see later, the practical relevance of this argument is rather limited.

In the following sections, we start with the more realistic assumption of correlated heterogeneity and introduce FE estimation (Sect. 5.1.2.1). We then proceed with RE estimation, which is more cumbersome to compute and partly builds on results from FE estimation (Sect. 5.1.2.2). After having a firm knowledge of both estimation techniques, we will present test procedures on how to decide which of them fits the data better (Sect. 5.1.2.3).

5.1.2.1 Correlated Heterogeneity: Fixed Effects Estimation

Estimating the discrete response probability model (5.19) with fixed effects is not as easy as in the linear case. $G(\cdot)$ could be either the normal or the logistic distribution function. Since both functions are non-linear, there is no simple data transformation that would eliminate the u_i from the model (like "time-demeaning" does in the case of the linear regression model for continuous dependent variables; see Sect. 4.1.2.1). Alternatively, one can use dummy variables in the linear case (see Sect. 4.1.2.1). Is this also a feasible option for categorical variables? Technically, it is no problem with nowadays computers, although maximizing a likelihood function with respect to several thousand parameters is a quite demanding task. Besides the regression coefficients, $(n-1)$ fixed effects have to be estimated (one for each unit $i = 1, \ldots, n$ except the reference unit). But do these estimates have nice statistical properties? In other words, are they unbiased and efficient?

At this point, things become complicated. As already mentioned, ML estimates are "only" consistent under the assumptions specified in Textbox 5.1, and in order to prove this assertion, statisticians study the behavior of the estimators with increasing sample size ($n \to \infty$), while holding the number of parameters constant. The latter condition is not feasible when estimating fixed effects with dummy variables. When $n \to \infty$, by definition, the number of parameters increases as the number of fixed effects and hence, the number of dummy variables increases. In the statistics literature, this problem is referred to as the *incidental parameters problem*. Early work on binary response models including fixed effects showed that parameter estimates are heavily biased away from zero when applying the dummy variable approach to short panels (see Greene (2004a) and the literature cited therein). But it has been debated whether the bias decreases with increasing panel length and whether it exists for all kinds of panel regression model that can be applied to categorical data. More recent work by Greene (2004a,b) showed that "the fixed effects estimator shows a large positive finite sample bias in discrete choice models when T is very small. [...] (T)his general result for the probit model is mimicked by the binomial logit and the ordered probit models. The bias is persistent, but it does drop off rapidly as T increases to 3 and more" (Greene, 2004b, 144). Moreover, Greene also showed that standard errors are underestimated and hence, statistical tests will often show significant effects when in fact there are none. Certainly, more work needs to be done to understand the behavior of FE estimates found by using dummy variables, but the available evidence is not very optimistic and hence, we do not recommend this approach without taking specific precautions that are beyond the scope of this textbook.

Fortunately, one can apply a slightly different estimation method within the logistic regression framework that—similar to the linear case and time-demeaning—controls for the overall probability of observing the category of interest within each unit's panel history.[14] The FE logistic regression model is defined as follows:

$$\Pr(y_{it}=1) = \frac{\exp(\beta_0(t) + \beta_1 x_{1it} + \cdots + \beta_k x_{kit} + \gamma_1 z_{1i} + \cdots + \gamma_j z_{ji} + u_i)}{1 + \exp(\beta_0(t) + \beta_1 x_{1it} + \cdots + \beta_k x_{kit} + \gamma_1 z_{1i} + \cdots + \gamma_j z_{ji} + u_i)} \quad (5.23)$$

---

[14] This is not possible within the probit regression framework. Surely, there are other approaches to estimate fixed effects with probit regression models, which, however, go beyond the scope of this introductory textbook. See the section on p. 245 for a simple method.

5.1 Modeling the Level of Y: Discrete Response Models for Panel Data

Instead of using the full likelihood, one maximizes a conditional likelihood that conditions for each unit i on the overall probability of observing $y_{it} = 1$ within the observation period. As we will shortly see, by using such a conditional likelihood, one gets rid of the u_i.

The overall probability of observing $y_{it} = 1$ within the observation period for unit i is measured by the number of ones observed over time.[15] The larger the probability of being in state $y_{it} = 1$ for a given unit, the more ones we will observe in its sequence of ones and zeros in the observation period. By conditioning on the number of ones, we are controlling for the "average" level of the dependent categorical variable. This is very similar to the continuous case, where we "time-demeaned" the data to control for the between-unit variance. In a way, time-demeaning in the continuous case is also a conditional estimation technique: FE regression coefficients β are estimated *given* the unit-specific arithmetic means (see the section on p. 133).

The basic idea of conditional maximum likelihood (CML) is most easily explained with a simple example. We start with a two-wave panel and use the wagepan data on union membership (see Example 3.1). Table 3.2 in Sect. 3.2 cross-tabulates membership status 1985 with membership status 1984. The binary dependent variable indicates whether a person is member of a union ($y_{it} = 1$) or not ($y_{it} = 0$). With only $T = 2$ measurements over time, there are only four different sequences of zeros and ones possible: 00 (no member in both years, $f_{22} = 386$), 01 (became a member in 1985, $f_{21} = 22$), 10 (left the union in 1985, $f_{12} = 37$), 11 (member in both years, $f_{11} = 100$).[16] The four sequences include zero, one, and two ones, respectively. Instead of maximizing the full likelihood:

$$L = \prod_{i=1}^{386} \left[\Pr(y_{i1}=0) \cdot \Pr(y_{i2}=0)\right] \cdot \prod_{i=387}^{408} \left[\Pr(y_{i1}=0) \cdot \Pr(y_{i2}=1)\right]$$

386 units with sequence 00 (characteristic: $\Sigma_{t=1}^{2} y_{it}=0$)
22 units with sequence 01 (characteristic: $\Sigma_{t=1}^{2} y_{it}=1$)

$$\cdot \prod_{i=409}^{445} \left[\Pr(y_{i1}=1) \cdot \Pr(y_{i2}=0)\right] \cdot \prod_{i=446}^{545} \left[\Pr(y_{i1}=1) \cdot \Pr(y_{i2}=1)\right]$$

37 units with sequence 10 (characteristic: $\Sigma_{t=1}^{2} y_{it}=1$)
100 units with sequence 11 (characteristic: $\Sigma_{t=1}^{2} y_{it}=2$)

we are now using a conditional likelihood, which focuses on the conditional probability of observing the respective sequence for a unit given that the unit belongs to the group of units sharing the characteristic of having the same number of ones over time:

---

[15] In statistical terms, the sum of the positive outcomes ($y_{it} = 1$) over time for unit i is a *minimal sufficient statistic* for the unit-specific effect u_i.

[16] The frequencies f_{ij} ($i = 1, 2; j = 1, 2$) can be computed from Table 3.2 using the absolute numbers of members and non-members in 1984 and the transition probabilities for the transition between 1984 and 1985.

$$CL = \prod_{i=1}^{386} \frac{\Pr(y_{i1}=0) \cdot \Pr(y_{i2}=0)}{\Pr(\Sigma_{t=1}^2 y_{it}=0)} \cdot \prod_{i=387}^{408} \frac{\Pr(y_{i1}=0) \cdot \Pr(y_{i2}=1)}{\Pr(\Sigma_{t=1}^2 y_{it}=1)}$$

$$\cdot \prod_{i=409}^{445} \frac{\Pr(y_{i1}=1) \cdot \Pr(y_{i2}=0)}{\Pr(\Sigma_{t=1}^2 y_{it}=1)} \cdot \prod_{i=446}^{545} \frac{\Pr(y_{i1}=1) \cdot \Pr(y_{i2}=1)}{\Pr(\Sigma_{t=1}^2 y_{it}=2)}$$

The corresponding probabilities of observing zero, one, or two ones over time are computed as follows:

$$\Pr(\Sigma_{t=1}^2 y_{it}=0) = \Pr(y_{i1}=0) \cdot \Pr(y_{i2}=0)$$

$$\Pr(\Sigma_{t=1}^2 y_{it}=1) = \Pr(y_{i1}=0) \cdot \Pr(y_{i2}=1) + \Pr(y_{i1}=1) \cdot \Pr(y_{i2}=0)$$

$$\Pr(\Sigma_{t=1}^2 y_{it}=2) = \Pr(y_{i1}=1) \cdot \Pr(y_{i2}=1)$$

If we plug these probabilities into the conditional likelihood function, we see that the first and the last multiplicand pertaining to sequences of either all zeros or all ones are practically irrelevant, since the corresponding ratios equal 1. Obviously, units without change of the dependent variable do not contribute to the conditional likelihood. The conditional likelihood uses only those units with a change in Y:

$$CL = \prod_{i=387}^{408} \frac{\Pr(y_{i1}=0) \cdot \Pr(y_{i2}=1)}{\Pr(y_{i1}=0) \cdot \Pr(y_{i2}=1) + \Pr(y_{i1}=1) \cdot \Pr(y_{i2}=0)}$$

$$\cdot \prod_{i=409}^{445} \frac{\Pr(y_{i1}=1) \cdot \Pr(y_{i2}=0)}{\Pr(y_{i1}=0) \cdot \Pr(y_{i2}=1) + \Pr(y_{i1}=1) \cdot \Pr(y_{i2}=0)}$$

In other words: Similar to the continuous case, CML uses only the within variation of Y.

Now let us assume a simple regression model including an error term u_i for unobserved heterogeneity at the unit level. For ease of exposition, we use the same linear trend model for the probability of union membership that we already used in Sect. 3.6.1.3:

$$\Pr(y_{it}=1) = \frac{\exp(\alpha_0 + \alpha_1 t_{it} + u_i)}{1 + \exp(\alpha_0 + \alpha_1 t_{it} + u_i)} \Leftrightarrow \ln \frac{\Pr(y_{it}=1)}{1 - \Pr(y_{it}=1)} = \alpha_0 + \alpha_1 t_{it} + u_i$$

Using this model in the conditional likelihood function, we arrive at the following equation:

$$CL = \prod_{i=387}^{408} \frac{(1 - \frac{\exp(\alpha_0+\alpha_1 t_{i1}+u_i)}{1+\exp(\alpha_0+\alpha_1 t_{i1}+u_i)}) \cdot (\frac{\exp(\alpha_0+\alpha_1 t_{i2}+u_i)}{1+\exp(\alpha_0+\alpha_1 t_{i2}+u_i)})}{A}$$

$$\cdot \prod_{i=409}^{445} \frac{(\frac{\exp(\alpha_0+\alpha_1 t_{i1}+u_i)}{1+\exp(\alpha_0+\alpha_1 t_{i1}+u_i)}) \cdot (1 - \frac{\exp(\alpha_0+\alpha_1 t_{i2}+u_i)}{1+\exp(\alpha_0+\alpha_1 t_{i2}+u_i)})}{A}$$

5.1 Modeling the Level of Y: Discrete Response Models for Panel Data

with $A = \left(1 - \dfrac{\exp(\alpha_0 + \alpha_1 t_{i1} + u_i)}{1 + \exp(\alpha_0 + \alpha_1 t_{i1} + u_i)}\right) \cdot \left(\dfrac{\exp(\alpha_0 + \alpha_1 t_{i2} + u_i)}{1 + \exp(\alpha_0 + \alpha_1 t_{i2} + u_i)}\right)$

$+ \left(\dfrac{\exp(\alpha_0 + \alpha_1 t_{i1} + u_i)}{1 + \exp(\alpha_0 + \alpha_1 t_{i1} + u_i)}\right) \cdot \left(1 - \dfrac{\exp(\alpha_0 + \alpha_1 t_{i2} + u_i)}{1 + \exp(\alpha_0 + \alpha_1 t_{i2} + u_i)}\right)$

After some algebraic transformations, it can be simplified as follows:

$$CL = \prod_{i=387}^{408} \frac{\exp(\alpha_0 + \alpha_1 t_{i2} + u_i)}{\exp(\alpha_0 + \alpha_1 t_{i1} + u_i) + \exp(\alpha_0 + \alpha_1 t_{i2} + u_i)}$$

$$\cdot \prod_{i=409}^{445} \frac{\exp(\alpha_0 + \alpha_1 t_{i1} + u_i)}{\exp(\alpha_0 + \alpha_1 t_{i1} + u_i) + \exp(\alpha_0 + \alpha_1 t_{i2} + u_i)}$$

and the unit-specific error term u_i can be eliminated:

$$CL = \prod_{i=387}^{408} \frac{\exp(\alpha_1 t_{i2})}{\exp(\alpha_1 t_{i1}) + \exp(\alpha_1 t_{i2})} \cdot \prod_{i=409}^{445} \frac{\exp(\alpha_1 t_{i1})}{\exp(\alpha_1 t_{i1}) + \exp(\alpha_1 t_{i2})} \quad (5.24)$$

Note also that the regression constant α_0 drops out of the equation.

The explanatory variable T measures calendar time and has the values $t_{i1} = 1984$ and $t_{i2} = 1985$. If we maximize this function with respect to the trend parameter α_1, we arrive at the estimate $\hat{\alpha}_1 = -0.5199$ that was already mentioned in Sect. 3.6.1.3. It tells us that the logit of union membership decreases by about 0.52 units within one year; or even better, that the odds of being a union member decrease by a factor of 0.5946 ($= \exp(-0.5199)$) or by roughly 41 % within one year. Several points are noteworthy with this CML approach:

1. The derivation of the likelihood is pretty straightforward. We only had to apply the well-known formulas for probabilities of combined events (a product when combined by an "and", a sum when combined by an "or"). Moreover, when using a regression model for the probabilities, we are implicitly applying the product rule for dependent events.[17] The regression model specifies the probabilities *conditional* on the values of the explanatory variables T, X, Z and U (in the aforementioned simple model, we only used T and U). We assume that all observations are independent of each other once we have controlled for all these variables.

2. By using a conditional likelihood, we are able to control for unobserved heterogeneity without having to estimate it. Conditioning is the equivalent to time-demeaning in the case of continuous dependent variables.

3. This virtue, however, comes at a price. Estimates are based only on the units with a change of Y, which is always a smaller sample and thus results in less precise estimates having larger standard errors.

---

[17] The product rule for independent events, $\Pr(A \cap B) = \Pr(A) \cdot \Pr(B)$, states that you can simply multiply the single probabilities, while the product rule for dependent events, $\Pr(A \cap B) = \Pr(A) \cdot \Pr(B|A)$, states that you should use the conditional probability $\Pr(B|A)$.

4. Conditioning not only eliminates the unit-specific error term u_i from the likelihood function, but also the effects of all time-constant explanatory variables Z (and the regression constant). In other words, CML controls for observed *and* unobserved heterogeneity at the unit level, but does not provide estimates of their effects.
5. CML estimates only the effects of the time-varying explanatory variables X. CML estimates of these effects are consistent and asymptotically normal distributed under similar conditions like the usual ML estimates.
6. Finally, there is an interesting connection to pooled logistic regression in the case of two-wave panels: Similar to FD estimates with continuous dependent variables, which are equivalent to FE estimates in two-wave panels (see Textbox 4.7), CML estimates based on two-wave panels are equivalent to logistic regression estimates based on "differenced" data.

In order to show this, we divide the numerator and denominator of both ratios in (5.24) by $\exp(\alpha_1 t_{i1})$, which does not change the result on the right-hand side of the equation. This transforms the conditional likelihood into the (unconditional) likelihood of a pooled logistic regression model:

$$CL = \prod_{i=387}^{408} \frac{\exp(\alpha_1 t_{i2} - \alpha_1 t_{i1})}{1 + \exp(\alpha_1 t_{i2} - \alpha_1 t_{i1})} \cdot \prod_{i=409}^{445} \frac{1}{1 + \exp(\alpha_1 t_{i2} - \alpha_1 t_{i1})}$$

$$= \prod_{i=387}^{408} \frac{\exp(\alpha_1 t_{i2} - \alpha_1 t_{i1})}{1 + \exp(\alpha_1 t_{i2} - \alpha_1 t_{i1})} \cdot \prod_{i=409}^{445} \left(1 - \frac{\exp(\alpha_1 t_{i2} - \alpha_1 t_{i1})}{1 + \exp(\alpha_1 t_{i2} - \alpha_1 t_{i1})}\right)$$

$$= \prod_{i=387}^{445} \left(\frac{\exp(\alpha_1 \cdot (t_{i2} - t_{i1}))}{1 + \exp(\alpha_1 \cdot (t_{i2} - t_{i1}))}\right)^{s_{it}} \cdot \left(1 - \frac{\exp(\alpha_1 \cdot (t_{i2} - t_{i1}))}{1 + \exp(\alpha_1 \cdot (t_{i2} - t_{i1}))}\right)^{(1-s_{it})}$$

(5.25)

s_{it} is a dummy variable indicating whether the corresponding unit belongs the group of units with the sequence 01 ($s_{it} = 1$) or the group of units with the sequence 10 ($s_{it} = 0$). Equation (5.25) is structurally equivalent to the logistic likelihood function (5.10), which tells us that in the case of two-wave panels, CML estimates can be found by using a logistic regression model with the binary dependent variable S (observing sequence 01 rather than sequence 10) regressed on the *differenced* explanatory variables.[18] However, with $T > 2$, this identity does not hold anymore. Thus, in the case of longer running panels ($T \geq 3$), we need specialized software to find CML estimates.

---

[18] Note that S is also a differenced variable. Define $\Delta y_{i2} = y_{i2} - y_{i1}$. $S = 1$ when $\Delta y_{i2} = +1$ (due to sequence 01) and $S = 0$ when $\Delta y_{i2} = -1$ (due to sequence 10). Observations with $\Delta y_{i2} = 0$ (due to sequences 00 and 11) are excluded from the analysis. Differencing the data results (i) in a loss of observations and (ii) eliminates time-constant explanatory variables Z from the model. More specifically, for each unit, we lose the first observation because we cannot compute the difference ($y_{i1} - y_{i0}$). Time-constant variables Z, when differenced, will be 0 for all units and thus show no variation and drop out of the model.

5.1 Modeling the Level of Y: Discrete Response Models for Panel Data

Table 5.9 Determinants of secondary job holding (pooled and FE logistic model)

Variable	Logit model		Logit model		FE logit model	
	Estimate	Std. Err.	Estimate	Robust S. E.	Estimate	Std. Err.
Desire to work more hours	0.3678	0.1375	0.3678	0.1738	0.8493	0.2580
Desire to work fewer hours	−0.0156	0.0819	−0.0156	0.1160	−0.0253	0.1663
Weekly working hours in first job	−0.0158	0.0047	−0.0158	0.0091	−0.0318	0.0119
Is not satisfied with pay	0.1865	0.0799	0.1865	0.1049	0.0477	0.1557
Is not satisfied with work itself	0.0395	0.0924	0.0395	0.1277	0.2939	0.1818
Logarithm of first job wage	−0.3553	0.1165	−0.3553	0.2307	−1.6483	0.4001
Logarithm of non-labor income	0.0334	0.0277	0.0334	0.0427	−0.0605	0.0583
Has public employer	0.3513	0.0960	0.3513	0.1751	−0.1908	0.3324
Logarithm of job tenure	0.1625	0.0427	0.1625	0.0710	0.1008	0.1034
Age	0.0474	0.0305	0.0474	0.0516	0.2951	0.1095
Age squared	−0.0009	0.0004	−0.0009	0.0007	−0.0025	0.0013
Has children under 15	0.2250	0.0828	0.2250	0.1258	0.3532	0.2207
Being a service class worker	0.2786	0.0853	0.2786	0.1522	−0.5928	0.2540
Constant	−2.3621	0.5770	−2.3621	0.9472		
Pseudo R^2	0.0211		0.0211			
LR/X_1^2	121.23		49.44		52.83	
df	13		13		13	
ln L	−2,807.4795		−2,807.4795		−566.4602	
N	10,687		10,687		1,607	
n			2,127		275	
T_{min}, T_{max}			1	10	2	10

Source: `heineck-schwarze` data (see Example 5.1)

To conclude the discussion of FE estimation, we apply this technique to our moonlighting example. Table 5.9 shows the estimates of a FE logistic regression model with the `heineck-schwarze` data. The results are again in line with the assumption that men who desire to work more hours are more likely to be dual job holders. Also, for a one hour increase of the first job's weekly working hours, we anticipate a decline in the odds of moonlighting by a factor of $\exp(-0.032) = 0.969$ or $(\exp(-0.032) - 1) \cdot 100 = -3.1$ %. Thus, the interpretation of the estimated coefficients is identical to the interpretation of the familiar pooled logistic regression model.

Now compare the estimates with those obtained from the pooled logistic regression models using either theoretical or robust standard errors. If you look at the standard errors, you will discover substantial differences between the FE model and the two pooled logistic regression models. The FE estimates have larger standard errors than the pooled logistic regression model with robust standard errors and the latter model has larger standard errors than those obtained from simple pooled logistic regression using theoretical standard errors. This is due to the fact that—when estimating FE—many pieces of information are lost. As the derivation of the conditional likelihood showed, units without change of the dependent variable do not contribute to the likelihood. In other words, whenever a man holds a side job at all interview dates and hence, there is no variation over time to explain, observations are ignored. The same holds true for those persons who never moonlight during the observation period. In the moonlighting example, serial correlation is quite substantial and thus, the number of units available for analysis is much smaller than in the pooled analysis. 1,852 men (contributing altogether 9,080 observations) had to be dropped from the analysis because of all positive or all negative outcomes on the dependent variable. Consequently, the FE estimates are based on "only" 275 men and 1,607 observations. However, by focusing on the changing units, FE controls for unobserved heterogeneity at the unit level and the possible omitted variable bias when important time-constant variables are ignored that correlate with the variables in the model. Therefore, from a theoretical point of view, an important finding is the significant positive effect of the desire to work more hours in all three regression models, providing support for the hours-constraints motive. At the same time, as the number of weekly working hours increases, the odds of moonlighting decrease. The findings for the heterogeneous-jobs motive are mixed, however. Once the panel structure is taken into account, there is no significant difference between those being satisfied and those being dissatisfied with their pay in the first job (dissatisfaction with the work itself had no significant effect at all).

Before concluding this section, a brief mention of another consequence of FE logistic regression is necessary. Like in the case of continuous dependent variables, the effects of time-constant explanatory variables cannot be quantified. This limitation did not show up in our analysis of the `heineck-schwarze` data because all explanatory variables used in our analysis are time-varying (see Example 5.1). But if, for instance, we had used the full data set of Heineck and Schwarze (2004) and had included variables for gender and nationality, it would not be possible to quantify the differences between men and women or between German and British employees using FE logistic regression. Some people think that this is a clear disadvantage of FE, but it should be stressed that time-constant explanatory variables are controlled for in the term u_i, although the procedure does not provide a numerical estimate of their effects.

5.1.2.2 Uncorrelated Heterogeneity: Random Effects Estimation

If you are interested in numerical estimates of the effects of time-constant explanatory variables Z on the response probability, you can use RE estimation, which is feasible both with the logistic and probit regression model. However, this advantage comes at a price: We have to assume that unobserved heterogeneity u_i at the unit level is independent of the variables in the model. So for instance, if the neglected time-constant variable in our moonlighting example is health status, we have to assume that it is independent of all the variables used in our former regression models. As already discussed, this may not be a realistic assumption for health status, because health status quite probably also influences preferences for working time. In the case that you suspect unobserved heterogeneity to be correlated with the variables in your model, it is better to stick with FE estimation. If, however, the independence assumption is true, RE effects will be more efficient than FE because RE uses both the longitudinal and the cross-sectional information in the data (i.e., its estimates are based on a much larger sample than FE estimates).

Some people also question why one should use RE estimation at all if the assumption is that unobserved heterogeneity is uncorrelated with the variables in the model. They propose to use pooled probit or logistic regression, which also assume independent error terms and which are computationally much more easy to handle. However, all pooled models erroneously assume independent observations and ignore the fact that the data include several measurements for each unit. As a consequence, pooled models will provide us with standard errors that are too small. Hence, from a statistical point of view, it is always better to use an estimation technique that accounts for the panel character of the data (like FE and RE). Nevertheless, one could ask whether the serial correlation in the data is large enough to justify the use of these panel techniques. We will come back to this question later in this section.

In order to understand RE estimation, we switch to a more general notation that is less tedious and can be applied to any kind of discrete response model (including probit and logistic regression). Let $\Pr(y_{it}|\boldsymbol{\beta}, \mathbf{x}_{it}, \boldsymbol{\gamma}, \mathbf{z}_i, u_i)$ be the probability of observing a certain value, say $Y_{it} = 1$, of the binary dependent variable at time point t for unit i. The vertical line "|" indicates that this probability is supposed to depend on time-varying (X) and time-constant (Z) explanatory variables, on their respective effects $\boldsymbol{\beta}$ and $\boldsymbol{\gamma}$, and on unobserved heterogeneity U. To shorten the notation, we use vectors (indicated by bold letters). So instead of naming all time-varying variables $x_{1it}, x_{2it}, \ldots, x_{kit}$, we talk about the vector $\mathbf{x}_{it}$ of all time-varying variables at time point t for unit i. The same applies for all time-constant variables, which are collected in the vector $\mathbf{z}_i$, and the regression coefficients that are collected in the vectors $\boldsymbol{\beta}$ and $\boldsymbol{\gamma}$. At the moment, $\Pr(y_{it}|\boldsymbol{\beta}, \mathbf{x}_{it}, \boldsymbol{\gamma}, \mathbf{z}_i, u_i)$ simply states that the response probability is a function of these effects and variables, but it does not specify the functional form of how they are related. As we will see, knowledge of its functional form is not necessary to understand the basic idea of RE estimation. But obviously, both the logistic and the probit regression model are special cases of this more general specification.

Now let us consider the full likelihood of all observations in our sample. Again, to make things simple, we focus on the individual likelihood that each unit i contributes to the overall likelihood. It is a product of the probabilities for all the $t = 1, \ldots, T$ observations of unit i over time:

$$L_i = \Pr(y_{i1}|\boldsymbol{\beta}, \mathbf{x}_{i1}, \boldsymbol{\gamma}, \mathbf{z}_i, u_i) \cdot \Pr(y_{i2}|\boldsymbol{\beta}, \mathbf{x}_{i2}, \boldsymbol{\gamma}, \mathbf{z}_i, u_i) \cdot \ldots \cdot \Pr(y_{iT}|\boldsymbol{\beta}, \mathbf{x}_{iT}, \boldsymbol{\gamma}, \mathbf{z}_i, u_i)$$
$$= L_{i1} \cdot L_{i2} \cdot \ldots \cdot L_{iT}$$

The full likelihood is then the product over the likelihood contributions L_i of all $i = 1, \ldots, n$ units. Note also that the formula applies to balanced and unbalanced panels. In the latter case, the number of observations varies between units and hence, the upper limit T of the product has to be individualized ($T = T_i$).

The problem with maximizing this likelihood is the term u_i, which we do not know. In the previous section, we showed that it is impossible to obtain consistent estimates of both the regression coefficients and the fixed effects u_i (the so-called incidental parameters problem). But what if we think of u_i as a random variable U_i and make a distributional assumption about it? For example, if we assume that $l = 1, \ldots, L$ different realizations of U_i are possible: $u_{i1}, u_{i2}, \ldots, u_{iL}$, each being observed with a different probability $\Pr(u_{il})$, then we could plug the assumed u_{il} into the regression equation for $\Pr(y_{it}|\boldsymbol{\beta}, \mathbf{x}_{it}, \boldsymbol{\gamma}, \mathbf{z}_i, u_i)$ and compute a weighted sum of the resulting probabilities (using the probability of observing each u_{il} as a weight for $\Pr(y_{it}|\boldsymbol{\beta}, \mathbf{x}_{it}, \boldsymbol{\gamma}, \mathbf{z}_i, u_{il}))$.[19] For example, if we assume that three different realizations $u_{i1} = -1, u_{i2} = 0$, and $u_{i3} = +1$ are possible at $t = 1$, which are observed with probability 0.25, 0.5, and 0.25, respectively, then the likelihood contribution of the first measurement $t = 1$ of unit i equals

$$L_{i1} = \Pr(y_{i1}|\boldsymbol{\beta}, \mathbf{x}_{i1}, \boldsymbol{\gamma}, \mathbf{z}_i, -1) \cdot 0.25$$
$$+ \Pr(y_{i1}|\boldsymbol{\beta}, \mathbf{x}_{i1}, \boldsymbol{\gamma}, \mathbf{z}_i, 0) \cdot 0.5 + \Pr(y_{i1}|\boldsymbol{\beta}, \mathbf{x}_{i1}, \boldsymbol{\gamma}, \mathbf{z}_i, +1) \cdot 0.25$$

For any number L larger than three, this formula looks like this:

$$L_{i1} = \sum_{l=1}^{L} \Pr(y_{i1}|\boldsymbol{\beta}, \mathbf{x}_{i1}, \boldsymbol{\gamma}, \mathbf{z}_i, u_{il}) \cdot \Pr(u_{il}|\mathbf{x}_{i1}, \mathbf{z}_i)$$

In this formula, we have used $\Pr(u_{il}|\mathbf{x}_{i1}, \mathbf{z}_i)$ instead of $\Pr(u_{il})$ because in the most general case, the probability of observing a certain u_{il} will depend on the characteristics X and Z of the unit. If we now think about the whole likelihood contribution of unit i, this results in a quite complicated expression:

$$L_i = \left(\sum_{l=1}^{L} \Pr(y_{i1}|\boldsymbol{\beta}, \mathbf{x}_{i1}, \boldsymbol{\gamma}, \mathbf{z}_i, u_{il}) \cdot \Pr(u_{il}|\mathbf{x}_{i1}, \mathbf{z}_i) \right)$$
$$\cdot \left(\sum_{l=1}^{L} \Pr(y_{i2}|\boldsymbol{\beta}, \mathbf{x}_{i2}, \boldsymbol{\gamma}, \mathbf{z}_i, u_{il}) \cdot \Pr(u_{il}|\mathbf{x}_{i2}, \mathbf{z}_i) \right)$$
$$\cdots \left(\sum_{l=1}^{L} \Pr(y_{iT}|\boldsymbol{\beta}, \mathbf{x}_{iT}, \boldsymbol{\gamma}, \mathbf{z}_i, u_{il}) \cdot \Pr(u_{il}|\mathbf{x}_{iT}, \mathbf{z}_i) \right)$$

However, using the assumption of independent unobserved heterogeneity, in which case $\Pr(u_{il}|\mathbf{x}_{i1}, \mathbf{z}_i)$ reduces again to $\Pr(u_{il})$, we can factor $\Pr(u_{il})$ out and simplify the expression to

---

[19] Remember that the probability of a combined event that results from an or-combination of the single events equals the sum of the corresponding probabilities.

5.1 Modeling the Level of Y: Discrete Response Models for Panel Data

$$L_i = \sum_{l=1}^{L} \{[\Pr(y_{i1}|\boldsymbol{\beta}, \mathbf{x}_{i1}, \boldsymbol{\gamma}, \mathbf{z}_i, u_{il}) \cdot \ldots \cdot \Pr(y_{iT}|\boldsymbol{\beta}, \mathbf{x}_{iT}, \boldsymbol{\gamma}, \mathbf{z}_i, u_{il})] \cdot \Pr(u_{il})\} \quad (5.26)$$

In square brackets, you find a product term that is already known to you from pooled logistic (or probit) regression (compare (5.10)). First, this product is evaluated for one of the possible realizations of U_i, say u_{i1}. The result equals the probability of observing the sequence of observations $y_{i1}, y_{i2}, \ldots, y_{iT}$ for unit i, given that unobserved heterogeneity equals u_{i1}. Then, similar calculations are done for all the other possible realizations of U_i. Finally, a weighted sum of all these probabilities is computed because unobserved heterogeneity may be equal to either u_{i1} or $u_{i2} \ldots$ or u_{iL}. The weights measure the (assumed) probability that the corresponding realization of U_i is the true one.

Naturally, the assumption that unobserved heterogeneity can take on only a few ($l = 1, \ldots, L$) discrete values is not very realistic. If it were, one would always ask: Why these ones (e.g., why -1, 0 and $+1$)? Hence, it is much more meaningful to assume that unobserved heterogeneity is measured by a continuous random variable U that ranges from minus to plus infinity. In that case, we have to define a density function that describes the distribution of U and we need a more general mathematical tool to "add" the now infinite number of possibilities in (5.26). If $f(u)$ is the density function of unobserved heterogeneity, then integrating over the range of possible realizations of U is the continuous equivalent to summation over a finite number of discrete values. Correspondingly, the summation in (5.26) has to be exchanged by an integral:

$$L_i = \int_{-\infty}^{+\infty} \{[\Pr(y_{i1}|\boldsymbol{\beta}, \mathbf{x}_{i1}, \boldsymbol{\gamma}, \mathbf{z}_i, u_{il}) \cdot \ldots \cdot \Pr(y_{iT}|\boldsymbol{\beta}, \mathbf{x}_{iT}, \boldsymbol{\gamma}, \mathbf{z}_i, u_{il})] \cdot f(u) \cdot du\}$$

Finally, the full likelihood is computed by multiplying the likelihood contributions L_i of all $i = 1, \ldots, n$ units in the sample:

$$L_i = \prod_{i=1}^{n} \left(\int_{-\infty}^{+\infty} \{[\Pr(y_{i1}|\boldsymbol{\beta}, \mathbf{x}_{i1}, \boldsymbol{\gamma}, \mathbf{z}_i, u_{il}) \cdot \ldots \cdot \Pr(y_{iT}|\boldsymbol{\beta}, \mathbf{x}_{iT}, \boldsymbol{\gamma}, \mathbf{z}_i, u_{il})] \cdot f(u) \cdot du\} \right)$$
(5.27)

Equation (5.27) is the general format of an RE response probability model. It includes several special cases depending on the distributional assumptions about unobserved heterogeneity and on the choice of the regression function that links the response probabilities to the explanatory variables X and Z. A common assumption is that unobserved heterogeneity is normally distributed with mean 0 and variance σ_u^2: $u_i \sim N(0, \sigma_u^2)$. The RE probit regression model then assumes the response probabilities in (5.27) to be equal to

$$\Pr(y_{it}|\boldsymbol{\beta}, \mathbf{x}_{it}, \boldsymbol{\gamma}, \mathbf{z}_i, u_i)$$
$$= \Phi\left(\beta_0(t) + \beta_1 x_{1it} + \cdots + \beta_k x_{kit} + \gamma_1 z_{1i} + \cdots + \gamma_j z_{ji} + u_i\right) \quad (5.28)$$

while the RE logistic regression model assumes them to be equal to

$$\Pr(y_{it}|\boldsymbol{\beta}, \mathbf{x}_{it}, \boldsymbol{\gamma}, \mathbf{z}_i, u_i)$$
$$= \frac{\exp(\beta_0(t) + \beta_1 x_{1it} + \cdots + \beta_k x_{kit} + \gamma_1 z_{1i} + \cdots + \gamma_j z_{ji} + u_i)}{1 + \exp(\beta_0(t) + \beta_1 x_{1it} + \cdots + \beta_k x_{kit} + \gamma_1 z_{1i} + \cdots + \gamma_j z_{ji} + u_i)} \quad (5.29)$$

If we plug either (5.28) or (5.29) into (5.27) and specify $f(u)$ as a normal density with mean 0 and variance σ_u^2,

$$f(u) = \frac{1}{\sqrt{2\pi \sigma_u^2}} \exp\left(-\frac{u^2}{2\sigma_u^2}\right)$$

then the likelihood function is completely specified and can be maximized with respect to the unknown regression coefficients $\boldsymbol{\beta}$ and $\boldsymbol{\gamma}$.

The likelihood of either the probit or the logistic RE regression model is far more complicated than the logistic FE model. Maximization is difficult because of the integral in (5.27). There is neither an analytical solution for the integral nor for the overall maximization problem. Hence, we need numerical optimization *and* numerical integration techniques to find the ML estimates of $\boldsymbol{\beta}$ and $\boldsymbol{\gamma}$.[20] This explains why ML estimation of RE models for categorical dependent variables is usually much more costly in terms of computing time than GLS estimation of RE models for continuous dependent variables (see Sect. 4.1.2.2) or ML estimation of pooled regression models for categorical dependent variables (see Sects. 5.1.1.2 and 5.1.1.3).[21]

To exemplify how to estimate an RE model with categorical dependent variables, we consider once more the determinants of secondary job holding. Table 5.10 summarizes the results of this exercise both for the RE probit and the RE logistic regression model. Let us first address the question of whether unobserved heterogeneity plays a significant role in this data set. This can be decided on the basis of the estimated variance $\hat{\sigma}_u^2$ of unobserved heterogeneity U, which is shown at the bottom of Table 5.10. A descriptive measure of the importance of unobserved heterogeneity is the proportion of total error variance that is contributed by unobserved heterogeneity. If unobserved heterogeneity is independent of the variables in the model and the idiosyncratic error implied by the discrete response model, the variance of the composite error ε_{it} equals the sum of both variances: $\text{Var}(\varepsilon_{it}) = \sigma_u^2 + \sigma_e^2$

---

[20] A popular numerical integration technique is *Gauss–Hermite quadrature*. As a rule, numerical integration approximates the integral by a polynomial function. The polynomial function is chosen in the way that it is identical to the function to be integrated at least for a discrete number of points on the abscissa. These points are termed *quadrature points* and obviously, the precision of the approximation depends on how many points we choose and where they lie on the abscissa. Many statistical programs for RE estimation provide routines to check the sensitivity of the quadrature approximation. Note also that computing time for RE estimation is inversely related to the number of quadrature points.

[21] Note in passing that RE models for continuous dependent variables could also be estimated with ML. In that case, you will encounter similar problems with respect to computing time.

5.1 Modeling the Level of Y: Discrete Response Models for Panel Data

Table 5.10 Determinants of secondary job holding (RE logistic and RE probit model)

Variable	RE logit model		RE probit model	
	Estimate	Std. Err.	Estimate	Std. Err.
Desire to work more hours	0.8616	0.2341	0.4664	0.1316
Desire to work fewer hours	0.0085	0.1451	−0.0074	0.0798
Weekly working hours in first job	−0.0267	0.0094	−0.0140	0.0051
Is not satisfied with pay	0.1184	0.1378	0.0634	0.0762
Is not satisfied with work itself	0.2200	0.1608	0.1228	0.0892
Logarithm of first job wage	−0.7900	0.2461	−0.4399	0.1342
Logarithm of non-labor income	−0.0263	0.0505	−0.0191	0.0277
Has public employer	0.4378	0.2332	0.2459	0.1289
Logarithm of job tenure	0.0799	0.0828	0.0384	0.0454
Age	0.1379	0.0691	0.0757	0.0377
Age squared	−0.0020	0.0009	−0.0011	0.0005
Has children under 15	0.2564	0.1719	0.1391	0.0947
Being a service class worker	−0.0272	0.1860	−0.0191	0.1009
Constant	−5.2796	1.2605	−3.0708	0.6913
X_1^2	59.17		60.13	
df	13		13	
ln L	−1,963.9930		−1,961.1574	
σ_u	3.4466	0.1504	2.0167	0.0818
ρ	0.7831	0.0148	0.8027	0.0129
N	10,687		10,687	
n	2,127		2,127	
T_{min}, T_{max}	1	10	1	10

Source: `heineck-schwarze` data (see Example 5.1)

(see Textbox 3.1). The proportion ρ of total error variance that is due to unobserved heterogeneity is then defined as

$$\rho = \frac{\sigma_u^2}{\sigma_u^2 + \sigma_e^2} \qquad (5.30)$$

Note that the idiosyncratic error variance equals $\sigma_e^2 = 1$ when using a probit regression model and $\sigma_e^2 = \pi^2/3$ when using a logistic regression model (see Textbox 5.2). Furthermore, (5.30) is also a measure of the serial correlation in the dependent categorical variable that is left over after controlling for the explanatory variables X and Z in the model (see the discussion in Sect. 4.1.2.2 and (4.21)). The reason for this (remaining) serial correlation is unobserved heterogeneity u_i, which causes different observations for one unit to have something in common and hence to be correlated.

For the probit regression model, ρ is estimated as $\hat{\rho} = 0.803$ $(= 2.017^2/(2.017^2 + 1))$. The logistic regression model estimates $\hat{\rho} = 0.783$ $(= 3.447^2/(3.447^2 + \pi^2/3))$. Thus, a fair amount of the total error (78–80 %) is due to unobserved heterogeneity at the unit level. Therefore, in this example, there are good reasons to prefer RE models over pooled logistic or pooled probit regression. This conclusion can be tested more formally with a likelihood ratio test, because the corresponding pooled and RE models are hierarchically nested (in the pooled models U is restricted to be 0). If we compute twice the difference of the log likelihoods of both the pooled and the RE model, we arrive at highly significant test statistics for both the logistic ($LR = 1{,}686.97$, $df = 1$, $p < 0.000$) and the probit regression model ($LR = 1{,}693.49$, $df = 1$, $p < 0.000$). We therefore conclude that unobserved heterogeneity is important and panel estimation methods are needed. You should note, however, that this test operates under the additional assumption that unobserved heterogeneity is uncorrelated with the independent variables in the model. Therefore, it does not tell us whether RE is the appropriate panel estimation method. We still need to test whether the assumption of uncorrelated unobserved heterogeneity is true, because in the opposite case, RE estimation is not the appropriate method either. This point will be discussed in greater detail in the next section.

Let us now turn to the other estimates in Table 5.10. The results, in terms of the direction of the coefficients, are similar to what we have obtained above in FE estimation. More particularly, a striking finding of the FE logit model, the RE logit model and RE probit model is support for the hours-constraints motive. At the same time, the results of these models suggest that the heterogeneous-jobs motive does not seem to interfere with moonlighting. Hence, the findings for our variables of interest are quite robust with regard to different estimation procedures. The same does not hold true for our control variables, however. Such deviations in size of effects and standard errors raise the question of whether the assumption of uncorrelated heterogeneity is violated. The next section touches upon this issue.

5.1.2.3 Choosing Between Pooled, Fixed, and Random Effects Estimation

Having introduced three different estimation procedures for categorical dependent variables—pooled, FE, and RE estimation—we can now discuss the question: When do we apply which method?

When Do We Apply Pooled, FE, and RE Estimation? Although the simplest technique, pooled (probit or logistic) estimation is also the most demanding procedure among the three. It assumes that all unit-specific heterogeneity can be controlled by the independent variables in the model so that the remaining unexplained variance is simply "white noise". Social science theories and data hardly fulfill this assumption. Therefore, pooled estimates are often biased or at least inefficient. Consider our indicator for the heterogeneous-jobs motive in the `heineck-schwarze` data as an example. While the simple pooled logistic model shows some support for the heterogeneous-jobs motive, satisfaction with pay loses its significance once the

5.1 Modeling the Level of Y: Discrete Response Models for Panel Data

specific nature of panel data is taken into account. Pooled estimation can be a starting point, but seldom will it provide the final estimates. Hence, in most panel applications, a choice has to be made between RE and FE estimation. It seems as if from a substantive point of view, RE estimation provides the most comprehensive conclusions from the data because it allows one to estimate effects of both time-constant and time-varying variables. But RE estimation—like pooled analysis—makes restrictive assumptions, too. It admits unobserved unit-specific heterogeneity, but assumes it to be independent of the explanatory variables X and Z. From the examples discussed for the continuous case, you should be aware that failure of this assumption results in biased estimates (see Fig. 4.4 in Sect. 4.1.2.3).

So how can we decide whether RE estimates are biased? Similar to the continuous case, the overall rule is that RE estimation is possibly biased, if FE and RE estimates differ substantially. If that is the case, some people give up using RE estimation and stick with FE estimation only. Nevertheless, there are two reasons why RE estimation is quite popular in panel research: (i) The effects of time-constant explanatory variables Z can be estimated, and (ii) FE estimates focus only on units where Y changes over time, which results in a loss of degrees of freedom and correspondingly less precise estimates, especially if there is also not much over time variation in the explanatory variables. In Sect. 4.1.2.3, we argued that both arguments are not very convincing. Since the number of units, n, is usually very large in micro panels, it is often no practical problem to find a sufficient number of units with changing Y. If there is insufficient within variation both of the dependent and the independent variables, then it does not make much sense to conduct an expensive panel survey anyway. With respect to the first argument, one could at least argue that although there are no estimates of the Z effects, these time-constant characteristics are controlled for—both the observed (Z) and the unobserved ones (U)—and this is done under much more general conditions than RE assumes. Only if we want to have an estimate of their effect size are we lost with FE estimation.

Nevertheless, it would be helpful to have a formal test procedure that tells us whether FE and RE effects differ significantly. Fortunately, Hausman's (Hausman, 1978) specification test can also be applied to discrete response models for panel data. As already mentioned in Sect. 4.1.2.3, the general test idea is the following: If two estimators are consistent under a given set of assumptions, their estimates should not differ significantly. Let us call this set of assumptions A. Under a different set of assumptions, say B, this may not be true. If, in this case, only one of the two estimators provides consistent estimates, the estimates from both estimators should differ significantly. Hausman showed that the standard error of these differences is a simple function of the variance-covariance matrices of both estimators. In our case, A equals a model in which unobserved heterogeneity is uncorrelated with the independent variables in the model. In this situation, both RE and FE estimates are consistent, with RE estimates being more efficient than FE estimates. B pertains to a model with correlated unobserved heterogeneity, in which RE estimation provides biased results, while FE estimation is still consistent.

We illustrate the Hausman test with results from the FE and RE logistic regression models applied to the `heineck-schwarze` data (see Tables 5.9 and 5.10).

For example, there seems to be no large difference between the FE estimate of the desire to work more hours effect ($\hat{\beta}^{FE} = 0.8493$) and the RE estimate ($\hat{\beta}^{RE} = 0.8616$). According to Hausman, the standard error of the difference between both estimates, $\hat{\sigma}_{(\hat{\beta}^{FE} - \hat{\beta}^{RE})}$, can be calculated from the standard errors of both estimates, $\hat{\sigma}_{\hat{\beta}^{FE}}$ and $\hat{\sigma}_{\hat{\beta}^{RE}}$, as follows:

$$\hat{\sigma}_{(\hat{\beta}^{FE} - \hat{\beta}^{RE})} = \sqrt{\hat{\sigma}^2_{\hat{\beta}^{FE}} - \hat{\sigma}^2_{\hat{\beta}^{RE}}}$$

This result can be used to compute a simple T test. However, statisticians usually compute a Wald test, because ML estimates are (only) consistent and tests are preferred that rely on large sample theory. The square of the test statistic t,

$$X_1^2 = \left[\frac{(\hat{\beta}^{FE} - \hat{\beta}^{RE}) - 0}{\hat{\sigma}_{(\hat{\beta}^{FE} - \hat{\beta}^{RE})}} \right]^2 \tag{5.31}$$

equals the Wald statistic[22] and is distributed as χ^2 with $df = 1$ degree of freedom. Using the estimated standard errors of both desire to work effects, $\hat{\sigma}_{\hat{\beta}^{FE}} = 0.2580$ and $\hat{\sigma}_{\hat{\beta}^{RE}} = 0.2341$, we arrive at a value of $X_1^2 = 0.013$, which is not significant when compared to a χ^2 distribution with $df = 1$ degree of freedom ($p = 0.910$). However, if we apply the same test to the estimated effects of the respondent's wage in the first job, $\hat{\beta}^{FE} = -1.6483$ and $\hat{\beta}^{RE} = -0.7900$, a significant test statistic of $X_1^2 = 7.403$ appears ($p = 0.007$).

In the general case, when testing the joint significance of the differences between *all* RE and FE estimates, one uses the following Wald test:

$$X_2^2 = \left(\hat{\boldsymbol{\beta}}^{FE} - \hat{\boldsymbol{\beta}}^{RE} \right)' \cdot \left(\hat{\boldsymbol{\Psi}}^{FE} - \hat{\boldsymbol{\Psi}}^{RE} \right)^{-1} \cdot \left(\hat{\boldsymbol{\beta}}^{FE} - \hat{\boldsymbol{\beta}}^{RE} \right) \tag{5.32}$$

which is a generalization of the former test procedure using matrix algebra. The vectors $\hat{\boldsymbol{\beta}}^{FE}$ and $\hat{\boldsymbol{\beta}}^{RE}$ represent the parameter estimates from FE and RE estimation, while the matrices $\hat{\boldsymbol{\Psi}}^{FE}$ and $\hat{\boldsymbol{\Psi}}^{RE}$ include the corresponding estimated variances and covariances of the estimates. X_2^2 is distributed as χ^2 with $df = l$ degrees of freedom (l being the number of coefficients tested). As already mentioned in Sect. 4.1.2.3, you do not need matrix algebra to apply the overall Hausman test, because many software packages include pre-programmed versions of (5.32). Applying this formula to the thirteen (comparable) estimates from our FE and RE models (excluding the constant) results in a test statistic of $X_2^2 = 38.33$, which is highly significant with $df = 13$ degrees of freedom ($p = 0.0003$). Hence, according to the Hausman test, FE estimates should be preferred over RE estimates.

---

[22] Like in Chap. 4, we use X^2 to symbolize the test statistic of a Wald test. Again, since we discuss different Wald tests, we distinguish the various test statistics by indices (X_1^2, X_2^2, etc.). Note, however, that numbering starts again in this chapter and X_1^2 from this and the former chapter are not identical.

5.1 Modeling the Level of Y: Discrete Response Models for Panel Data

A Hybrid Model In Sect. 4.1.2.3, we also introduced a hybrid model for the continuous case that somehow was a compromise between RE and FE models. Since FE estimation uses only the within variation of Y and X (Z, by definition, has none), the basic idea of this hybrid model was to estimate an RE model and include time-varying explanatory variables X in such a way that their within- and between-unit effects are separated. Therefore, besides time-constant explanatory variables Z, the hybrid model uses unit-specific means $(\bar{x}_{1i.}, \ldots, \bar{x}_{ki.})$ and time-demeaned values $(x_{1it} - \bar{x}_{1i.}, \ldots, x_{kit} - \bar{x}_{ki.})$ instead of the original time-varying explanatory variables X. In the continuous case, or more specifically, with linear models, this hybrid model replicates the FE estimates *and* in addition also estimates effects for time-constant explanatory variables Z. Such a hybrid model can also be specified for categorical dependent variables. However, because of its non-linearity, the hybrid response probability model does not exactly replicate the FE estimates, but the estimates come very close to them. Again, the attraction of this hybrid model is that it allows us to test differences between FE and RE estimates in a very general way. Furthermore, it allows us to find FE estimates for other link functions besides the logit (e.g., for probit regression models).

To illustrate how to specify this hybrid model, we use the logistic regression model: $\Pr(y_{it} = 1) = \exp(\eta)/(1 + \exp(\eta))$. The linear predictor η of this model is now defined as

$$\eta = \beta_0 + \beta_1 \cdot (x_{1it} - \bar{x}_{1i.}) + \varphi_1 \bar{x}_{1i.} + \cdots + \beta_k \cdot (x_{kit} - \bar{x}_{ki.}) + \varphi_k \bar{x}_{ki.}$$
$$+ \gamma_1 z_{1i} + \cdots + \gamma_j z_{ji} + u_i \quad (5.33)$$

For ease of exposition, we have not assumed a time trend (i.e., $\beta_0(t) = \beta_0$). Otherwise, we would also have to decompose the time trend into its within- and between-unit components. In this hybrid model, the regression coefficients $\beta_1, \ldots, \beta_k$ measure within-unit effects of the time-varying explanatory variables X. They should be nearly identical to the regression coefficients of the corresponding FE model. The regression coefficients $\varphi_1, \ldots, \varphi_k$ measure the between-unit effect of the time-varying explanatory variables X. As discussed in Sect. 4.1.2.3, they are hardly of interest. Including mean values of the time-varying explanatory variables X into the model simply controls unobserved heterogeneity to that extent that we can replicate the FE estimates with $\beta_1, \ldots, \beta_k$. Furthermore, $\beta_1, \ldots, \beta_k$ and $\varphi_1, \ldots, \varphi_k$ should be identical, if unobserved heterogeneity is uncorrelated with the variables in the model, and hence, RE and FE estimates should only differ randomly. By using single parameter tests or by using linear restrictions on several parameters, we can test whether this assumption is true. Finally, the regression coefficients $\gamma_1, \ldots, \gamma_k$ measure the effects of the time-constant explanatory variables Z, which—compared to the simple logistic RE model—now control for correlated unobserved heterogeneity as far as it is measured by the means $\bar{x}_{1i.}, \ldots, \bar{x}_{ki.}$.

Table 5.11 displays the results of the hybrid model for the heineck-schwarze data and reveals that the coefficients for the deviations from unit-specific means are

Table 5.11 Determinants of secondary job holding (RE logistic hybrid model)

Variable	Estimate	Std. Err.
Unit-specific means		
Desire to work more hours	1.0879	0.6838
Desire to work fewer hours	0.2249	0.3699
Weekly working hours in first job	−0.0283	0.0164
Is not satisfied with pay	0.6637	0.3780
Is not satisfied with work itself	0.0005	0.4346
Logarithm of first job wage	−0.9665	0.4255
Logarithm of non-labor income	0.0538	0.1046
Has public employer	1.0535	0.3445
Logarithm of job tenure	0.1960	0.1601
Age	0.0777	0.0945
Age squared	−0.0016	0.0012
Has children under 15	0.3631	0.3067
Being a service class worker	0.7887	0.3155
Deviations from unit-specific means		
Desire to work more hours	0.8873	0.2582
Desire to work fewer hours	−0.0486	0.1586
Weekly working hours in first job	−0.0367	0.0118
Is not satisfied with pay	0.0637	0.1503
Is not satisfied with work itself	0.2351	0.1771
Logarithm of first job wage	−1.7637	0.3749
Logarithm of non-labor income	−0.0623	0.0580
Has public employer	−0.1763	0.3156
Logarithm of job tenure	0.1020	0.0995
Age	0.2863	0.1068
Age squared	−0.0024	0.0013
Has children under 15	0.2733	0.2115
Being a service class worker	−0.5395	0.2425
Constant	−4.6131	1.8088
X_1^2	92.2	
df	26	
ln L	−1,942.3418	
σ_u	3.4929	0.1486
ρ	0.7876	0.0142
N	10,687	
n	2,127	
T	10	

Source: `heineck-schwarze` data (see Example 5.1)

quite close to the corresponding coefficients in the FE model. By applying a Wald test, we can test whether the coefficients for deviations from unit-specific means are the same as coefficients for unit-specific means. For example, for the desire to work more hours and the respondent's wage, the Wald test gives you values of $X_3^2 = 0.07$ and $X_3^2 = 1.98$, respectively, which are not significant when compared to a χ^2 distribution with $df = 1$ degree of freedom ($p = 0.788$ resp. $p = 0.159$). However, performing an overall test returns a test statistic of $X_3^2 = 38.69$, which is highly significant ($df = 13$, $p = 0.0002$) and again suggests that FE estimates should be preferred over RE estimates.

5.1.3 Extensions

As discussed in Sect. 3.4.2, serial correlation may result from spurious and true state dependence. While we have controlled for spurious state dependence due to unobserved heterogeneity in the FE (and RE) regression model, we have not yet dealt with the second source of serial correlation. The estimation models that we have considered to this point rest upon the assumption that there is no state, or what is also called structural dependence: the probability of moonlighting at time t is not determined by the propensity to hold a side job at time $t - 1$. If there is true state dependence, then you would use a dynamic model rather than a static model. This is achieved by plugging a new variable $y_{i,t-1}$ into the model, which allows you to capture the degree to which the lagged dependent variable $y_{i,t-1}$ determines the discrete response y_{it} at time t.[23] There may not only be feedback effects from your lagged dependent variable on the response probability at time t. There are also instances in which your lagged dependent variable determines the explanatory variables at time t. Readers interested in such extensions of our general approach presented above are referred to Honoré (2002), Maddala (1987), Heckman (1981b) and Heckman (1981a). More advanced treatments can be found in Honoré and Kyriazidou (2000), Arellano and Carrasco (2003) and Honoré and Lewbel (2002).

The above discussion of discrete response models is restricted to dichotomous variables. However, in several applications, the dependent variable has more than two distinct values. The monograph by Long (1997) as well as the more recent textbooks by Long and Freese (2006), Agresti (2002) and Hosmer and Lemeshow (2000) deal with regression models for such ordinal, nominal, and count outcomes. These books include various practical examples with cross-section data and are written at an introductory level.

At a more advanced level, various articles and textbooks detail specific methods and review their refinements for the analysis of panel data. An example for multinomial, unordered, dependent variables is the analysis of different labor market states,

---

[23] Notice that this is a special case of the Markov model discussed in Sect. 3.5.2.2.

such as high-wage, low-wage, unemployed, and economically inactive. An introduction to regression models specifically designed for the analysis of such unordered variables is given in Borooah (2001). Multinomial regression models are also covered in greater detail in some standard textbooks such as Hosmer and Lemeshow (2000). Chamberlain (1980) shows how to extend the CML approach to the multinomial logit model. Sophisticated treatment of omitted variable bias in multinomial logit models can be found in Lee (1982).

For variables that show an inherent ordering, such as satisfaction with pay or job security, specific techniques have been developed, namely ordinal logit or probit analysis. Readers interested in introductions to ordinal logit or probit analysis are referred to the classical article by McKelvey and Zavoina (1975). A gentle introduction to the analysis of ordered responses is given in Daykin and Moffatt (2002). The textbook by Agresti (2010) covers the specific techniques for ordinal variables in greater detail. Various two-step approaches have been proposed to allow for fixed effects in ordered response models (Winkelmann and Winkelmann, 1998; Ferrer-i-Carbonell and Frijters, 2004). First, the ordinal logit model is simplified to a binary logit model (e.g., by collapsing the categorical responses or dichotomizing the dependent variable at the individual mean). In the second step, fixed effects estimation can be applied, as demonstrated in Sect. 5.1.2.1. Ferrer-i-Carbonell (2005) gives an example of how to estimate a RE ordered probit model.

Count variables can take on only a limited number of values greater or equal to zero. For instance, count variables arise in demographic analysis (e.g., the number of children ever born to a women), but we can also think of application examples in econometric analysis (e.g., number of job changes in a given time interval). If you look at the distribution of such variables, you will see that these discrete variables are often highly skewed. We therefore need a method appropriate for such dependent variables: the Poisson model. An insightful review on estimation models for count variables can be found in Winkelmann and Zimmermann (1986), Winkelmann (2003), and Cameron and Trivedi (1986, 1998). Hausman et al. (1984) illustrate how to estimate RE and FE Poisson models. The issue of individual effects and dynamics in count data models is raised in a more recent paper by Blundell et al. (2002).

5.2 Modeling the Change of Y: Discrete-Time Event History Models for Panel Data

In this section, we use discrete-time event history models to analyze the change of a categorical dependent variable Y. Event history modeling is a very comprehensive tool to analyze time-to-event data. In this textbook, we will use only part of this methodology; more specifically, procedures that apply to research designs when the state of a unit i over time is only known at discrete points in time $(t_1, t_2, \ldots, t_T)$, as is usually the case with panel data. Nevertheless, to make full use of these discrete-time methods, it is necessary to be familiar with the basic terminology of event

5.2 Modeling the Change of Y: Discrete-Time Event History Models for Panel Data

history modeling, which we introduce in Sect. 5.2.1. A nice side effect of focusing on discrete-time methods is that we can use statistical tools (logistic regression) that are already familiar to us to estimate the parameters of the corresponding event history models.

However, these techniques have to be adopted to a missing data problem that is specific to event history data and which is called *censoring*. Units that do not experience a change of Y during the observation period are called censored observations. It is not quite obvious how to treat them—whether they should be ignored or whether they should be included in the model and if so, how this could be done. In Sect. 5.2.2, we will show that simple techniques using traditional models for continuous or categorical variables lead to seriously biased estimates. Only maximum likelihood estimation is able to treat the specific information of censored (no event until end of observation) and uncensored observations (event at t_i) correctly. As we show in Sect. 5.2.2.3, the likelihood function of a discrete-time event history model is identical to the likelihood of a logistic regression model and hence, logistic regression can be used to estimate the parameters of model (5.2).

In Sect. 5.2.3, we will exemplify how to apply a discrete-time logistic model to panel data using an analysis of women's retirement decisions. This example addresses the change of a dependent categorical variable. It looks at women's labor market status and analyzes the change from being active to being retired.

Example 5.2 (hank data)   The example focuses on the effects of early life family events on women's late life labor market behavior. Prior research on married women's labor force participation has intensively focused on the interplay between a woman's reproductive history and their spouses' employment decisions. In recent years, multifaceted transformations have had large scale consequences for the work and family life of women. Aspects of change embrace declining birth rates, extended educational enrollment and labor force participation, and transformation of the gender division of work. So far, little is known about the relationship between childbearing and retirement. Using panel data from the first 18 waves of the German Socio-Economic Panel Study (SOEP) covering 1984–2001, Hank (2004) concentrates attention on ever-married West German women aged 50–69 and addresses the research question of whether mothers will withdraw from the workforce earlier than childless women. Or, will they prolong their working life to make up for employment interruptions during their reproductive years?

One supposition is that having children goes along with an earlier retreat from the labor market since mothers not only have a weaker labor market orientation over the life course, but they are also more likely to have a male breadwinner in the household who is expected to provide economic security throughout, rendering the mothers' employment career less important. At the same time, and relative to men, women experience more fragmentation, more part-time employment, more employment in the lower segments of the labor market, and lower returns to work. These factors affect pension rights and incomes and hence women's retirement behavior. Therefore, mothers might decide to work longer than childless women in order to make up for losses in their own personal income and retirement benefits.

Table 5.12 gives a short description of the variables included in this data set. The variables were measured annually. Most explanatory variables are time-varying except number of children, education, years of labor force experience, and number of employment spells. All in all, the sample consists of $n = 837$ women that have been observed between $T_{min} = 1$ and $T_{max} = 18$ years (altogether $N = 5,765$ observations).

Unfortunately, some of the variables include quite a lot of missing data. For education and the partner's employment status, Hank created indicator variables with the value 1 indicating that the information for the corresponding variable is missing (0 otherwise). There are also cases for which the information on income is lacking. Hank used a simple imputation technique, which appears adequate for his specific research purpose since income is only a control variable. He used the median income in this sample (3,700 DM) to fill gaps. Note that this is not explained in the journal article; the results for the indicator variables are not shown.

Table 5.12 Measures used in the analysis of the hank data

Variable	Description	
Variable of interest		
Number of children	number of children at age 50	time-constant
Control variables		
Age group	six dummies for: 50–53 years = reference, 54–57 years, 58–59 years, 60–61 years, 62–63 years, 64–69 years	time-varying
Educational attainment	four dummies for: no degree = reference, vocational degree, university degree, imputation flag	time-constant
Labor force career	number of years in the labor force at age 50	time-constant
	number of employment spells at age 50	
Partner	four dummies for: no partner = reference, partner retired, partner not retired, imputation flag	time-varying
Income	monthly household income (in DM)/100	time-varying
	monthly household income squared	
Home owner	dummy for home owner	time-varying
Health status	dummy for poor health	time-varying
Person needing care	dummy for person needing care in the household	time-varying

Throughout most parts of this section, we assume—like in the case of the hank data—that the dependent variable is dichotomous. But the concluding Sect. 5.2.4 discusses some extensions to more complicated change processes, including categorical variables having more than two values or variables that change several times during the observation period.

5.2.1 Basic Terminology

A lifetime is characterized by a sequence of events. *Event* is a byword for an individual's transition from one discrete state to one (of several) other discrete state(s) within a well-defined interval of time. Examples are entry into unemployment in labor market analysis, partnership formation in demographic studies, and mortality in medical research. We can also observe sequences of events for other research units, such as warfare of nations in political science research or formation of companies in economics. An event history documents whether, and if so, when, such events occur. With techniques of event history modeling, one can deal with time-to-event data and analyze the hazard rate of experiencing a particular event at time t.

In the technical language of event history modeling, the time span a unit of analysis spends in a specific state is called an *episode* or a *spell*. Individuals in a specific discrete state who have a chance to switch to a specific outcome state during this period of time belong to the *risk set*. In turn, the time period that a person is at risk for event occurrence is defined as the *risk period*.

In simple event history models, one considers only two states: one origin and one destination state. As an illustration, in the analysis of mortality, the underlying process starts with birth and ends with death. This is also called a *single event*. In other cases, there are multiple destinations. For instance, a job can end with a change of employer, unemployment, economic inactivity, or further education. This is called a *multiple event*.

Sometimes we are not interested in one single spell but in multiple episodes. For instance, individuals may hold various jobs during their economically active life. Event history modeling allows the analysis of such *repeatable events* with models for multiple episodes. Our Example 5.2 analyzes a single event that does not repeat over time.

Often we do not observe the whole sequence of events. Depending on the data collection design, time is observed with varying precision, and time units may be too large to represent the underlying processes that generate event sequences. Accordingly, in the literature on event history analysis, one distinguishes between techniques for *continuous-time* and *discrete-time data*. Both terms refer to the mode of data collection and the resulting dependent variable T measuring spell duration ($t =$ end date $-$ start date). Here, continuous time means that time can take on any nonnegative value ($t \geq 0$). By contrast, in discrete-time methods, time can take on only positive integer values ($t = 1, 2, 3, \ldots$). Panel data often yield only discrete-time data (see the annual data in Example 5.2). Given the special focus of this textbook, we will concentrate on methods for discrete-time event history data.

Fig. 5.6 Different types of censoring

Besides the discrete or continuous-time nature of the data, there is another more fundamental data collection problem typical for event histories: The period of observation is always limited and hence, we observe episodes for which we do not know the start date, and for others, there is no end date. There are also episodes that are completely unobserved. This problem is called *censoring*. Figure 5.6 illustrates the censoring problem with an example from labor market analysis. It shows a selection of employment spells for a sample of six individuals (time T is measured continuously since each individual's entry into the labor market). The episodes of person 1 and person 4 are examples for observations that continue until time t_i, at which point an event occurs (marked by a *). Since all of this happens during the observation period (indicated by the observation window), start and end dates (including origin and destination states) are known for both episodes. The other episodes in Fig. 5.6 are censored, however. They are distinguished as completely right censored (person 6), right censored (person 3), left censored (person 2), and completely left censored (person 5). Censoring means that the information that describes an episode (start date, origin state, end date, destination state) is either partially or completely missing. A central assumption that is usually made in event history analysis is that the driving force behind censoring mechanisms is independent of the change process being analyzed. This assumption is often true, if one chooses the observation period independent of the process under study (as is the case in Example 5.2).

Obviously, completely censored episodes cannot be used in the analysis. But what about partially censored episodes? Quite generally speaking, left censoring poses more problems than right censoring. Event history analysis has developed specific statistical techniques to deal with the latter problem and we will discuss them in Sect. 5.2.2. Statistical techniques for left censored data are much more complicated (if they exist at all), but sometimes the missing information can be collected

5.2 Modeling the Change of Y: Discrete-Time Event History Models for Panel Data

by retrospective interviewing.[24] Therefore, one should ignore left censored episodes and focus the analysis on the uncensored and right censored episodes.

Left censored episodes are also prevalent in the hank data: Since the study focuses on ever-married West German women aged 50–69, the majority of them ($n = 507$) have been gainfully employed for a long time when the observation period starts at age 50. Ignoring all of these episodes would drastically decrease the sample size. Alternatively, Hank (2004) controls for the previous employment history by using years of labor force experience and number of employment spells as explanatory variables in the model.

Having introduced the concepts of discrete time and censoring, we come now to the fundamental statistical concepts of event history analysis. In discrete-time analysis, we observe the event history of n independent units ($i = 1, \ldots, n$), beginning at some natural starting point $t = 0$. In Sect. 3.3, we introduced the concept of the discrete-time hazard rate $h_p(t)$. It is defined as the conditional probability that an individual will experience an event (a transition from state p to another state) at a particular time t, given that this event did not occur before time t and the individual still was a member of the risk set.[25] The discrete-time hazard rate is given as

$$h_p(t) = \Pr(y_{it} \neq p | y_{i1} = \cdots = y_{i,t-1} = p) \tag{5.34}$$

One can also study the survival probability (i.e., the probability that an individual will not experience an event at or before time t and thus, remain in the origin state p):

$$S_p(t) = \Pr(T \geq t) \tag{5.35}$$

To derive this probability, one simply accumulates the period-by-period risks of event non-occurrence (see the example in Sect. 3.3, especially Table 3.4). The discrete-time survival probability is given as the product of probabilities of not experiencing an event in each of the t intervals up to and including the current one:

$$S_p(t) = \left(1 - h_p(1)\right) \cdot \left(1 - h_p(2)\right) \cdot \ldots \cdot \left(1 - h_p(t)\right) = \prod_{l=1}^{t} \left(1 - h_p(l)\right) \tag{5.36}$$

Naturally enough, changes in the hazard rate bring about changes in the survivor function. The survival probability declines swiftly with higher hazard rates and, vice versa, it declines slowly when the discrete-time hazard rate is low.

---

[24] Obviously, no similar data collection strategy exists for right censored data (up until now, social science research is not able to look in the future).

[25] Note here that originally the hazard rate is not defined as a probability but as a rate that summarizes the instantaneous transition intensity. As such, the hazard rate is a continuous-time quantity, a case that is left unconsidered in this textbook. Therefore, from a statistical point of view, it would be more accurate only to talk about conditional transition probabilities in case of discrete-time event histories. In this text, we are a bit more liberal and use the term discrete-time *hazard* as a synonym for conditional transition probability.

By focusing on the conditional transition probability resp. the discrete-time hazard rate, we have made the implicit assumption that the underlying process allows change to happen only at discrete points in time. This may be acceptable as an approximation, but in reality, nearly all change processes operate in continuous time. For example, working people can retire each month of the year and not only in yearly intervals. Therefore, from a theoretical point of view, it is much more realistic to assume a process operating in continuous time that—due to the panel design—has been observed imprecisely. In that case, you should model the continuous-time hazard rate and adapt the statistical model to the imprecise measurements of T. We will come back to this alternative when discussing several extensions of our basic event history model (see Sect. 5.2.3.5).

Nevertheless, the assumption of an underlying process operating in discrete time with discretely measured spell durations makes statistical modeling very easy. You only need to define how the discrete-time hazard rate $h_{ip}(t)$ depends on T and your explanatory variables X and Z (see (5.2); in that case we also need an index i because the hazard rate now depends on individual characteristics X and Z):

$$h_{ip}(t) = G\big(\beta_0(t) + \beta_1 x_{1it} + \cdots + \beta_k x_{kit} + \gamma_1 z_{1i} + \cdots + \gamma_j z_{ji}\big)$$

$G(\cdot)$ is a suitable transformation function to ensure that the right-hand side of the equation provides values that are within the proper limits of probabilities (i.e., $0 \leq h_{ip}(t) \leq 1$). A common choice in discrete-time event history modeling is the logistic distribution function, which models the discrete-time hazard as follows:

$$h_{ip}(t) = \frac{\exp(\beta_0(t) + \beta_1 x_{1it} + \cdots + \beta_k x_{kit} + \gamma_1 z_{1i} + \cdots + \gamma_j z_{ji})}{1 + \exp(\beta_0(t) + \beta_1 x_{1it} + \cdots + \beta_k x_{kit} + \gamma_1 z_{1i} + \cdots + \gamma_j z_{ji})} \quad (5.37)$$

You may want to start with the simplest model, where the hazard rate varies between each pair of panel waves (modeled with a series of dummies for the $\beta_0(t)$ term), but does not depend on explanatory variables. You can then introduce more complexity by allowing the hazard rate to depend on time-constant (Z) and time-varying (X) explanatory variables. As we will see in Sect. 5.2.3.5, the more realistic assumption of an imprecisely measured continuous-time process provides very similar estimates of the effects of the explanatory variables.

5.2.2 How to Estimate a Discrete-Time Hazard Model

How, then, can we derive an estimate of the hazard rate, when it is not directly observable? With survey data, we can only observe whether, and if so, when, transitions from a given origin state to a specific destination state occur. Looking back to the statistical methods discussed so far, at least two possible analytical strategies come to mind. First, one could measure the time until event occurrence, t_i, for each individual i and consider this duration variable as a dependent variable in a linear model. Second, thinking of the methods used to analyze categorical dependent

5.2.2.1 OLS Regression

Let us first consider the case of a linear regression model. Statistically, in linear regression, the dependent variable is assumed to be continuous and varying between negative and positive infinity. Often, it is also assumed to be normally distributed. With time-to-event data, however, the dependent variable T is restricted to be greater than zero and in most cases has a right skewed distribution. This deficiency can easily be solved by taking the natural logarithm of T. This transformation will make the distribution more symmetric and spell durations lower than 1 will result in negative values since $\ln(a) < 0$ if $0 < a < 1$. Therefore, a simple regression model for duration data would be (see (4.3)):

$$\ln(t_i) = \alpha_0 + \beta_1 x_{1it} + \cdots + \beta_k x_{kit} + \gamma_1 z_{1i} + \cdots + \gamma_j z_{ji} + \varepsilon_{it} \qquad (5.38)$$

If one is analyzing non-repeatable events (like in Example 5.2), then there is only one observation for each unit i (including the dependent variable T and the explanatory variables X and Z) and one can use OLS to estimate the parameters. Note that this is a model on spell *durations*, which are inversely related to the hazard rate: The longer the spell duration, the lower the hazard rate and vice versa. Hence, explanatory variables having a positive (negative) effect on the hazard rate should have a negative (positive) effect in (5.38). This model is easily applied and parameter estimates have a well-known interpretation. There are, however, important insufficiencies of this simple linear model.

First, OLS is not able to deal with censored observations that are typical of event history data. The study by Hank (2004) is an example of a situation where time-to-event data is incomplete for some units. Not all women retire during the observation period: 451 out of $n = 837$ women contribute a right censored spell to the data set. How can these spells, for which we do not observe the total spell duration, enter our analysis?

A simple approach could be to construct the dependent variable in such a way that it measures the maximum duration the observations were under observation and hence, consider these spells as if they had ended at the last point of observation. Obviously, this is not a very meaningful procedure. The assumption that the true survival time t_i^* is equal to the length of the observation period τ is not met for right censored observations: $t_i^* > \tau$.[26] Depending on the number of censored cases, summary statistics, such as the median survival time (in the origin state) and estimated parameters, will be heavily biased.[27] Median survival time will be underestimated,

---

[26] If time of observation were different for each unit, τ should have an index i, too. But this is not the case in Example 5.2.

[27] The estimated median lifetime equals that value of T, at which 50 % of the units at risk have already experienced the transition under study (i.e., that time point when the estimated survivor function is 0.5).

if t_i is set equal to τ, albeit the true survival time t_i^* is larger than τ. Using this approach, women in the hank data will appear to retire earlier than they actually do. Moreover, if mothers work longer, as hypothesized by some authors, then mothers have a higher probability of being censored during a limited observation period. The simple approach for censored observations would make their spell durations shorter than they actually are and hence, would decrease the differences between mothers and other women. The corresponding parameter estimate would underestimate the true effect of motherhood.

Another approach would be to simply ignore censored observations and exclude the corresponding records from the analysis. This would result in a very selective sample since longer spell durations with a correspondingly high probability of censoring are systematically excluded from the analysis. Again, this would result in biased parameter estimates and summary statistics.

Second, OLS estimation assumes that the dependent variable (or equivalently, the error term in the regression model) is normally distributed (see Textbox 4.1). As already mentioned, using $\ln(t_i)$ instead of t_i itself often solves the distributional problems. But from a theoretical point of view, using data transformations such as the natural logarithm is not very convincing, because it always looks like data fitting. Moreover, if space would allow us to learn more about continuous-time event history analysis, we would know that spell durations by definition are often *not* normally distributed. Depending on the trend of the hazard rate over time, whether it is constant over time, monotonically increasing, or u-shaped, we could even specify the theoretical distribution of the spell durations.[28] These distributions would not be identical with the normal (or log-normal) distribution.

Finally, it turns out that the simple model (5.38) is not as flexible as it looks at first glance. For example, you may have noticed that it includes only a regression constant α_0 and not the general term $\beta_0(t)$ that has been used in earlier models to operationalize changes over time. But with time T as the dependent variable, it is not possible to include a term on the right-hand side that is itself a function of process time t.[29] This is a clear limitation because, as already mentioned, we might have a clear hypothesis that the process itself changes over time. As we will see later, this problem does not arise if we model the hazard rate (and not the spell duration).[30] Another limitation of (5.38) appears with the time-varying explanatory variables X. How can one incorporate such variables in the simple model (5.38)? Each unit is represented only *once* in the data set, with one single dependent variable (duration) and you can provide only one value for each explanatory variable. For women who retire, you would have to decide which value of a time-varying explanatory variable is associated with a change in the hazard rate—the one at the beginning of the spell

---

[28] Therefore, continuous-time event history models are often called *duration models*.

[29] This does not preclude that the model includes variables that use other definitions of time (e.g., in the hank data, year of birth for each women, i.e., cohort membership).

[30] Continuous-time event history models (duration models) solve this problem by deriving a distributional assumption for spell durations from hypotheses concerning the trend of the hazard (see previous paragraph).

5.2 Modeling the Change of Y: Discrete-Time Event History Models for Panel Data

or the value at the end or some value in between? Ideally, we would like to use *all* of the values of X that are observed during spell duration, but that is not possible with only one data record for each unit. Hence, the simple model (5.38) does not allow a proper incorporation of time-varying explanatory variables.

5.2.2.2 Logistic Regression

Consider now a binary response model where the dependent variable is equal to 1 if we observe an event and equal to 0 if the observation is censored. The discrete response model (5.7) can be used to analyze the explanatory variables on which these events depend:

$$\ln\left(\frac{\Pr(y_{it}=1)}{1-\Pr(y_{it}=1)}\right) = \beta_0(t) + \beta_1 x_{1it} + \cdots + \beta_k x_{kit} + \gamma_1 z_{1i} + \cdots + \gamma_j z_{ji} \quad (5.39)$$

Two cautionary notes are in order here. One obvious consequence of this analytical strategy is a large loss of important information. All units are lumped together even though they experience the target event at many different points in time; event occurrence data do not yield information on the timing of event occurrence. As a remedy, some people include time (to event or censoring) as an explanatory variable on the right-hand side of the equation (by a suitable specification of $\beta_0(t)$). Although this looks like a good solution, it is not what we are looking for. We want to know the probability of an event at each point in time, given that the unit is still at risk of experiencing the event (i.e., the *conditional* transition probability). While the number of units still at risk (the risk set) diminishes over time as more and more units have experienced the event, (5.39) focuses on the *unconditional* probability of experiencing the event, even if we differentiate this probability for different points in time by specifying $\beta_0(t)$. In other words, (5.39) relates event occurrences always to the whole sample size and not to the number of units that are left over at each point in time. Therefore, also from a theoretical point of view, the discrete response models are not a good choice since they do no rest on the analytical strategy we have in mind when working with time-to-event data.

5.2.2.3 Logistic Discrete-Time Hazard Model

Although the former two approaches have their limitations, they show us what a more suitable estimation method should look at. First, it should use all the available information in the data. Besides analyzing events, it should make proper use of the censored observations, which tell us that up to the end of the observation window *no* event occurred, although the corresponding unit has been *at risk* of experiencing the event of interest. Second, the estimation method should recognize the timing of events and censoring. While OLS regression can cope with the timing, it obviously has problems with censored observations. Logistic regression, on the other hand, can distinguish between events and censored observations, but has problems in modeling how the process changes over time. What we are now looking for, is an estimation method that combines the virtues of both techniques and additionally, provides a sensible model of the underlying change process.

As discussed in Sect. 3.5.2.2, the conditional transition probability, or for short, the discrete-time hazard rate, is a meaningful measure of this change process and thus, $h_{ip}(t)$ should be the dependent variable of our regression models. If unit i is observed t_i (discrete) time units and has not experienced an event (i.e., if it is a censored observation), (5.36) shows us how we can use $h_{ip}(t)$ to compute the probability of surviving up to time point t_i:

$$L_i^{censored} = \prod_{t=1}^{t_i} \left(1 - h_{ip}(t)\right) \tag{5.40}$$

Similarly, we can compute the probability that unit i experiences an event at time point t_i, given that i is still member of the risk set. This necessitates that unit i has survived $t_i - 1$ time units:

$$L_i^{event} = \left(\prod_{t=1}^{t_i-1} \left(1 - h_{ip}(t)\right)\right) h_{ip}(t_i) \tag{5.41}$$

Now, if we make up a new data set that includes $N = \sum_{i=1}^{n} t_i$ records (i.e., each unit is represented as often in this data set as it is observed at discrete points in time), then it is very easy to specify the likelihood of the whole sample. We only have to record in a dummy variable for each of the t_i records of unit i, whether the record ended with an event ($event_{it} = 1$) or did not end with an event ($event_{it} = 0$). Using this dummy variable, we can combine the former two (5.40) and (5.41) and by multiplying over all $i = 1, \ldots, n$ units, derive the full likelihood:

$$L = \prod_{i=1}^{n} \prod_{t=1}^{t_i} \left(1 - h_{ip}(t)\right)^{(1-event_{it})} \cdot h_{ip}(t)^{event_{it}} \tag{5.42}$$

As an example, consider the following hypothetical data set including $n = 3$ individuals (see Table 5.13). Person 1 experiences an event after one year ($t_1 = 1$), person 2 after four years ($t_2 = 4$), and person 3 is observed four years without experiencing an event (a censored observation at $t_3 = 4$).

Altogether, the data set includes $N = \sum_{i=1}^{n} t_i = 9$ records. The dummy variable event indicates whether the corresponding record ends with an event (event = 1) or not (event = 0). In the likelihood function (5.42), you can think about this dummy variable as a switch that turns on the right terms depending on whether the record indicates an event or a survival. For example, for person 2, who experiences an event at $t_2 = 4$, the following likelihood contribution that equals the general format of likelihood contributions for events (5.41) results:

$$L_2 = \left[\left(1 - h_{2p}(1)\right)^{(1-0)} \cdot h_{2p}(1)^0\right] \cdot \left[\left(1 - h_{2p}(2)\right)^{(1-0)} \cdot h_{2p}(2)^0\right]$$
$$\cdot \left[\left(1 - h_{2p}(3)\right)^{(1-0)} \cdot h_{2p}(3)^0\right] \cdot \left[\left(1 - h_{2p}(4)\right)^{(1-1)} \cdot h_{2p}(4)^1\right]$$
$$= \left(1 - h_{2p}(1)\right) \cdot \left(1 - h_{2p}(2)\right) \cdot \left(1 - h_{2p}(3)\right) \cdot h_{2p}(4)$$

5.2 Modeling the Change of Y: Discrete-Time Event History Models for Panel Data

Table 5.13 Hypothetical data set for estimating discrete-time event history data

ID	Wave	Duration	Event	Number of children	Labor force experience	Age group	Poor health
1	1	1	1	0	30	60–61	no
2	1	4	0	2	15	58–59	no
2	2	4	0	2	15	60–61	yes
2	3	4	0	2	15	60–61	yes
2	4	4	1	2	15	62–63	yes
3	1	4	0	1	27	54–57	no
3	2	4	0	1	27	58–59	yes
3	3	4	0	1	27	58–59	no
3	4	4	0	1	27	60–61	no

For person 3, a unit without an event, the likelihood contribution equals the general format of likelihood contributions for censored observations (5.40):

$$L_3 = \left[(1 - h_{3p}(1))^{(1-0)} \cdot h_{3p}(1)^0\right] \cdot \left[(1 - h_{3p}(2))^{(1-0)} \cdot h_{3p}(2)^0\right]$$
$$\cdot \left[(1 - h_{3p}(3))^{(1-0)} \cdot h_{3p}(3)^0\right] \cdot \left[(1 - h_{3p}(4))^{(1-0)} \cdot h_{3p}(4)^0\right]$$
$$= (1 - h_{3p}(1)) \cdot (1 - h_{3p}(2)) \cdot (1 - h_{3p}(3)) \cdot (1 - h_{3p}(4))$$

Hence, the basic idea of organizing discrete-time event histories is to split each spell into as many data records as there are discrete observations over time. With panel data that are organized in long format (see Sect. 2.1), this is an easy task. You only have to delete all measurements after the event of interest has occurred. So, for example, let us assume that the hypothetical data from Table 5.13 come from a four-year panel. Person 1 retired after one year and hence, all measurements following the first measurement are dropped from the discrete-time event history data set. Note also that this format allows us to use all the different values that are observed for time-varying explanatory variables during the observation period (see the variables "age group" and "poor health" in Table 5.13).

By inserting the regression model (5.37) for the hazard rates into the likelihood function

$$L = \prod_{i=1}^{n}\prod_{t=1}^{t_i}\left(1 - \frac{\exp(\beta_0(t) + \beta_1 x_{1it} + \cdots + \beta_k x_{kit} + \gamma_1 z_{1i} + \cdots + \gamma_j z_{ji})}{1 + \exp(\beta_0(t) + \beta_1 x_{1it} + \cdots + \beta_k x_{kit} + \gamma_1 z_{1i} + \cdots + \gamma_j z_{ji})}\right)^{(1-event_{it})}$$
$$\cdot \left(\frac{\exp(\beta_0(t) + \beta_1 x_{1it} + \cdots + \beta_k x_{kit} + \gamma_1 z_{1i} + \cdots + \gamma_j z_{ji})}{1 + \exp(\beta_0(t) + \beta_1 x_{1it} + \cdots + \beta_k x_{kit} + \gamma_1 z_{1i} + \cdots + \gamma_j z_{ji})}\right)^{event_{it}} \quad (5.43)$$

we can use explanatory variables X and Z and functions of time $\beta_0(t)$ to model how the hazard varies between units with different characteristics X and Z and over time T. Depending on the hypotheses about time dependence, one may either introduce duration as a continuous variable (e.g., a linear duration effect) or include a set of dummy variables (discontinuous duration effect). For instance, you could use $\beta_0(t) = \alpha_0 \ln(t)$ if you assume that the hazard rate is linearly falling or linearly increasing with process time T.[31] Alternatively, you could use dummy variables for different time periods during the observation period: $\beta_0(t) = \alpha_0 + \alpha_1 d_1 + \cdots + \alpha_l d_l$. Within these time periods, the hazard does not change, while it can be different between time periods. Do not forget that for model identification, you have to exclude one time period (the reference period, during which the hazard rate equals α_0). It is also important that each time period includes at least one event, otherwise you will not be able to estimate the effect of the corresponding dummy. Quite generally, the decision of the functional form is up to the analyst who has to find good theoretical arguments as to why he or she expects a certain functional form, describing the process under study.

Estimates of the model parameters are found by maximizing the likelihood function (5.43) (or its natural logarithm, which is easier) with respect to the parameters α, β, and γ. This can be done by any kind of statistical program that offers routines for maximizing user-defined likelihood functions. This usually implies a little bit of programming for the user. However, a closer look at (5.43) shows that this cumbersome solution is not necessary. If we compare it with the likelihood (5.10) for the logistic regression model, we see that both functions are structurally equivalent, except that the latter one includes the unconditional probability of observing a certain category of the dependent variable, while the former one includes the conditional transition probability.[32] While the likelihood for the logistic regression model uses the (categorical) dependent variable itself to switch on and off the correct terms in the likelihood function, we now need an additional indicator variable event to do this job for us. But obviously, we can use any logistic regression program to estimate the parameters of our discrete-time hazard model, if we organize the data like in our small hypothetical data set (see Table 5.13) and use an indicator variable that measures events as the dependent variable.

But beware! We are using a program that has been made for other purposes. We are taking advantage of its numerical capabilities, but we should not uncritically accept all terms in its output. For example, if it names parameter estimates "odds ratios", this is fine for the logistic regression model, but we are focusing on hazard rates and hazard ratios (see the discussion below). You should note that parameter estimates now tell us the effect of the corresponding explanatory variable on the conditional transition probability (resp. hazard rate) and not on the overall probability of observing an event.

---

[31] Because the regression model is non-linear, use $\ln(t)$ and not t as the independent variable to estimate a linear effect of process time.

[32] The fact that (5.10) includes a fixed number of observations over time, T, which is identical for all units, while (5.42) includes a varying number of observations, t_i, which is different for each unit i is no real difference. In the case of an unbalanced panel, T in (5.10) would have to be individualized as well (see footnote 8).

5.2 Modeling the Change of Y: Discrete-Time Event History Models for Panel Data

More specifically, let us define $h_{ip}(t)/(1-h_{ip}(t))$ as the odds of experiencing the event at time point t versus "surviving" time point t without an event (for short, the *odds of experiencing an event* at t). Estimates resulting from a logistic discrete-time hazard model:

$$\ln\left(\frac{h_{ip}(t)}{1-h_{ip}(t)}\right) = \beta_0(t) + \beta_1 x_{1it} + \cdots + \beta_k x_{kit} + \gamma_1 z_{1i} + \cdots + \gamma_j z_{ji} \quad (5.44)$$

provide the change in the *log-odds of experiencing an event* at time point t. This is a quantity, which is even more difficult to communicate than the logit in logistic regression. If you want to make statements about the hazard rate, you have to insert your estimates into (5.37) and make forecasts using values of the independent variables X, Z, and T. These forecasts can be nicely illustrated in conditional effect plots (see Fig. 5.7 on p. 266). More specifically, the marginal effect of a time-varying variable X_k on the hazard rate at time point t equals

$$\frac{\partial h_{ip}(t)}{\partial x_{ki}} = \beta_k \cdot h_{ip}(t) \cdot (1 - h_{ip}(t)) \quad (5.45)$$

It is a function of the corresponding parameter estimate β_k and the values of all variables in the model at time point t, which enter into the equation through $h_{ip}(t)$ (the marginal effects of Z and T are computed similarly). From (5.45), we conclude that a negative sign of a coefficient implies that the hazard rate decreases with rising values of the corresponding explanatory variable. In turn, a positive sign indicates that the hazard increases. If the coefficient is equal to zero, the explanatory variable has no effect on the hazard.

Alternatively, you can focus on the odds of experiencing an event:

$$\frac{h_{ip}(t)}{1-h_{ip}(t)} = \exp(\beta_0(t) + \beta_1 x_{1it} + \cdots + \beta_k x_{kit} + \gamma_1 z_{1i} + \cdots + \gamma_j z_{ji})$$

$$= \exp(\beta_0(t)) \cdot \exp(\beta_1)^{x_{1it}} \cdot \ldots \cdot \exp(\beta_k)^{x_{kit}} \cdot \exp(\gamma_1)^{z_{1i}} \cdot \ldots \cdot \exp(\gamma_j)^{z_{ji}} \quad (5.46)$$

by computing the antilogarithms $\exp(\beta_0(t))$, $\exp(\beta_1)$, ..., $\exp(\gamma_j)$ of the model parameters. Similar to odds ratios (see Sect. 5.1.1.2), they are interpreted as *hazard ratios*. They measure how much the hazard changes if the corresponding explanatory variable increases by one unit. Note again that it is a multiplicative model and hence, change is measured as multiplicative change (the multiplicant equals the antilogarithm, say $\exp(\beta_1)$) or, equivalently, percentage change: $(\exp(\beta_1) - 1) \cdot 100$. An antilogarithm equal to 1 means that the corresponding explanatory variable has no effect.

Finally, there is one last subtle but important difference to the logistic panel regression model. The unit of analysis in event history analysis is the spell and not the repeated measurements over time. From this point of view, the hypothetical data in Table 5.13 include only three cases. The fact that we split them up into small pieces of one year does not matter here. This has been done to model changes of the hazard during the duration of the spell (with the side effect of being able to include all values of time-varying explanatory variables that have been measured during the duration of the spell). The likelihood is still based on all (censored and non-censored) spells and when their contribution to the likelihood is computed the small pieces of information are again put together (see (5.40) and (5.41)).

This has important implications about how we think about the single records in the data set. While they represented possibly serially correlated, but separate measurements in the case of logistic panel regression models, they now represent consecutive information about an underlying unit of analysis (i.e., the spell). Necessarily, they belong together and hence, we do not have to control for their serial correlation. This would be different if we were to analyze repeatable events. Let us assume that each individual contributes $s = 1, \ldots, S_i$ spells to the analysis (e.g., in an analysis of job durations). The likelihood would then have to be evaluated over all individuals and all spells:

$$L = \prod_{i=1}^{n} \prod_{s=1}^{S_i} \prod_{t=1}^{t_{si}} \left(1 - h_{ip}(t)\right)^{(1-event_{ist})} \cdot h_{ip}(t)^{event_{ist}} \tag{5.47}$$

Possibly, spells belonging to one individual are not independent of each other. For example, if a person is unqualified, he faces more labor market risks than other individuals and hence, spell durations will be comparatively short for that person. The qualification of a person is easily controlled for in the corresponding discrete-time hazard model, but probably other unknown factors exist at the level of individuals that contribute to the serial correlation of the spells within each individual's event history. In this case, we need more specialized models that control for the serial correlation due to unknown variables at the level of individuals (see also Sect. 5.2.3.4). We repeat, however, that it is a serial correlation between spells and not between yearly measurements.

5.2.3 Applying the Discrete-Time Event History Model

In this section, we apply the discrete-time event history model to real data. It is a model developed for the analysis of singular events like the change from employment to retirement in the hank data (see Example 5.2). We start our discussion in Sect. 5.2.3.1 with the simple case of non-repeatable singular events. Retirement is obviously such a non-repeatable event (usually, you retire once and not several times). Then we take up again the question of unobserved heterogeneity, which—as it turns out—in event history models, results in biased estimates, even if unobserved heterogeneity is uncorrelated with the variables in the model (Sect. 5.2.3.2). Following this discussion, we introduce random and fixed effects into our discrete-time event history model (Sects. 5.2.3.3 and 5.2.3.4). We discuss whether and how both extensions deal with the adverse effects of unobserved heterogeneity. Finally, since the assumption of a discrete change process makes sense for discrete (panel)

5.2 Modeling the Change of Y: Discrete-Time Event History Models for Panel Data

measurements, but not for the underlying process itself, we ask how we can adapt continuous-time event history models to data that—due to the panel design—have been measured only very imprecisely (Sect. 5.2.3.5).

5.2.3.1 Non-repeatable Singular Events

Consider again the research example by Hank (2004) and suppose we wish to use the data from Example 5.2 to derive his reported estimation results. For this illustration, we begin by requesting the estimates for a parsimonious specification that controls for individual characteristics (education and employment history), household characteristics, and the reproductive history of the women. The conditional probability for the transition into retirement in year t, $h_{ip}(t)$, is assumed to take the form

$$\ln\left(\frac{h_{ip}(t)}{1-h_{ip}(t)}\right) = \alpha_0 + \alpha_1 d_{i1} + \cdots + \alpha_6 d_{i6} + \beta_1 x_{1it} + \cdots + \beta_k x_{kit} + \gamma_1 z_{1i} + \cdots + \gamma_j z_{ji} \tag{5.48}$$

where the set of α's captures the underlying process time (age). We use a set of age dummies and assume that the transition rates are constant in each of the age intervals but can change between them. The results of this discrete-time event history model are summarized in Table 5.14. Columns 2 and 4 show the coefficients and hazard ratios, while standard errors are given in Columns 3 and 5.

There is evidence for the impact of a woman's reproductive history, with women having children retiring later. The parameter estimate for the variable number of children is -0.098. You can take the antilogarithm of this estimate, $\exp(-0.098) = 0.907$, and interpret the resulting hazard ratio. As discussed in Sect. 5.2.2.3, it is then possible to look at the percentage change in the ratio. For our example, we get $(\exp(-0.098) - 1) \cdot 100 = -9.335$. This tells you that with each additional child, the hazard of retiring decreases by a factor of 0.907 or about 9.3 %. In addition, the transition into retirement is also influenced by age, health status, marital status, and labor force experience. The other explanatory variables, however, do not play a significant role in the retirement decision.

Let us now look at the result of some other specifications and see how different models can be tested against each other. Our goal is to replace the explanatory variable number of children with other measures that grasp different aspects of women's reproductive history. All other control variables are left unchanged. In Hank (2004), the results for four additional specifications are reported. Specification 1 controls for the number of children (see Table 5.14). Specification 2 includes instead a binary indicator of whether a woman ever had a child. Specification 3 captures the effect of age at first birth, while specification 4 focuses on mother's labor force participation. Finally, the last specification differs from specifications 1 to 4 in that it controls for various interaction effects.

Which one of the alternative models fits the data better? When models are nested, you can use likelihood ratio tests that compare the likelihood of a current model, which controls for explanatory variables whose effects you would like to evaluate, and the likelihood of a restricted model, which does not control for these variables (see Sect. 7.2.2).

Table 5.14 Determinants of female retirement (discrete-time logistic hazard model)

Variable	Estimate	Std. Err.	Hazard Ratio	Std. Err.
No. of children	−0.0976	0.0481	0.9070	0.0436
Age group				
54–57 years	1.1275	0.2698	3.0880	0.8332
58–59 years	1.5948	0.2871	4.9276	1.4146
60–61 years	3.9589	0.2570	52.4005	13.4680
62–63 years	2.7524	0.3075	15.6795	4.8222
64–69 years	3.2638	0.2905	26.1478	7.5948
Educational attainment				
Vocational degree	−0.1280	0.1267	0.8799	0.1115
University degree	−0.4709	0.2924	0.6244	0.1826
Imputation flag: Education	0.1729	0.4931	1.1888	0.5862
Labor force career				
Years in labor force at age 50	0.0192	0.0067	1.0194	0.0068
No. of employment spells	0.0494	0.0551	1.0507	0.0579
Partner				
Partner, not retired	−1.1598	0.1890	0.3136	0.0592
Partner, retired	−0.3349	0.1691	0.7154	0.1210
Imputation flag: Partner	0.8165	0.2447	2.2625	0.5537
Income				
Household income	0.0032	0.0098	1.0032	0.0098
Squared household income	0.0000	0.0000	1.0000	0.0000
Imputation flag: Income	0.1951	0.2519	1.2154	0.3061
Other personal characteristics				
Home owner	−0.0850	0.1278	0.9185	0.1174
Poor health	0.3699	0.1681	1.4476	0.2433
Person needing care	0.1284	0.2505	1.1370	0.2849
Constant	−4.3016	0.4008		
LR			755.65	
df			20	
ln L			−1,038.5887	
N			5,765	
n			837	
T			18	
Events			386	

Source: hank data (see Example 5.2)

5.2 Modeling the Change of Y: Discrete-Time Event History Models for Panel Data

Table 5.15 Determinants of female retirement (results of competing model specifications)

Variable	Model 1	Model 2	Model 3	Model 4	Model 5
Number of children	−0.098∗				
Ever had a child		−0.457∗			
Early first birth			−0.339		
Late first birth			−0.533∗		
Child young, mother employed				−0.558∗	
Child young, mother not employed				−0.411	
Early 1st birth/employed					−0.484
Early 1st birth/not employed					−0.260
Late 1st birth/employed					−0.618∗
Late 1st birth/not employed					−0.495∗
N	5,765	5,765	5,765	5,765	5,765
n	837	837	837	837	837
ln L	−1,038.589	−1,038.361	−1,037.287	−1,037.915	−1,036.640
AIC	2,119.177	2,118.723	2,118.575	2,119.831	2,121.279
BIC	2,259.028	2,258.574	2,265.085	2,266.341	2,281.109

Notes: ∗ $p < 0.05$; ∗∗ $p < 0.01$; ∗∗∗ $p < 0.001$
Each model controls for age, educational attainment, labor force career, partner, income and other personal characteristics
Source: hank data (see Example 5.2)

In this example, the models are not nested, however, and so the likelihood ratio test cannot be used. A common approach to this problem is to use either Akaike's information criterion (*AIC*) or the Bayesian Information Criterion (*BIC*) (see Sect. 7.2.2). In order to demonstrate the application of these statistics, we calculate both information criteria. The results of this procedure are summarized in Table 5.15. Based on these findings, we conclude that model 2 should be preferred over the other specifications because it has the smallest *BIC* and the second smallest *AIC* value. Differences between the measures of fit are small, however.

Sometimes researchers are also interested in making predictions. For instance, it is possible to predict the conditional transition probability for each person by using (5.48). As an illustration, consider the case of a woman who is 60 years of age, with a university degree, 27 years of labor force experience (at age 50), two employment spells, a monthly net household income of 720 DM, and a child. She is a single person, has no health problems, and lives in a rented apartment without persons needing care (i.e., all dummies are assumed to be zero). Substituting variables and parameter estimates in (5.48), the predicted conditional transition probability (resp. hazard rate) for this woman is 0.432. Now imagine that this woman was younger, say 55 years old, and predict the hazard rate using the same explanatory variables as were used before. In what way do the results change? The fitted hazard rate is now

Fig. 5.7 Predicted hazard rate of female retirement (using estimates of model 1)

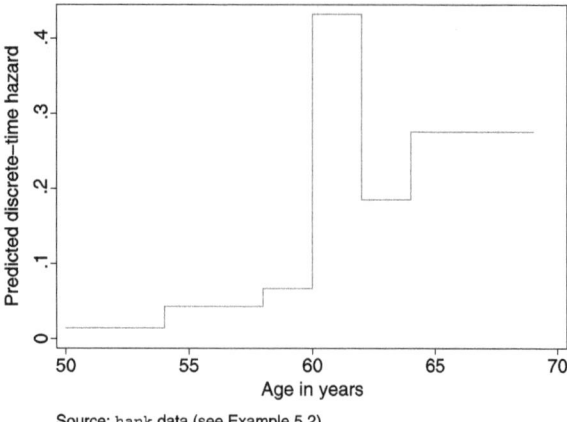

Source: hank data (see Example 5.2)

much smaller: 0.043. Repeat this exercise, but this time, assume that the 60-year-old woman has three children. The predicted hazard rate for this constellation is 0.385. The latter result is in line with the hypothesis that mothers decide to work longer, while the former result indicates that the conditional probability to retire increases with age.

We can also use this simple example to show how to plot the hazard rate in order to characterize the shape of the hazard function. Reconsider the case of a single woman with a university degree, 27 years of labor force experience (at age 50), two employment spells, a monthly net household income of 720 DM, and one child, without health problems and living in a rented apartment without persons needing care. Figure 5.7 shows how the predicted hazard rate varies with respect to age and the process time. According to our specification of a semi-parametric model with age dummies, hazard rates are constant in each of the six age intervals, but can change between them. It can be easily seen that hazard rates increase slowly between age 50 and 60, peak at the ages of 60–61 due to early retirement pathways, decline thereafter, and increase again at about the official retirement age.

5.2.3.2 Unobserved Heterogeneity in Event History Models

Until now, the discussion of the discrete-time logistic hazard model has assumed that all relevant explanatory variables are included in the regression equation and measurement error is not present. In practice, however, this assumption is rarely met. As an example, the regression model presented by Hank (2004) leaves out some variables of interest, such as occupational status, branch of industry, gender roles, or intergenerational financial transfers. Similar to the panel methods focusing on the level of the dependent variable, neglecting relevant explanatory variables in change models such as event history models will result in unobserved heterogeneity. Omitted variable bias is a special case of unobserved heterogeneity when there is correlation between the explanatory variables included in the regression model and the neglected explanatory variables. Section 4.1.2 has elaborated the consequences of omitted variable bias in the analysis of continuous panel data. In event history

5.2 Modeling the Change of Y: Discrete-Time Event History Models for Panel Data

analysis, the problem of unobserved heterogeneity is even more serious: Neglecting relevant explanatory variables may distort the results of your estimation model irrespective of whether there is correlation between the observed and unobserved explanatory variables or not. More precisely, ignoring unobserved heterogeneity in event history modeling may have different ramifications, such as downward bias in duration dependence, distorted effects of the explanatory variables, dependent censoring, dependent multiple events, and dependent repeatable events. This section focuses on the consequences of unobserved heterogeneity on the hazard rate and the effects of the explanatory variables. More detailed discussion on the topics can be found at the end of this chapter in Sect. 5.2.4. In order to exemplify the consequences of unobserved heterogeneity, consider the following simple illustrative example.

Example 5.3 (cancer data) Assume a data set resulting from the following simple experiment in medicine. Cancer patients are randomly assigned to a control group and a treatment group. The two groups of individuals suffering from cancer can be distinguished by a variable Z_1, with $z_1 = 0$ for patients receiving a conventional cancer treatment and $z_1 = 1$ for cancer patients receiving a new cancer-directed therapy. At the beginning of the experiment ($t = 0$), the proportions of these two groups in the total sample are 0.5 and the experiment runs over a period of $T = 60$ months. Our interest lies in the analysis of the mortality hazard and, more particularly, the effectivity of the new cancer treatment.

Suppose now that the true hazard rates for these two groups are constant over time, but patients with $z_1 = 0$ show a two times larger hazard of dying than individuals with $z_1 = 1$:

$$\frac{h_{ip}(t|z_1=0)}{h_{ip}(t|z_1=1)} = 2$$

Furthermore, let us as assume that there is also variation within the control group and the treatment group since the sample includes patients with recurrent cancer. Consider a second variable Z_2, which captures this information and assume that the proportions of patients with and without recurrent cancer are 0.5 within the control group and the treatment group, respectively. The hazard rate of dying is still constant over time. Yet, regardless of whether patients receive a conventional cancer therapy or the new therapy, the hazard rate of dying is two times higher for patients with recurrent cancer ($z_2 = 1$):

$$\frac{h_{ip}(t|z_1, z_2=1)}{h_{ip}(t|z_1, z_2=0)} = 2$$

All in all, the cancer data consist of four equally sized subpopulations and Table 5.16 shows the true hazard rates in each group. We now want to show how

Table 5.16 True hazard rates

	Control group ($z_1 = 0$)		Treatment group ($z_1 = 1$)	
	First-time cancer ($z_2 = 0$)	Recurrent cancer ($z_2 = 1$)	First-time cancer ($z_2 = 0$)	Recurrent cancer ($z_2 = 1$)
$h_{ip}(t)$	0.02	0.04	0.01	0.02

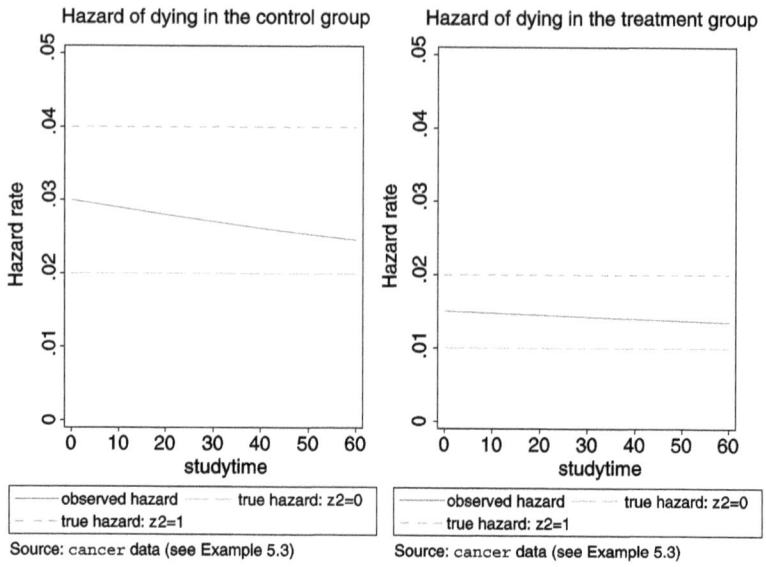

Fig. 5.8 True and observed hazard rate of dying for the control and treatment group

unobserved heterogeneity may bias the observed hazard rate. To this end, suppose that we only know Z_1 (i.e., whether patients receive a conventional cancer therapy or rather a new treatment). But we do not know that the mortality risk of these patients also depends on recurrence of cancer.

Figure 5.8 gives you a picture of the resulting hazard rates for patients belonging to the control group and treatment group, respectively. The observed hazard rates in these two groups differ from the true hazard rates in the subpopulations: The hazard rates decline even though the hazard of dying is constant within each group.[33] This shape of the hazard rates is simply due to unobserved heterogeneity. The reason for this is that the proportion of patients at risk of dying of cancer changes over time, since patients with recurrent cancer die earlier than patients who suffer from cancer for the first time. For example, one year after the start of the experiment at $t = 12$, the survival probability for a patient who receives a conventional cancer treatment and suffers from cancer for the first time is

---

[33] Since we are assuming a discrete-time process, the observed hazard functions in Figs. 5.8 and 5.9 should be decreasing step functions. To simplify both figures, we have used continuously falling functions that approximate the respective step functions.

5.2 Modeling the Change of Y: Discrete-Time Event History Models for Panel Data

$$S(t = 12|z_1 = 0, z_2 = 0) = \left(1 - h_{ip}(1)\right) \cdot \left(1 - h_{ip}(2)\right) \ldots \left(1 - h_{ip}(12)\right) = 0.7847$$

Similarly, for patients with recurrent cancer, we get

$$S(t = 12|z_1 = 0, z_2 = 1) = \left(1 - h_{ip}(1)\right) \cdot \left(1 - h_{ip}(2)\right) \ldots \left(1 - h_{ip}(12)\right) = 0.6127$$

To put it into words, $(1 - 0.6127) \cdot 100 = 38.73$ % of the patients with recurrent cancer have already died within the first 12 months of the experiment, whereas only $(1 - 0.7866) \cdot 100 = 21.34$ % of the patients who suffer from cancer for the first time died during this period. But this means that the risk set at time $t = 12$ includes significantly more patients who suffer from cancer for the first time and hence, have a lower hazard of dying. To put it in figures, suppose there are 50 patients with recurrent cancer and 50 patients with first-time cancer in the control group at the beginning of the experiment. Then, 19 and 11 patients already died within the first year, respectively. So 70 patients are still at risk of dying, but the proportion of patients with and without recurrent cancer has changed and the risk set consists of 44 % patients with recurrent cancer and 56 % who felt ill with cancer for the first time. Hence, as time passes, the sample increasingly consists of first-time cancer patients, who have a half as large hazard rate compared to the recurrent cancer patients. As a consequence of this selection effect, an event history model results in a decreasing hazard rate even if the true individual hazards in both groups are constant. The same downward bias of the hazard rate can also be observed in the treatment group (see right panel of Fig. 5.8).

In our example, it is assumed that the true hazard rates in the control and treatment group are constant. In other applications, the true hazard rates in the subpopulations might also be monotonically increasing, falling, or at first rising and then falling, etc. Neglecting unobserved heterogeneity leads the hazard rate that is estimated for the total population to be a mixture of different (unobserved) hazard rates in subpopulations. From what we have just learned, we know that the potential bias from unobserved heterogeneity depends on the selection effect in the risk population. The composition of the risk set changes continuously over time and subpopulations with a higher risk of event occurrence leave the risk set earlier than subpopulations with a lower hazard rate. If our model fails to control for these different hazard rates in subpopulations, the risk set will be more and more composed of the low-risk populations. As a consequence, models suffering from unobserved heterogeneity will overestimate negative duration dependence and underestimate positive duration dependence.

A second lesson from the example just cited above is that the effects of the explanatory variables will also be biased. Recall from above that the true hazard ratio for patients in the control and treatment group is constant and equal to $h_{ip}(t|z_1 = 0)/h_{ip}(t|z_1 = 1) = 2$. Figure 5.9 illustrates that the ratio of the observed (estimated) hazard rates for patients in these two groups is not constant, but decreases over time. At the beginning of the experiment, when the population proportions in the four subgroups are of equal size, the ratio of the estimated hazards amounts to $\hat{h}_{ip}(t = 0|z_1 = 0)/\hat{h}_{ip}(t = 0|z_1 = 1) = 2$. But due to selective mortality, one year after we obtain $\hat{h}_{ip}(t = 12|z_1 = 0)/\hat{h}_{ip}(t = 12|z_1 = 1) = 1.96$, and at

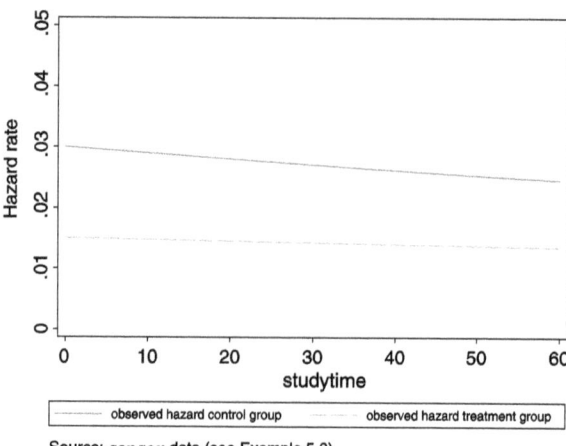

Fig. 5.9 Observed hazard rate of dying for the treatment and control group

the end of the experiment, we obtain $\hat{h}_{ip}(t=60|z_1=0)/\hat{h}_{ip}(t=60|z_1=1) = 1.81$. Thus, if the regression model fails to control for unobserved heterogeneity, we will underestimate the true effect of Z_1.

There are more issues involved here which merit mention before concluding this section. As illustrated above, if our model fails to control for the fact that there are different hazard rates in subpopulations, time dependence can be considered an expression of unobserved heterogeneity. But this means that changing the set of explanatory variables in the event history model will also lead to a change in duration dependence. At the same time, including additional explanatory variables is also likely to result in a change of the parameter estimates of the variables already in the regression model, even if the new variables are uncorrelated with these variables (a phenomenon that is different from traditional regression models). Furthermore, if the values of a time-dependent explanatory variable and the dependent process itself are influenced by the same unobserved individual factors, the analysis will yield spurious effects. A final consequence of unobserved heterogeneity is dependent censoring. Section 5.2.1 introduced different types of censoring and emphasized that the driving mechanisms behind censoring need to be independent of the process of interest. It may well be that some unobserved variables affect the process under study and the censoring process. In such cases, censoring is non-ignorable and consequently, the likelihood function presented in Sect. 5.2.2.3 is invalid.

An important guideline to remedy the problem of unobserved heterogeneity is using reliable and comprehensive panel data for model building. However, in various applications, we are not able to control for all relevant factors. In other applications, our theoretical model does not include all important factors. Section 5.1.2 showed how RE and FE estimation help to sort out unobserved heterogeneity when modeling the level of Y. The following two sections elaborate upon the question of whether similar approaches can be used in event history modeling.

5.2.3.3 Uncorrelated Heterogeneity: Random Effects Event History Models

Let us start with the more simple case in which unobserved heterogeneity is assumed to be uncorrelated with the variables in the regression model. Reconsider the logistic discrete-time hazard model and re-formulate (5.44), now allowing for unobserved heterogeneity in such a way that u_i captures unobserved unit-specific risk factors. The model would then be:

$$\ln\left(\frac{h_{ip}(t)}{1-h_{ip}(t)}\right) = \beta_0(t) + \beta_1 x_{1it} + \cdots + \beta_k x_{kit} + \gamma_1 z_{1i} + \cdots + \gamma_j z_{ji} + u_i \quad (5.49)$$

Again, the unit-specific effects u_i are assumed to be random and independent of X and Z. You have to assume a certain distribution for the u_i. One way to do this is to make the assumption that the u_i are normally distributed. This is the RE logistic regression model developed in Sect. 5.1.2.2.

Let us return to our Example 5.2. Again, we fit the basic regression model shown above, but this time we control for unobserved heterogeneity, assuming normally distributed random effects (i.e., instead of a standard logistic regression, apply an RE logistic regression to the event history data set). Table 5.17 presents the results. To check whether the model controlling for unobserved heterogeneity should be preferred to the reference model, we look at the estimate of the standard deviation of unobserved heterogeneity, $\hat{\sigma}_u$, and the estimated proportion of total variance contributed by unobserved heterogeneity $\hat{\rho}$ (compare (5.30)). When $\hat{\rho}$ is zero, unobserved heterogeneity is unimportant. The logistic hazard model and the RE logistic hazard model are hierarchically nested since U is restricted to be 0 in the former model (and hence, $\rho = 0$). This allows us to use the likelihood ratio test to test the null hypothesis that ρ is zero. For the `hank` data, we can calculate the test statistic as two times the difference of the log likelihoods ($LR = 8.74$ with $df = 1$). Given a significance level of 0.01, we conclude that the null hypothesis should be rejected. In other words, there is evidence for unobserved heterogeneity in the `hank` data.

What are the implications for our parameter estimates? First, let us start with the estimates of duration dependence. Theoretically, one would expect retirement to increase as employees gradually approach the official retirement age. Hence, one would expect some kind of positive duration dependence. In the last section, we concluded that models suffering from unobserved heterogeneity will overestimate negative duration dependence and underestimate positive duration dependence. In other words, if the expectation is positive duration dependence, the positive effect should be larger in models controlling for unobserved heterogeneity than in models that ignore unobserved heterogeneity. As the estimates of the age intervals in Table 5.17 show, in both specifications, the predicted hazard rate increases with age (the process time), but the positive duration dependence is somewhat larger in the RE model than in the reference model. Reconsider our example used to generate the conditional effects plot presented in Fig. 5.7. We now repeat this exercise with the results obtained from the RE model. Figure 5.10 compares the predicted hazard from our reference model and the RE model and visualizes that differences between

Table 5.17 Determinants of female retirement (logistic and RE logistic hazard model)

Variable	Logistic model		RE logistic model	
	Estimate	Std. Err.	Estimate	Std. Err.
Number of children	−0.0976	0.0481	−0.1531	0.0694
Age group				
54–57 years	1.1275	0.2698	1.2922	0.3033
58–59 years	1.5948	0.2871	1.9267	0.3666
60–61 years	3.9589	0.2570	4.8289	0.5345
62–63 years	2.7524	0.3075	3.8055	0.6446
64–69 years	3.2638	0.2905	4.6743	0.7782
Educational attainment				
Vocational degree	−0.1280	0.1267	−0.1659	0.1750
University degree	−0.4709	0.2924	−0.7077	0.4090
Imputation flag: Education	0.1729	0.4931	0.2983	0.6781
Labor force career				
Years in labor force at age 50	0.0192	0.0067	0.0223	0.0094
Number of employment spells	0.0494	0.0551	0.0564	0.0754
Partner				
Partner, not retired	−1.1598	0.1890	−1.5950	0.3112
Partner, retired	−0.3349	0.1691	−0.5702	0.2445
Imputation flag: Partner	0.8165	0.2447	1.1733	0.3542
Income				
Household income	0.0032	0.0098	0.0020	0.0112
Squared household income	0.0000	0.0000	0.0000	0.0000
Imputation flag: Income	0.1951	0.2519	0.1861	0.3021
Other personal characteristics				
Home owner	−0.0850	0.1278	−0.1374	0.1727
Poor health	0.3699	0.1681	0.4559	0.1969
Person needing care	0.1284	0.2505	0.2941	0.3148
Constant	−4.3016	0.4008	−4.6477	0.5571
LR or X_1^2	755.65		177.08	
df	20		20	
ln L	−1,038.5887		−1,034.2207	
σ_u			1.1741	0.3491
ρ			0.2953	0.1238
N	5,765		5,765	
n			837	
T	18		18	
Events	386		386	

Source: hank data (see Example 5.2)

5.2 Modeling the Change of Y: Discrete-Time Event History Models for Panel Data

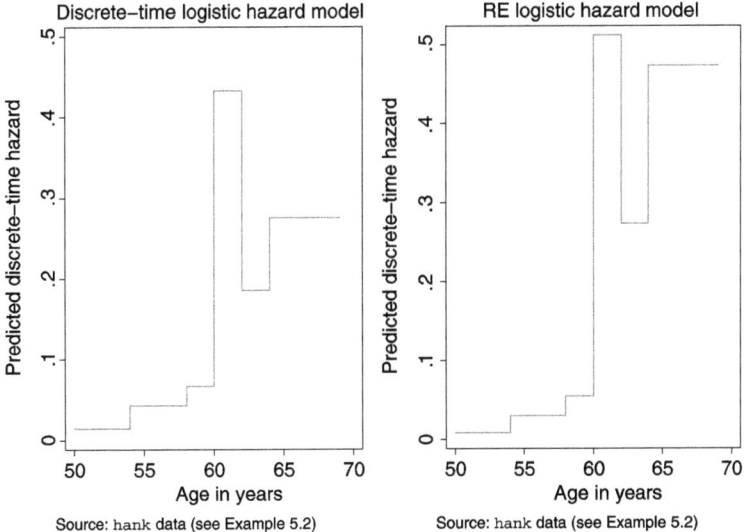

Fig. 5.10 Predicted hazard of retiring

the logistic hazard model and the RE logistic hazard model appear to be particularly large at older ages.

Second, the estimated coefficient of the explanatory variable number of children is larger in magnitude than the corresponding coefficient in the reference model. In the RE logistic hazard model, we obtain $(\exp(-0.1531) - 1) \cdot 100 = -14.196$ as compared to $(\exp(-0.098) - 1) \cdot 100 = -9.335$ in the logistic model that does not control for unobserved heterogeneity.[34] This is also in line with the discussion in the former section, in which we concluded that unobserved heterogeneity possibly leads to downward biased estimates of the effects of X and Z. Similarly, we observe the same downward bias for all other explanatory variables, with the only exception being the variables controlling for income. All in all, the RE estimates give you a clear example of how unobserved heterogeneity may bias duration dependence and result in underestimated regression coefficients.

The problem of RE event history models is the assumption that unobserved heterogeneity is independent of the variables included in the regression model. Hence,

---

[34] Notice, however, that the rules of interpretation differ somewhat for RE models. To exemplify this, consider once more our variable of interest, namely, number of children. In the logistic hazard model, the parameter estimate for this variable can be interpreted as the population averaged effect. It shows you to what extent the hazard of retiring differs for two randomly selected women, with one woman having one more child than the other woman, holding constant all other explanatory variables in the model. In the RE model, we control for unobserved unit-specific risk factors. When interpreting the results of the RE logistic hazard model, it is therefore important to keep in mind that the parameter estimates are now unit-specific. In other words, we would then conclude that for two women with the same random effect, the hazard of retiring decreases by a factor of 14 % with an additional child.

if it is reasonable to assume that your model suffers from correlated unobserved heterogeneity, RE event history models are not the appropriate strategy to tackle this problem. The next section raises the question of whether there are FE models available for event history analysis.

5.2.3.4 Correlated Heterogeneity: Fixed Effects Event History Models

As the previous chapter has demonstrated, an attractive feature of the FE model is that it allows to control for all time-constant, unit-specific effects. Hence, omitting time-constant variables cannot distort the results from FE regression. Quite the opposite is true in the random effects model, which presumes that all unit-specific effects are uncorrelated with the explanatory variables and thus cannot control for unmeasured, stable, unit-specific characteristics that are correlated with the variables in the model. The precondition for FE regression is, however, variation within observations, necessitating at least two measurements per unit of analysis, which, in the case of event histories, means at least two spells.[35] This is the case when analyzing multi-episode data from individuals who experience repeatedly events over the observation period (see Sect. 5.2.4). FE regression is not feasible when analyzing non-repeatable events (e.g., retirement or death) or when confining the analysis to first events only (e.g., the duration of the first job). Since each unit experiences only one single event, within-individual comparisons are rendered impossible. Hence, if only one spell is available, special techniques are necessary, such as the case-crossover and the case-time-control design. Both designs have been used in epidemiological research to control for unobserved characteristics in case-control studies (Maclure, 1991; Suissa, 1995). Allison and Christakis (2006) have applied these techniques to non-repeatable events. As the following discussion will show, both designs are applicable, but they have important pitfalls.

Let us exemplify the *case-crossover design* by once again turning to the hank data (see Example 5.2). Select all women who retire during the observation period (the "cases") and compare each woman with herself at different points in time (i.e., each woman acts as her own "control"). We then ask: Why did a woman retire in one particular year and not in any of the other years? In case of the hank data, women can retire in one of the 18 years under study. Think of a woman who reports in the fourth panel wave that she is retired. Technically, the former research question can also be posed in terms of sequences: What is the conditional probability of observing for this woman the sequence 0001 rather than any of the sequences observed in the data (1, 01, 001, 00001, etc.)? This is basically what the CML approach does. It conditions on all sequences that include the same number of ones as the sequence in question (see Sect. 5.1.2.1). However, compared to the analysis on levels, with event history data, these sequences are very specific: The ones appear only at the end of the sequence (and not somewhere in the middle). This has pros and cons.

---

[35] Up until now, we have used the term "measurement" to denote the repeated measurements (panel waves) over time. Here, the term "measurement" refers to the different spells that are part of an event history. For example, in an analysis of employment spells, an employee may have participated (and been observed) in three consecutive waves of a panel, but may have not quit his job and hence, is observed still in his first job.

5.2 Modeling the Change of Y: Discrete-Time Event History Models for Panel Data

On the one hand, each sequence indicates that a change occurred at some point in time (indicated by the one) after nothing has changed for several (panel) measurements over time (as indicated by the preceding zeros). Hence, to describe this and other sequences, we need the conditional transition probability $h_{ip}(t)$ and obviously, CML is the best approach to answer our research question. We specify again a logistic regression model for the conditional transition probability and add a term u_i:

$$\ln\left(\frac{h_{ip}(t)}{1-h_{ip}(t)}\right) = \beta_0(t) + \beta_1 x_{1it} + \cdots + \beta_k x_{kit} + \gamma_1 z_{1i} + \cdots + \gamma_j z_{ji} + u_i \quad (5.50)$$

u_i captures all unmeasured unit-specific effects that are stable over time. This implies that the conditional transition probability $h_{ip}(t)$ is a logistic function of the explanatory variables X and Z and unobserved heterogeneity U:

$$h_{ip}(t) = \frac{\exp(\beta_0(t) + \beta_1 x_{1it} + \cdots + \beta_k x_{kit} + \gamma_1 z_{1i} + \cdots + \gamma_j z_{ji} + u_i)}{1 + \exp(\beta_0(t) + \beta_1 x_{1it} + \cdots + \beta_k x_{kit} + \gamma_1 z_{1i} + \cdots + \gamma_j z_{ji} + u_i)} \quad (5.51)$$

This probability is used to construct the various sequences included in the conditional likelihood. As the discussion in Sect. 5.1.2.1 showed, the unit-specific error term u_i and the effects of all time-constant explanatory variables Z cancel out of the likelihood function. In other words, CML estimates the effects of the time-varying explanatory variables X while controlling for all (observed and unobserved) time-constant characteristics, even if they are correlated with X.

In sum, we can use a normal FE regression program and apply it to our discrete-time event histories, although we focus on only one spell per unit of analysis. Similar to all the other FE models, the analysis focuses only on units that change over time (i.e., experience an event) and numerical estimates of the effects of time-constant explanatory variables Z are not available (these variables are only controlled for). In the hank data, this results in a loss of 451 out of 837 women and there are no numerical estimates of the effects of the key explanatory variables describing women's reproductive history (e.g., the number of children). The time-constant effects of educational attainment and labor force career will also be absorbed into the error term. Similarly, a lack of variation in time-varying variables over time implies that observations also do not add information to the conditional likelihood function. Take home ownership in the hank data as an example for a time-varying variable with very low within-unit variation during the observation period.

On the other hand, the fact that the one always occurs at the end of the sequence severely limits the kinds of time-varying explanatory variable X that can be included in the model. If the ones appear only at the end, then the probability of observing a specific sequence is a function of spell duration and all other characteristics associated with spell duration. Including any monotonic function of time in the regression model will result in convergence failure when maximizing the likelihood, since event occurrence in a woman's sequence is then predicted perfectly (a phenomenon known from traditional logistic regression models as *complete separation*). Yet, dropping the variable time from the model and assuming no duration dependence is also no alternative. For example, the hank data builds upon data from

Table 5.18 Women's retirement by home ownership (in percent)

	Retired	Homeowner	
		No ($X=0$)	Yes ($X=1$)
	No ($Y=0$)	92.79	93.60
	Yes ($Y=1$)	7.21	6.40
	Total	100	100

Source: hank data (see Example 5.2)

18 panel waves and in this and many other applications with long observation periods, we have good reason to assume that the discrete-time hazard changes over time (e.g., the hazard of retiring should increase). In a similar vein, if explanatory variables can only change in one direction over time (i.e., only increase or decrease), the discrete-time hazard will also be completely determined. An example for a variable in the hank data that is likely to only increase over time is the indicator for retirement of a woman's partner. Of course, neglecting time dependence and explanatory variables that are correlated with time is not a solution, since this will likely result in biased estimates (see also Sect. 5.2.3.2). Taken together with the major limitation that the key time-invariant explanatory variables describing a woman's reproductive history are absorbed into the error term, the case-crossover design cannot be recommended for the analysis of the hank data. We therefore abstain from presenting any results derived from the case-crossover design.

Obviously, the limitations of the case-crossover design are due to the very specific sequences that result from event history data. If it would be possible to model sequences that include ones not necessarily at the end of the sequence, this would make FE models for non-repeatable events much easier. This idea is the starting point of the *case-time-control design*. It builds upon the fact that odds ratios are symmetric measures of statistical association. Hence, it does not make a difference whether you regress a dichotomous variable Y on a dichotomous variable X in a logistic regression model or whether you do it the other way around; in both cases, you will arrive at the same parameter estimates of the explanatory variable X (or Y).

To verify this specific feature of logistic regression, consider two binary variables in the hank data: women's retirement (Y) and home ownership (X). Table 5.18 displays the relative frequencies in a two-way table. Knowing this, we can calculate the odds ratio for retirement given X:

$$\frac{(92.79/7.21)}{(93.60/6.40)} = \frac{(92.79 \cdot 6.40)}{(93.60 \cdot 7.21)} = 0.88$$

You will obtain the same odds ratio when using the conditional probabilities of being a homeowner given Y, as shown in Table 5.19:

$$\frac{(36.38/63.62)}{(39.38/60.62)} = \frac{(36.38 \cdot 60.62)}{(39.38 \cdot 63.62)} = 0.88$$

Notice, however, that this nice feature of symmetry only approximately holds in multivariate logistic regression (unless you are specifying a completely saturated model without continuous variables).

5.2 Modeling the Change of Y: Discrete-Time Event History Models for Panel Data

Table 5.19 Home ownership by women's retirement (in percent)

Source: hank data (see Example 5.2)

Homeowner	Retirement	
	No ($Y = 0$)	Yes ($Y = 1$)
No ($X = 0$)	36.38	39.38
Yes ($X = 1$)	63.62	60.62
Total	100	100

The idea of the case-time-control design is now to exchange the (dichotomous) dependent variable (measuring event occurrence) with one of the dichotomous explanatory variables X so that the resulting sequence of the X values is not a simple function of spell duration (including a one only at the end). Assume, for example, that we want to establish whether women's transition to retirement depends on their partner's retirement. We use censored as well as uncensored observations in the hank data. In setting up a case-time-control design, the (unconditional) probability $\Pr(y_{it} = 1)$ that the partner is retired at year t becomes the dependent variable and X_1 is a dummy variable for women's retirement:

$$\ln\left(\frac{\Pr(y_{it}=1)}{1-\Pr(y_{it}=1)}\right) = \beta_0(t) + \beta_1 x_{1it} + u_i \tag{5.52}$$

Again, CML is used to derive estimates while controlling for all stable characteristics. There needs to be within-unit variation on the dependent variable and hence, unpartnered women and women whose partner did not retire during the observation period do not contribute to the conditional likelihood. Even though the dependent and independent variables have been exchanged, the resulting odds ratio is interpreted the other way around, given the symmetry of the odds ratio. Hence, β_1 is interpreted as the effect of partner's retirement on the woman's conditional transition probability into retirement.

One great advantage of this design is that time dependence can be included as an explanatory variable. This is because the sequences of zeros and ones indicating whether a woman's partner retires in a specific year differ from the woman's own sequence, since the partner's transition into retirement may occur before the woman's retirement and hence, may include a one in the middle of the sequence. Recall from above that when specifying a case-crossover design, all sequences end with event occurrence and hence, monotonic functions of time cannot be included in the regression model.

An obvious constraint is, however, that the case-time-control design cannot be used for continuous explanatory variables. This is especially unfavorable for the analysis of our example since the key indicator in the hank data is the number of children. Over and above, this variable is time-constant and hence lacks within-group variance. Even if the symmetry argument would also apply to continuous variables, FE estimation is not able to estimate effects of time-constant variables. Another disadvantage of this approach lies in the fact that the symmetry feature applies only approximately in multivariate logistic regression models.

In our application example, however, we want to add additional explanatory variables to obtain unbiased estimates of our key variables. Estimates of the effects of these additional explanatory variables will remain biased in a case-time-control design.

In sum, there is no convincing FE approach to analyze non-repeatable events (single episode data). Both the case-crossover and the case-time-control design are still rather unusual approaches and suffer from various limitations. Both techniques may be of help in some applications, but are not appropriate FE methods for the analysis of data similar to our Example 5.2. It seems as if for a serious application of FE models, you need multi-episode data, in which the event of interest occurs repeatedly and hence, each unit of analysis contributes several spells to the data.

5.2.3.5 Applying Continuous-Time Event History Models Within a Panel Design

Discrete-time event history analysis is very popular in panel research. The advantages of this model are all too obvious: It is the most popular and convenient model and many empirical researchers have a familiar ring with the terminology, estimation procedure, and interpretation of coefficients. Nevertheless, the discrete-time logistic hazard model is based on the assumption that events can only occur at discrete points in time. Such discrete-time data result from intrinsically discrete transition processes, such as changes in voter turnout from one election to the next. In many applications, however, it is more realistic to assume that the process under study is continuous, but the time variable is measured imprecisely. Event history data are considered to be interval-censored when the precise dates of transitions are unknown and event occurrences are observed only within given time intervals. For instance, in our Example 5.2, women may retire at any month during the observation period, but we only know whether a transition into retirement has occurred between two consecutive panel waves. An alternative for discrete-time analysis, which is better suited when the data used are generated by a continuous-time process, is the complementary log–log hazard model.

To understand the reason behind this, it is necessary to identify the relationship between the conditional transition probability $h_{ip}(t)$ in discrete time and the instantaneous hazard rate of event occurrence $r_{ip}(t)$ in continuous time. In discrete time, we assume that change can happen only at $t = 1, \ldots, T$ discrete points in time. In an imprecisely measured continuous-time process, T is a real-valued random variable that can take on every value between 0 and $+\infty$. However, the exact value of T is not observed. We only know whether T falls into a particular interval $(t_{l-1}, t_l]$, say, the interval from shortly after the previous panel wave $l - 1$ to and including the time point of the present panel wave l: $t_{l-1} < T \leq t_l$. Recall from Sect. 5.2.1 that the discrete-time hazard rate $h_{ip}(t)$ is defined as the conditional probability that an individual will experience a transition from state p to another state at the end of interval l, given that the individual still belongs to the risk set. There is a simple relationship between the survival probability in continuous time and the conditional

5.2 Modeling the Change of Y: Discrete-Time Event History Models for Panel Data

transition probability. The conditional transition probability at the lth panel wave (at the end of the lth interval) equals

$$h_{ip}(t_l) = \Pr(t_{l-1} < T \le t_l | T > t_{l-1}) = \frac{S_{ip}(t_{l-1}) - S_{ip}(t_l)}{S_{ip}(t_{l-1})} = 1 - \frac{S_{ip}(t_l)}{S_{ip}(t_{l-1})} \quad (5.53)$$

Obviously, the conditional transition probability in discrete time can be derived by comparing the probability of surviving between t_l and t_{l-1} in continuous time.[36] The difference of these two survival probabilities equals the probability of an event within interval l, which has to be related to the probability of surviving the previous interval $l-1$ to finally get the *conditional* transition probability.

In continuous-time event history analysis, the instantaneous hazard rate $r_{ip}(t)$ is defined as a conditional transition probability within an infinitesimal small interval:[37]

$$r_{ip}(t) = \lim_{\Delta t \to 0} \frac{\Pr(t \le T < t + \Delta t | T \ge t)}{\Delta t} \quad (5.54)$$

It is a continuous function of time allowing for increasing, decreasing, and even changing trends of event occurrence. Moreover, the hazard rate may depend on time-varying and time-constant variables X and Z. A basic hazard rate model assumes that the hazard rate can be written as

$$r_{ip}(t) = \exp\bigl(\beta_0(t) + \beta_1 x_{1it} + \cdots + \beta_k x_{kit} + \gamma_1 z_{1i} + \cdots + \gamma_j z_{ji}\bigr) \quad (5.55)$$

In this equation, time dependence is independent of the explanatory variables X and Z (i.e., there are no interactions between T and X resp. Z) so that the effects of these variables are constant over time. Hence, the estimated hazard rates of two units i and i' with different values of X and Z are proportional at any point in time:

$$\frac{r_{ip}(t)}{r_{i'p}(t)} = \exp\bigl[(x_{1it} - x_{1i't})\beta_1 + \cdots + (x_{kit} - x_{ki't})\beta_k$$
$$+ (z_{1i} - z_{1i'})\gamma_1 + \cdots + (z_{ji} - z_{ji'})\gamma_j\bigr] \quad (5.56)$$

Therefore, this model is also called a *proportional hazard rate model* (PHM). For this type of hazard models, the regression function can be split into two parts:

$$r_{ip}(t) = \exp\bigl(\beta_0(t)\bigr) \cdot \exp\bigl(\beta_1 x_{1it} + \cdots + \beta_k x_{kit} + \gamma_1 z_{1i} + \cdots + \gamma_j z_{ji}\bigr) = \lambda_0(t) \cdot f(x, z) \quad (5.57)$$

---

[36] The survival probability in continuous time is defined similar to (5.35). But now it is a continuous function of time that monotonically decreases as units experience an event and drop out of the risk set.

[37] In continuous time, intervals are usually defined as [*lower, upper*) (i.e., including the lower, but excluding the upper bound): *lower* $\le T <$ *upper*. This seems to be different from our present discussion of discrete-time processes, in which we defined change to happen at the *end* of the interval and correspondingly intervals as (*lower, upper*]: *lower* $< T \le$ *upper*. However, the probability that a continuous random variable T equals a particular value t is zero: $\Pr(T = t) = 0$. Hence, in continuous time: $\Pr(t \le T < t + \Delta t | T \ge t) = \Pr(t < T \le t + \Delta t | T > t)$.

with the first part $\lambda_0(t)$ describing the time dependence (the so-called baseline hazard) and the second part $f(x, z)$ describing the dependence on the explanatory variables.

Assuming $\Delta t \to 0$ in (5.54), turns the conditional probability into an *instantaneous* hazard rate (also called a transition intensity). Yet, by integrating over some time interval, we can again derive probability statements:[38]

$$\Pr(t \leq T < t + \Delta t | T \geq t) = \frac{S_{ip}(t) - S_{ip}(t + \Delta t)}{S_{ip}(t)} = 1 - \frac{S_{ip}(t + \Delta t)}{S_{ip}(t)}$$

$$= 1 - \exp\left[-\int_t^{t+\Delta t} r_{ip}(u) \, d(u)\right] \quad (5.58)$$

Applying this general formula to our former equation (5.53) for the discrete-time case shows how the discrete-time transition probability $h_{ip}(t_l)$ at the end of the lth interval is related to the continuous-time hazard rate $r_{ip}(t)$. Define $t = t_{l-1}$ and $t + \Delta t = t_l$ and you have

$$h_{ip}(t_l) = 1 - \exp\left[-\int_{t_{l-1}}^{t_l} r_{ip}(u) \, d(u)\right] \quad (5.59)$$

If both quantities are made functions of time and explanatory variables and if we assume a PHM, we can use (5.57) in the integral. Let us ignore the dependence on X and Z for a moment (i.e., $f(x, z) = 1$) and focus only on the time dependence $\lambda_0(t)$:

$$h_{ip}(t_l) = 1 - \exp\left[-\int_{t_{l-1}}^{t_l} \lambda_0(u) \, d(u)\right] \quad (5.60)$$

If we have no information about the time dependence, a simple assumption would be that the baseline hazard varies between intervals, but is constant within intervals: $\lambda_0(t) = \lambda_l$ for $l = 1, \ldots, L$ intervals. In that case, the integral can be solved as follows:

$$h_{ip}(t_l) = 1 - \exp\left[-\lambda_l \cdot (t_l - t_{l-1})\right] = 1 - \exp[-\lambda_l \cdot e_l] \quad (5.61)$$

Furthermore, as you can easily check with a hand calculator, λ_l and $h_{ip}(l)$ are almost identical, when both the interval length $e_l = (t_l - t_{l-1})$ and the instantaneous hazard rate λ_l are small. This implies that the discrete-time event history model approximates the continuous-time event history model quite well, when the (continuous-time) hazard of event occurrence is low (say, 0.1 or lower) and constant within intervals, which themselves are rather short. In that case, we will find very similar parameter estimates in both types of model.

If we would have used a PHM with explanatory variables, the expression would be as follows:

---

[38] You can think of the integral as some kind of summation over the many time points within the interval.

5.2 Modeling the Change of Y: Discrete-Time Event History Models for Panel Data

$$h_{ip}(t_l) = 1 - \left[\exp(-\lambda_l \cdot e_l)\right]^{\exp(\beta_1 x_{1it} + \cdots + \beta_k x_{kit} + \gamma_1 z_{1i} + \cdots + \gamma_j z_{ji})} \quad (5.62)$$

Taking logs, we arrive at the following linear-additive model, the so-called *complementary log–log model*:

$$\ln\left[-\ln(1 - h_{ip}(l))\right] = \beta_l + \beta_1 x_{1it} + \cdots + \beta_k x_{kit} + \gamma_1 z_{1i} + \cdots + \gamma_j z_{ji}$$
$$\text{with } \beta_l = \ln \lambda_l + \ln e_l \quad (5.63)$$

This model should include a constant β_l for each interval (panel wave) that controls for the interval-specific hazard rate λ_l and the interval length e_l. Such constants are easily modeled by dummy variables for the intervals.

Equation (5.63) shares the useful properties of the logit transformation, because it also transforms the conditional transition probability into a quantity between minus and plus infinity. However, there are also important differences. As shown above, the discrete-time estimates derived from the complementary log–log hazard model are at the same time estimates of the underlying continuous-time PHM. This is a major advantage over the discrete-time logistic hazard model, since the parameters of the complementary log–log model control for the time dependence of the underlying continuous-time process and hence, are also independent of the length e_l of each interval $l = 1, \ldots, L$. Moreover, each interval may be of different length and estimates of the interval-specific hazard rates can be derived from the estimated regression constants by the following formula: $\hat{\lambda}_l = \exp(\hat{\beta}_l - \ln e_l)$.[39] By contrast, the discrete-time logistic hazard model is sensitive to the length of the intervals and also must assume that all intervals have the same length.

Returning to our research example, we can now estimate a complementary log–log model for the hank data by inserting the complementary log–log hazard model (5.63) into the likelihood function (5.42). Table 5.20 summarizes the estimates of the base model. If you compare the estimates of this model with the ones shown in Table 5.17, you will largely arrive at similar conclusions about the effects of the explanatory variables. Given the former conclusion that the discrete-time model approximates the continuous-time model when the hazard of event occurrence is low, this comes as no surprise. Since $h_{ip}(t)$ is particularly small for all combinations of X and T in the hank data, the results of the complementary log–log hazard model do not significantly differ from the logistic hazard model.

5.2.4 Extensions

One extension of the discrete-time hazard model takes into account alternative destination states. In various social processes, destination states are competing and the occurrence of one event removes the individual from risk of the other events. To

---

[39] Although the complementary log–log model can handle unequal interval lengths, unequal time intervals are problematic when the continuous-time transition rate is not constant within intervals or when the model is not a proportional hazard model.

Table 5.20 Determinants of female retirement (complementary log–log hazard model)

Variable	Estimate	Std. Err.
Number of children	−0.0914	0.0428
Age group		
54–57 years	1.1228	0.2668
58–59 years	1.5821	0.2821
60–61 years	3.6984	0.2482
62–63 years	2.6668	0.2947
64–69 years	3.1303	0.2767
Educational attainment		
Vocational degree	−0.1063	0.1111
University degree	−0.4321	0.2621
Imputation flag: Education	0.0610	0.4287
Labor force career		
Years in labor force at age 50	0.0179	0.0059
Number of employment spells	0.0481	0.0476
Partner		
Partner, not retired	−0.9727	0.1700
Partner, retired	−0.2360	0.1459
Imputation flag: Partner	0.7061	0.2188
Income		
Household income	0.0040	0.0090
Squared household income	0.0000	0.0000
Imputation flag: Income	0.1506	0.2182
Other personal characteristics		
Home owner	−0.0762	0.1121
Poor health	0.2251	0.1435
Person needing care	0.0898	0.2167
Constant	−4.4006	0.3715
LR	749.90	
df	20	
ln L	−1,041.4641	
N	5,765	
n	837	
T	18	
Events	386	

Source: hank data (see Example 5.2)

simplify matters, reconsider our Example 5.3 of cancer-related death of patients. Some patients may also have died due to another health problem, such as cardiovascular disease. To measure the effectivity of the new cancer-directed therapy, we need to control for deaths due to other causes as competing risks of the death due to cancer. By contrast, in the analysis of retirement decisions in our Example 5.2, there are no competing risks.

Unlike in continuous-time event history models, the discrete-time likelihood function cannot be factored into a set of different hazard rates, each one for a different transition (Allison, 1982; Vermunt, 1997). Rather, to control for alternative outcomes in the multiple-risk model, multinomial logistic regressions must be estimated, in which the dependent variable is coded zero for non-occurrence and $1, \ldots, Q$ for event occurrence of the Q competing risks. There may be time-constant, unit-specific unobserved factors that affect each type of transition. For instance, in the analysis of first partnership formation, physical attractiveness and the desire to partner are examples of such unmeasured factors. This will result in dependence among risks and affect the observed hazard of an event for each of the alternative destination states. The problem with ignoring unobserved heterogeneity is therefore bias in the results obtained from multinomial discrete-time hazard models. A practical solution to allowing for shared unobserved risk factors is to include random effects.

It is also feasible to pursue an analysis for each destination state separately (Allison, 1995; see also Begg and Gray, 1984) to evidence the risk of a particular event in the absence of all other competing risks. In practice, researchers focus on one specific event type and treat all spells ending with an alternative event as if they were censored at that date. However, in contrast to the multinomial logistic estimator, this estimator is not fully efficient. Nonetheless, in practice, this approach leads to very similar results compared to those generated by multinomial logistic regression.

A more extensive survey on competing risks models is given by Hachen (1988). More sophisticated discussion of techniques for competing risks can be found in Hill et al. (1993), Goldstein et al. (2004), and Steele et al. (2004).

Another issue in event history analysis is repeated events. Examples for repeatable events include marriages or changes of employer. Multi-episode data provide two more pieces of information, namely, information on the sequence of events for each individual and information on event occurrence and non-occurrence in each of these episodes. A simple approach to analyze these data is to pursue a separate analysis for each event, eliminating from the pooled data set all records that pertain to the other events. This approach has the advantage that we can simply specify the likelihood function for single episode data, as established in Sect. 5.2.2.3. Remember that in the logistic discrete-time hazard model, the trick was to split each episode into as many records as there are discrete observations over time and to generate a dummy variable signaling whether these observations end with an event or not. However, it was only a technical tool that allows us to model possible time dependencies of the discrete-time hazard and to include the full time trajectory of the time-dependent explanatory variables X. The basic unit of analysis is still the single episode (spell) contributed by each individual (see the discussion at the end of Sect. 5.2.2.3).

When handling repeated events, the most important difference is that we may now observe more than one episode (spell) for each individual. So the question of whether each individual's sequence of events in multiple episode data can be treated as independent is crucial. Violation of the assumption of statistical independence may result from:

1. *Unobserved heterogeneity*: An important objection is the presence of unobserved unit-specific factors that are constant across all episodes and relevant to recurrent event occurrence in all episodes. As already highlighted in Sect. 5.2.3.2, the central question is then whether all common risk factors are observed and measured without error. If not, correlation between the durations of episodes from the same unit will result. In turn, this violation of the assumption of statistical independence will severely bias estimation results. You are therefore well advised to control for unobserved heterogeneity in the analysis of repeatable events. You may want to start with an RE model in which the random effects capture unobserved unit-specific factors that are common to all episodes, but independent of the variables X and Z in the model.
2. *Event dependence*: A simple model for repeated events may assume that the time at which a particular event occurs is independent of the previous event history. In practice, there is good reason to suspect that this assumption is violated. Event occurrence in one spell is likely to be influenced by event occurrence in previous spells. In other words, current change of the dependent variable Y is dependent on previous change of the dependent variable. When treating observations in multi-episode data as independent, the problem with correlated events is biased estimates of standard errors, comparable to the problem of autocorrelation in traditional regression models. One remedy is to include explanatory variables that capture the previous history of each event. The regression model should at least include a set of dummy variables representing the different events in the sequence.
3. *Episode-changing effects of X and Z*: A related issue is episode-constant and episode-changing effects of the independent variables. In various applications, there are some reasons to believe that the influence of explanatory variables is not constant across episodes, but rather is episode-specific. The problem, then, is once again the non-independence of observations in multi-episode data.

Different estimation models for recurrent events are discussed in more detail by Allison (1996), Yamaguchi (1986), Box-Steffensmeier et al. (2007), and Box-Steffensmeier and Zorn (2002).

5.3 Conclusion and Further Reading

This chapter aimed to give a brief introduction into panel models for categorical dependent variables. The first part of the chapter was devoted to the analysis of discrete responses with panel data. Our review was focused on models for binary dependent variables. Yet, references were made to the specific techniques required for other types of discrete variable, namely, polytomous nominal variables, ordinal variables,

5.3 Conclusion and Further Reading

and count variables. Before we end this discussion, we would like to recommend further reading.

More detailed introductory texts include Amemiya (1981) and Aldrich and Nelson (1984). The introductory level textbooks by Allison (2001), Agresti (2002), Hosmer and Lemeshow (2000), and Long and Freese (2006) show you how to implement the regression techniques discussed above in SAS and Stata. In Liang and Zeger (1986) and Pendergast et al. (1996), attention is given to methods for analyzing clustered binary response data. In order to gain a deeper understanding of the CML approach, consult the seminal article by Chamberlain (1980). To go into more details of FE and RE models, you are invited to look at the pathbreaking articles by Heckman (1981b) and Heckman and Willis (1976), as well as the reviews in Maddala (1987) and Hsiao (2003).

The second part of this chapter described how techniques of discrete-time event history analysis can be used to analyze the change of categorical dependent variables in panel data. We did not attempt an exhaustive review of all statistical roots here. For more in-depth reading, you may want to consult one of the following books and articles. The book by Singer and Willett (2003) includes chapters on discrete-time event history analysis that are valuable for understanding the approach just described. A gentle introduction and a set of guidelines on how to apply discrete-time event history analysis can be found in Allison (1982), Allison (1984) and Yamaguchi (1991). The chapter also raised the issue of omitted variable bias and unobserved heterogeneity. To what extent this pitfall affects the results of event history models is shown more extensively in Lancaster (1990). More discussion on correlated heterogeneity and FE event history models can be found in Allison (2009).

We often have the opportunity to analyze continuous-time data through the use of retrospective data included in most panel surveys. A sophisticated treatment of techniques for continuous-time data can be found in the pioneering early monographs by Kalbfleisch and Prentice (1980) and Cox and Oakes (1984). Other popular but sophisticated monographs are Lancaster (1990) and Blossfeld et al. (1989). Blossfeld and Rohwer (2002), Blossfeld et al. (2007), Cleves et al. (2002), and Allison (1995) take on different aspects in event history analysis using the software TDA, Stata and SAS, respectively.

How to Do Your Own Panel Analysis

We hope that our textbook has motivated you to do your own panel analysis. Many panel data are available for secondary analysis. Table 6.1 shows a selection of them and their characteristics. Ruspini (2002) gives more detailed information on these and other panel surveys and compares them with other longitudinal data such as pooled cross-sections and (retrospective) event histories. Textbooks on specific panel studies are rare (for an exception see Hill, 1997), because data collection and dissemination may change quite quickly during the course of the study. Hence, you should search for the most recent information on the corresponding web sites of the panel studies.

The chapters in Rose (2000) discuss the pros and cons of panel surveys and how to maintain the quality of panel data. The edited volume also includes some typical examples of panel analyses. A gentle introduction into the most common problems of panel analysis for social scientists can be found in Taris (2000). The edited volumes by Kasprzyk et al. (1989), Lynn (2009) and Menard (2008) provide more comprehensive methodological discussions on designing, collecting, and analyzing panel data. Finally, a recent research report from the Institute for Social and Economic Research in Essex (Lynn et al., 2005) provides a nice summary of our current knowledge about the design and implementation of longitudinal surveys and about the use of such data.

As already mentioned in the introductory chapters, it is difficult for the novice user to have a basic orientation to the various statistical methods of panel data analysis. This also has to do with the different disciplines using panel data, each of which has its own methodological tradition. This diversity has its pros and cons. An advantage is certainly that you find many good textbooks on panel data analysis in your discipline that tie in with your methodological training. But, at a certain point, you will come across (seemingly) different panel data methods and will have to ask yourself whether it is a new methodology or something that you already know. Often, you will have the feeling that it is only different people talking about the same problem, but using different terminology. This lack of interdisciplinary communication and generality is certainly a disadvantage.

Table 6.1 Characteristics of selected panel studies

Mnemonic	Name	Country	Main sample Units	Population	Measurements	Sampling design	Sample size	Additional samples
NLS	National Longitudinal Surveys	USA	individuals	US resident population	face-to-face interviews on an annual basis since 1966	national probability sample of civilians	original: 5,000 individuals per survey	old/young men, mature/young women, children and young adults, survey of youth
BHPS	British Household Panel Study	Great Britain	individuals within households	British resident population (aged 16 and older)	annual interview of every household member aged 16 and older since 1991	stratified clustered design (postcode address file)	original: 5,511 households (10,300 individuals)	Scotland, Wales (1999) and Northern Ireland (2001), children aged 11–15 (1994)
CNEF	Cross National Equivalent File	Australia (HILDA), Canada (SLID), Germany (SOEP), United Kingdom (BHPS), United States (PSID), Korea, Switzerland	individuals within households	private households	since 1980: information from panel surveys into a framework of comparably defined variables (mostly annual)	address and telephone random national samples (see corresponding panels)	58,000 households (188,000 individuals)	
ECHP	European Community Household Panel	European Community	individuals within households	resident population of the member states of the European Community (aged 16 and older)	on an annual basis from 1994–2001 (PAPI, CAPI, CATI)	multi-stage stratified random sampling (most samples use a two-stage design)	original: 60,819 households (130,000 individuals)	

6 How to Do Your Own Panel Analysis

Table 6.1 (Continued)

Mnemonic	Name	Country	Main sample Units	Population	Measurements	Sampling design	Sample size	Additional samples
EU-SILC	European Union Statistics of Income and Living Conditions	European Union	individuals within households	all members of private households of the EU Member States (aged 16 and older)	on an annual basis since 2003	frame, register or sample design ('complete' household or 'selected' respondents)	100,000 households (200,000 individuals)	
SOEP	German Socio-Economic Panel	Germany	individuals and households	German resident population	annual face-to-face interview of every household member aged 16 and older since 1984	multi-stage random samples (regionally clustered), random walk	original: 6,000 households, now: 11,000 households (20,000 individuals)	GDR, foreigners, immigrants, high income, refreshment
HILDA	Household, Income and Labor Dynamics in Australia Survey	Australia	individuals within households	all members of private dwellings in Australia	interviews with all household members aged 15 and older on an annual basis since 2001	multi-stage sampling (area, dwelling, household)	original: 7,682 households (19,914 individuals)	
IAB-EP	IAB Establishment Panel	Germany	enterprises	enterprises of all sectors with at least one employee (subject to social insurance contribution)	face-to-face interviews on an annual basis since 1993 in Western Germany, in Eastern Germany since 1996	random sample (based on a register with regard to industrial sector and size of enterprise)	16,000 enterprises	

Table 6.1 (Continued)

Mnemonic	Name	Country	Main sample Units	Population	Measurements	Sampling design	Sample size	Additional samples
KLIPS	Korea Labor Income Panel Study	Korea	individuals within households	Korean households living in urban areas aged 15 years and older	face-to-face interviews on an annual basis since 1998	two-stage stratified cluster sampling	original: 5,000 households (11,000 individuals)	
PSID	Panel Study of Income Dynamics	USA	individuals within households	US resident population	telephone interviews on an annual basis since 1968, beginning with face-to-face interviews	national equal probability sample	original: 4,800 families; 7,000 families in 2001	cross-sectional national sample, sample of low-income families, Latino (immigrant) households
SHP	Swiss Household Panel "Vivre en Suisse - Leben in der Schweiz"	Switzerland	individuals within households	all members of private households (aged 14 and older)	telephone interviews (CATI) on an annual basis since 1999	multi-stage probability sample	original: 5,074 households (12,931 individuals)	
SOCX	Social Expenditure Database	33 OECD countries	OECD countries	All OECD member states	updated every 2 years, data are available from 1980 onwards	data taken from OECD Health Data, OECD database on Labour Market Programmes, EUROSTAT	census of all member states	

Nevertheless, it is useful to have a look at the different methodological traditions and their associated literature. While it cannot claim to be comprehensive the following list names some of the most prominent traditions and mentions some introductory textbooks for each of them:
- Economics: Economics research traditionally focuses on continuous variables (often on the aggregate level of countries). Econometrics has a strong tradition in regression and time-series analysis for longitudinal data. First differences, fixed effects, and other estimation methods for the linear model can be found in every econometrics textbook (e.g., Cameron and Trivedi, 2005; Greene, 2008) or in specialized textbooks on panel data analysis (e.g., Baltagi, 2008; Hsiao, 2003; Wooldridge, 2010). However, with the advent of large socioeconomic household panels, economic research is increasingly focusing on categorical variables too (Cameron and Trivedi, 2005; Hsiao, 2003; Diggle et al., 2002). Frees (2004), Singer and Willett (2003), and Taris (2000) are textbooks that specifically focus on the social sciences.
- Psychology: Psychological research often uses experimental data, where the dependent variable is continuous and the independent variables are mostly categorical (e.g., the treatment) and sometimes continuous (e.g., the controls). Analysis of variance is perfectly suited to this type of data. Within this tradition, psychometrics has developed specific methods to analyze experimental data with repeated measurements using analysis of variance and variance components models (Fitzmaurice et al., 2011; Kirk, 1995). Furthermore, psychology, but also other social science disciplines, are interested in the development of individual characteristics over time. Growth models provide a perfect tool for these kinds of developmental research question and structural equation model (SEM) are a nice environment, in which one can test all kinds of hypothesis about inter- and intra-individual change (Bollen and Curran, 2006; Duncan et al., 2006). Furthermore, SEM also allow to model change that happens continuously in time (Oud and Delsing, 2010).
- Educational science, psychology, and increasingly internationally comparative survey researchers: These researchers often have to deal with grouped data, e.g., students within classes within schools, or respondents within regions within countries. For this kind of hierarchical data, they have developed multi-level modeling, which, as we have seen, can also be applied to panel data (Goldstein, 2011; Hox, 2010; Singer and Willett, 2003; Snijders and Bosker, 2011).
- Political science: Quantitative research in political science, especially in comparative politics, often uses continuous data at the aggregate level of countries. Since the number of countries, for which the necessary information is available, is often limited (e.g., to the OECD countries), there is a strong interest to increase the number of observations by using data from different years. Hence, political scientists have made important contributions to the analysis of pooled time-series cross-section data (macro panels) (Beck and Katz, 1995, 1996, 2011; Beck, 2001).
- Sociology and life sciences: Demography, epidemiology, and even engineering are the traditional users of survival analysis and duration data. Surveys of life histories have made these techniques interesting for social scientists too, among

whom they have become known as event history analysis (Blossfeld et al., 2007; Singer and Willett, 2003; Yamaguchi, 1991). Somewhat related to these methods is a technique called sequence analysis (Abbott and Tsay, 2000; Brzinsky-Fay and Kohler, 2010; MacIndoe and Abbott, 2004). While survival and event history analysis focus on single events, sequence analysis uses the whole sequence of spells and events that is observed for a unit in the observation period. It tries to find different types of sequence in the data by using optimal matching procedures.

- Sociology, political science, and marketing: The first social science panel studies were rather short, consisting of only 2–3 measurements over time (e.g., before, during, and after an election). The same has been true for research on consumer behavior (e.g., evaluating marketing campaigns). Within these analyses of short panels, processes of change (e.g., voter turnover, brand loyalty) have traditionally focused on transition matrices using Markov models. Applications of these kinds of model can be found in the context of categorical data analysis (e.g., van de Pol and Langeheine, 1990; Vermunt et al., 2008).

- Sociology and psychology: Finally, sociologists and psychologists with their interest in often unreliable attitude data have developed methods to account for random and systematic measurement error. The classical work by Blalock (1970), Heise (1969), and Wiley and Wiley (1970) discusses reliability assessment with panel data and how to separate true change from artificial change due to measurement error. All of these analyses can be undertaken in a much more general framework by using structural equation models (SEM). Gentle introductions into SEM are provided, among others, by Bollen (1989), Kline (2010), and Schumacker and Lomax (2004). The traditional SEM literature focuses on continuous latent variables. If you think that latent variables are categorical, you have to refer to latent class analysis (LCA). LCA is the methodological basis of structural equation models for categorical data. Bergsma et al. (2010) and Hagenaars (1990) show how to assess stability and change of possibly unreliable categorical dependent variables. Recent developments in panel analyses with latent variables are discussed by Little et al. (2007).

So, how to do your own panel analysis? Simply by following the steps that we have set for you in this book, and by building on the methodological strategies specific to your own discipline. If you still think that it is not simple at all, remember that the appetite comes with eating! It is only once you start doing your own analysis that you better understand how the various pieces of the puzzle fit together. What is more important, it is only by doing your own analysis that you get a taste of the statistical methods. You are now equipped with the concepts and the understanding of the mechanisms behind them. Just do it!

Useful Background Information 7

7.1 Functions of Random Variables

A random variable can take on a set of possible different values, each with an associated probability. Statisticians distinguish between continuous and discrete random variables. In the following, we focus on continuous random variables. A continuous random variable can be described by a statistical distribution (e.g., the normal distribution). Expected values, variances, and other statistics are used to measure certain characteristics of the distribution, e.g. its center and its spread (Greene, 2008, Appendix B). In this section, we discuss functions of random variables that are used at various places in the text. For example, if you have two normally distributed random variables X and Y, then their sum $Z = X + Y$ is also a random variable which is normally distributed. Table 7.1 shows how we can compute the expected value and the variance of Z as a function of X or Y or both of them. The formulas apply for any kind of distribution. Hence, X and Y do not need to be normally distributed.

According to the last line of Table 7.1, if we compute a function of two random variables that covary with each other, as we do when we look at the difference of two regressions coefficients (see (7.14)), then their respective variances ($\sigma^2_{\hat{\beta}_1}, \sigma^2_{\hat{\beta}_2}$) and their covariance ($\sigma_{\hat{\beta}_1, \hat{\beta}_2}$) has to be accounted for when computing the variance of the difference:

$$\sigma^2_{(\hat{\beta}_1 - \hat{\beta}_2)} = \sigma^2_{\hat{\beta}_1} + \sigma^2_{\hat{\beta}_2} - 2 \cdot \sigma_{\hat{\beta}_1, \hat{\beta}_2} \tag{7.1}$$

Another example is the error term of our panel regression models. It consists of two random variables: unobserved heterogeneity at the unit level (U) and idiosyncratic errors at the measurement level (E). However, we assumed the two error components to be independent of each other ($\sigma_{u,e} = 0$). If the covariance of U and E is zero, the variance of the composite error $\varepsilon_{it} = u_i + e_{it}$ equals the sum of both error variances:

$$\sigma^2_\varepsilon = \sigma^2_u + \sigma^2_e - 2 \cdot 0 = \sigma^2_u + \sigma^2_e \tag{7.2}$$

Table 7.1 Expected values and variances for functions of random variables

$Z = f(X, Y)$	E(Z)	Var(Z)
a	a	0
$b \cdot X$	$b \cdot E(X)$	$b^2 \cdot \text{Var}(X)$
$a + b \cdot X$	$a + b \cdot E(X)$	$b^2 \cdot \text{Var}(X)$
$a \cdot X + b \cdot Y$	$a \cdot E(X) + b \cdot E(Y)$	$a^2 \cdot \text{Var}(X) + b^2 \cdot \text{Var}(Y) + 2ab \cdot \text{Cov}(X, Y)$
$X + Y$	$E(X) + E(Y)$	$\text{Var}(X) + \text{Var}(X) + 2 \cdot \text{Cov}(X, Y)$
$X - Y$	$E(X) - E(Y)$	$\text{Var}(X) + \text{Var}(X) - 2 \cdot \text{Cov}(X, Y)$

Note: a and b are arbitrary constants

Similar formulas apply for covariances of random variables that are functions of other random variables. One of them is particularly important for us:

$$\text{Cov}(X + Y, Z) = \text{Cov}(X, Z) + \text{Cov}(Y, Z) \tag{7.3}$$

For example, when analyzing the serial correlations among the repeated observations y_{it} ($t = 1, \ldots, T$), we assumed that the serial correlations are due to the time-constant unit-specific effects u_i in the error term: $\varepsilon_{it} = u_i + e_{it}$. We also assumed that both error components U and E are independent of each other ($\text{Cov}(u_i, e_{it}) = \sigma_{u,e} = 0$; $t = 1, \ldots, T$) and that idiosyncratic errors are not serially correlated ($\text{Cov}(e_{it}, e_{is}) = \sigma_{e_t, e_s} = 0$; $t \neq s$). Using these assumptions and (7.3), we can determine the covariance between the error terms at two different time points t and s ($t \neq s$):

$$\begin{aligned}
\text{Cov}(\varepsilon_{it}, \varepsilon_{is}) &= \text{Cov}(u_i + e_{it}, \varepsilon_{is}) \\
&= \text{Cov}(u_i, \varepsilon_{is}) + \text{Cov}(e_{it}, \varepsilon_{is}) \\
&= \text{Cov}(u_i, u_i + e_{is}) + \text{Cov}(e_{it}, u_i + e_{is}) \\
&= \text{Cov}(u_i, u_i) + \text{Cov}(u_i, e_{is}) + \text{Cov}(e_{it}, u_i) + \text{Cov}(e_{it}, e_{is}) \\
&= \text{Cov}(u_i, u_i) + 0 + 0 + 0
\end{aligned} \tag{7.4}$$

Note that the covariance of a variable with itself equals its variance and hence, $\text{Cov}(\varepsilon_{it}, \varepsilon_{is}) = \text{Cov}(u_i, u_i) = \text{Var}(u_i) = \sigma_u^2$ ($t \neq s$).

Repeated application of (7.3) will also show that the correlation between the two observed variables Y_{80} and Y_{81} in path diagram (3.6) can be obtained by multiplying the two reliabilities and the stability of the latent variables Y_{80}^* and Y_{81}^*: $\text{Corr}(y_{i,81}, y_{i,80}) = 0.6 \cdot 0.454 \cdot 0.6 = 0.163$. When trying to do this proof yourself, you should remember that all variables in the path diagram are in standard form and hence, have variances (and standard deviations) equal to one. Since the Pearson correlation coefficient is defined as the covariance divided by the standard deviations of both variables that are correlated with each other (see (3.7)), the correlation of two standardized variables equals their covariance.

7.2 Estimation and Testing

7.2.1 Ordinary Least Squares

7.2.1.1 How to Compute a Regression Model Fitting the Data?
Figure 7.1 shows the statistical association between two continuous variables X and Y. The data come from a data set that we designed for this section.

> *Example 7.1* (sixcases data) We call these data the sixcases data, because they include only $n = 6$ cases. With these few data, all computations can be done manually, which makes all formulas easy to follow. Table 7.2 shows the values of both variables X and Y for each case $i = 1, \ldots, 6$.
>
> Furthermore, Table 7.2 shows some additional variables that have been derived from X and Y, among them $\hat{y}_i = 3 + 0.5 \cdot x_i$.

According to Fig. 7.1, the relationship is positive and approximately linear. We have indicated this with a straight line running from the lower left to the upper right corner of the graph. The line $\hat{y}_i = \beta_0 + \beta_1 x_i$ seems to fit the data quite well, although we have drawn it more or less freehand using some plausible values for

Table 7.2 Example with two continuous variables X and Y

i	x_i	y_i	$\hat{y}_i$	$(y_i - \hat{y}_i)$	$(y_i - \hat{y}_i)^2$	$|y_i - \hat{y}_i|$
1	0.5	3	3.25	−0.25	0.0625	0.25
2	1	2	3.5	−1.5	2.25	1.5
3	2	5	4	1	1	1
4	4	7	5	2	4	2
5	5	4	5.5	−1.5	2.25	1.5
6	5.5	5	5.75	−0.75	0.5625	0.75

Fig. 7.1 Example with two continuous variables X and Y

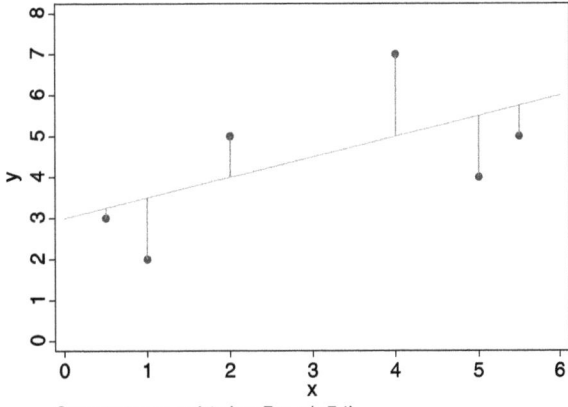

Source: sixcases data (see Example 7.1)

the intercept ($\ddot{\beta}_0 = 3$) and the slope ($\ddot{\beta}_1 = 0.5$). The points on the line, the predicted values $\hat{y}_i$, indicate our expectations about the level of Y given a particular level of X. How well the line actually fits the data cannot be answered until we have defined what we mean by "data fit". Broadly speaking, a "fitting" line should minimize the vertical distances (indicated by perpendicular lines) between the data points and the predicted values.

Unfortunately, minimizing the sum of the residuals ($y_i - \hat{y}_i$) does not provide a unique solution (there are different solutions, all providing $\Sigma_i(y_i - \hat{y}_i) = 0$ because positive and negative residuals cancel each other out). But minimizing either the squared $(y_i - \hat{y}_i)^2$ or the absolute residuals $|y_i - \hat{y}_i|$ is feasible. With the freehand parameters above, the sum of squared residuals amounts to $SSR = \Sigma_i(y_i - \hat{y}_i)^2 = 10.125$ and the sum of absolute residuals amounts to $SAR = \Sigma_i|y_i - \hat{y}_i| = 7$. By "playing" a little bit with the parameters both sums can be made even smaller: $SSR = 9.956$ for $\hat{\beta}_0 = 2.867$ and $\hat{\beta}_1 = 0.489$; $SAR = 6$ for $\tilde{\beta}_0 = 2.8$ and $\tilde{\beta}_1 = 0.4$. Modern spreadsheet programs can do this search for the "best" parameters by using some kind of numerical optimization algorithm (see the Excel file on the web site). Hence, you do not need to do it manually. Minimizing $SSR = \Sigma_i(y_i - \hat{y}_i)^2$ is the *computational basis* of ordinary least squares estimation (OLS), while minimizing $SAR = \Sigma_i|y_i - \hat{y}_i|$ is the equivalent for least absolute difference estimation (LAD). Both computational procedures are easily extended to multiple regression models including several X variables. Again, parameter values can be found by some numerical optimization procedure. For OLS, there is also an analytical solution which significantly simplifies the "search" for the right parameters. Regression programs routinely apply these formulas for the OLS estimates $\hat{\beta}$. However, as the examples of LAD and ML show (for ML see Sec. 7.2.2), analytical formulas are not always available and numerical optimization has to be used instead.

In case of OLS estimation, a descriptive measure of model fit can be derived by splitting the total variation $SST = \Sigma_i(y_i - \bar{y}.)^2$ of the dependent variable Y into two components: one being explained by the statistical model ($SSE = \Sigma_i(\hat{y}_i - \bar{y}.)^2$) and the other being not explained by the model ($SSR = \Sigma_i(y_i - \hat{y}_i)^2$). The *coefficient of determination* is defined as

$$R^2 = \frac{SSE}{SST} = \frac{\Sigma_i(\hat{y}_i - \bar{y}.)^2}{\Sigma_i(y_i - \bar{y}.)^2} = 1 - \frac{SSR}{SST} = 1 - \frac{\Sigma_i(y_i - \hat{y}_i)^2}{\Sigma_i(y_i - \bar{y}.)^2} \quad (7.5)$$

Hence, knowing the variance σ_y^2 of the dependent variable and the sum of squared residuals SSR of the regression model, one can easily compute the coefficient of determination as

$$R^2 = 1 - \frac{SSR}{(n-1) \cdot \sigma_y^2} \quad (7.6)$$

With $0 \leq R^2 \leq 1$, it is usually interpreted as the share of the total variance that is explained by the model. One can think of SSR as an absolute measure of fit that increases with the sample size and the variance of the dependent variable. The coefficient of determination turns it into a measure of relative fit. There is also an adjusted coefficient of determination

7.2 Estimation and Testing

$$R^2 = 1 - \frac{SSR/(n-k-1)}{SST/(n-1)} = 1 - \frac{SSR/(n-k-1)}{\sigma_y^2} \qquad (7.7)$$

that accounts for the complexity of the model relative to the sample size n, by controlling for the number of estimated regressions coefficients k and one estimated regression constant.

7.2.1.2 Sampling and Sampling Errors

Having reviewed the computational basis of both OLS and LAD, we can now discuss the issue of estimation and testing. *Estimation* is about drawing valid inferences from a sample about a larger population from which the units of the sample have been randomly selected. More precisely, one wants to compute estimates of population parameters (e.g., parameters of a regression model) that deviate as little as possible from the "true" population parameters. On average, i.e., across different samples, the estimates should be identical with the population parameters (criterion of *unbiasedness*) and their deviations from the "true" parameters, i.e., the sampling errors or more specifically, the variance of the sampling errors, should be as small as possible (criterion of *efficiency*). Given certain assumptions about the regression model (see Textbox 4.1), statisticians can prove that OLS estimates of the regression parameters are unbiased and vary less (i.e., are more efficient) than all other linear estimators (i.e., they are the best linear unbiased estimators and therefore, are called BLUE). Instead of repeating this proof (it can be found in every textbook on regression analysis), we want to motivate the general idea with some real data. In order to check the unbiasedness and efficiency of the estimates, one needs to know the "true" population parameters, which is usually not the case. Therefore, the following example analyzes the unrealistic case of a known population. Although there is no need to draw a sample in that case (everything is known about the population), selecting random samples and estimating population parameters with these known data shows what can happen when making inferences from random samples.

Example 7.2 (wpgen data)   The example uses SOEP data (see also Example 2.2). The file wpgen includes data on education and monthly gross labor income for all employees interviewed in 2006. Let us assume that log earnings (Y, measured in Euro) is only a function of schooling (X, measured in years of education). All other determinants of earnings are assumed to be random, zero on average and independent of schooling.[1] An OLS regression of

---

[1] This is not a very realistic assumption, as we discuss in Example 3.2. The return to education in this simple schooling model is possibly overestimating the real effect of schooling. But this is not our problem here. We want to see whether it is possible to replicate the regression coefficient that is found for employees with this simple model in the wpgen data (the population) with a random sample from that population.

the simple schooling model shows that one year of schooling increases earnings by about 11 %: $y_i = 6.1187 + 0.1108 \cdot x_i + e_i$, with e_i including all other determinants of earnings as well as errors of measurement of earnings. Let us assume that the $n = 6{,}026$ employees in the wpgen data represent our population. In the following, we want to analyze whether we can replicate this finding also in random samples of this population.

Drawing random samples from a data set is a routine task with statistical software programs. We selected randomly $n = 300$ employees. Regressing earnings on schooling provided the OLS estimates $\hat{\beta}_0 = 6.2111$ and $\hat{\beta}_1 = 0.1042$, which obviously are not identical with the population parameters $\beta_0 = 6.1187$ and $\beta_1 = 0.1108$, although the deviations (the *sampling errors*) are rather small. We wanted to see whether that is due to the one specific random sample and therefore, selected additional random samples (altogether thousand random samples), each time estimating the same schooling model with OLS. The upper panel of Fig. 7.2 shows the distribution of the $k = 1{,}000$ OLS estimates of the schooling effect β_1. This distribution is also called the *empirical sampling distribution* of β_1. The true population parameter $\beta_1 = 0.1108$ is indicated by a vertical line and we see that the thousand estimates vary symmetrically around this value with most estimates in the vicinity of $\beta_1 = 0.1108$. More specifically, the average of all the estimates ($\Sigma_k \hat{\beta}_{1k}/1000 = 0.1113$) is almost identical to the true population parameter. This indicates that OLS provides unbiased estimates. However, each single estimate is affected by a more or less strong sampling error. In fact, estimates vary from $\hat{\beta}_1 = 0.0288$ to $\hat{\beta}_1 = 0.1936$ with an overall standard deviation of $s_{\hat{\beta}_1} = 0.0229$.

This sampling experiment (also called a *simulation* study with thousand *replicates*) illustrates the statistical proof of unbiased OLS estimates (the issue of efficiency will be discussed later). Using the assumptions in Textbox 4.1, statisticians derive the *theoretical* sampling distribution of β_1 mathematically. According to this proof, OLS estimates of β_1 are normally distributed with a normal distribution centered at the population value of β_1. Empirical sampling distributions derived from simulation experiments like ours approximate this theoretical distribution. Furthermore, statisticians derive a formula for the standard deviation of the regression estimates (i.e., for the spread of the normal distribution) and show that this standard deviation $\sigma_{\hat{\beta}}$ is smaller than the standard deviation of the estimates of any other kind of linear estimator. More specifically, they show that the variance of the estimates of a particular regression coefficient β_j is a function of (i) the variance σ^2 of the error terms e_i, (ii) the variation SST_j of the corresponding independent variable X_j in the sample and (iii) how X_j (in the multivariate case) is related to the other independent variables in the model (as measured by the coefficient of determination R_j^2 when regressing X_j on all other independent variables in the model):

$$\sigma_{\hat{\beta}_j} = \sqrt{\frac{\sigma^2}{SST_j \cdot (1 - R_j^2)}} \quad \text{with } SST_j = \sum_{i=1}^{n}(x_{ij} - \bar{x}_{.j})^2 \quad (7.8)$$

7.2 Estimation and Testing

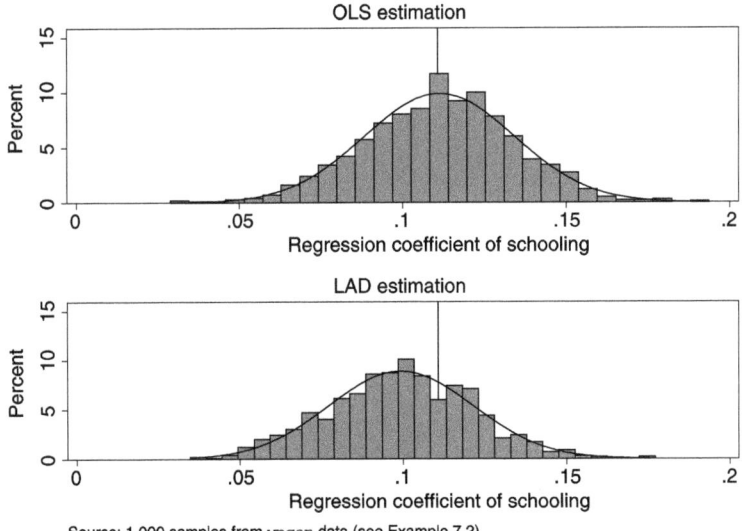

Fig. 7.2 Empirical sampling distribution of β_1

The standard deviation of a sample statistic (in our example, the estimated regression coefficient $\hat{\beta}_j$) across different samples is also called a standard error. Therefore, (7.8) shows the *standard error* of the regression coefficient β_j.

If we interpret $\sigma_{\hat{\beta}_j}$ as a measure of imprecision of the estimate, the components of the formula make intuitively sense: (i) The more determinants of Y that are not controlled for by the regression model and the less reliable Y is measured (hence, the larger σ^2), the less precise the estimates. (ii) Imprecision is lower, the more different conditions one observes for the effect in the sample, i.e., the more X_j varies (and hence, SST_j increases). For example, if earnings are assumed to be a function of schooling, testing this assumption with individuals with hardly differing levels of education (say, 10 and 12 years) provides less reliable estimates than performing this test with individuals having many different levels of education (say, 10, 12, 16 and 18 years). (iii) Finally, the lower is the imprecision, the less variable X_j correlates with the other independent variables in the model (and hence, the lower R_j^2) and consequentially, the easier it is to distinguish its effect from the effects of the other independent variables.

7.2.1.3 How to Choose Between Different Estimation Methods?

Very often different estimators are available to estimate the parameters of a statistical model. For example, instead of using OLS, the parameters of the linear regression model could be estimated with LAD. This raises the question, which estimation procedure is better and should be preferred in empirical work. To illustrate this decision, we have done a similar simulation experiment with LAD, the results of which are shown in the lower panel of Fig. 7.2. Without knowing what LAD is doing, one would clearly give up LAD in favor of OLS estimation, because the es-

Fig. 7.3 How to choose between different estimators

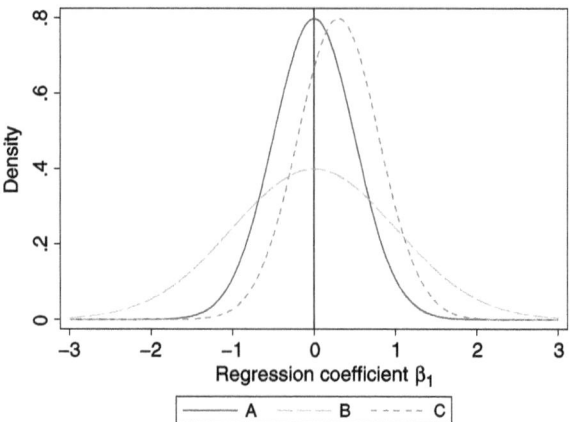

timates seem to be biased downward. As Fig. 7.2 shows, LAD estimates are again approximately normally distributed with about the same standard deviation as the OLS estimates, but on average they are smaller than the true population parameter ($\Sigma_k \tilde{\beta}_{1k}/1{,}000 = 0.0993$).

However, this seemingly unpleasant result comes as no surprise, if one takes into account what LAD estimation is doing. While OLS is estimating the expected *arithmetic mean* of Y for a given level of the independent variables, LAD is estimating the expected *median* of Y. Earnings distributions are known to be skewed (even after a logarithmic transformation). Obviously, in the wpgen data this skewness increases with rising levels of schooling resulting in a less steep increase in median compared to mean earnings. This explains the on average lower estimates of the schooling effect when using LAD.[2] Therefore, if we would have used an example with a symmetrically distributed dependent variable (where median and mean are identical), we would have achieved similar unbiased estimates as in the case of OLS estimation. However, the variance of the LAD estimates would be slightly larger than the variance of the OLS estimates (technically, LAD estimates are less efficient than OLS estimates). This is a reason to prefer OLS over LAD estimates. Although both of them are unbiased (i.e., correct on average), OLS estimates in single samples are expected to deviate less from the true population parameter than LAD estimates. This is clearly an advantage, if replicated sampling is not possible, as it is typical for the social sciences. Moreover, the fact that OLS estimates can be derived analytically makes them computationally easier than LAD estimates.

Figure 7.3 illustrates the choice between different estimators in a more general perspective. It shows the theoretical sampling distribution of the estimates $\hat{\beta}_1$, if one uses either method A, B or C to estimate the regression coefficient β_1. For

---

[2]Therefore, defining the LAD estimates in Fig. 7.2 as "biased" is not quite correct. Indeed, they measure the relationship between *median* earnings $\tilde{Y}$ and schooling X quite correctly. In the population this relationship is estimated as $\tilde{y}_i = 6.4135 + 0.1000 \cdot x_i + \tilde{e}_i$ and $\Sigma_k \tilde{\beta}_{1k}/1000 = 0.0993$ comes very close to the regression coefficient 0.1000 of schooling.

7.2 Estimation and Testing

simplicity we have assumed that each sampling distribution has the shape of a normal distribution. β_1 equals zero in the population (see the vertical line at $\beta_1 = 0$ in Fig. 7.3). Obviously, estimation methods A and B provide unbiased estimates, because the center of both sampling distributions is located at $\beta_1 = 1$. However, method B is less efficient than method A, because the spread of its sampling distribution is larger. When comparing OLS and LAD estimation, we have exactly the same situation: Both methods are unbiased, but OLS is more efficient than LAD. Finally, method C shows an example of a biased estimator. Usually, methods A and B would be preferred over C, because one would like to have estimators that provide the true population parameter, at least on average. But there are situations, where the choice is not that obvious and the focus on unbiasedness may be misleading.

As an example compare methods B and C. Although the average of the estimates provided by C does not equal $\beta_1 = 0$, the majority of the estimates (e.g., the 50 % of the estimates around the center of the distribution) is closer to the true population parameter than the middle 50 % of the estimates provided by B, simply because the sampling distribution of C is much more widespread than the sampling distribution of B (or, equivalently, because method B is less efficient than C). Hence, in a single sample, method C may provide an estimate that is much closer to the true population parameter than method B. This shows that the sole focus on unbiasedness may be misleading in some applications, in which unbiasedness and efficiency should be evaluated together.[3]

This was the case when comparing FE and RE estimation (see Sect. 4.1.2.4). FE estimates are unbiased under more general conditions than RE estimates, which assume that unobserved heterogeneity is uncorrelated with the variables in the model. But RE estimation provides less widespread (more efficient) estimates than FE regression and therefore, in a single sample and under certain conditions, may provide estimates closer to the true population parameter than FE estimation. As noted in Sect. 4.1.2.4, these conditions should be researched in greater detail by simulation studies.

7.2.1.4 How to Estimate the Parameters of an Unknown Population with a Sample of Data?

Our simulation study started from a known population and illustrated the statistical proof of unbiased and efficient OLS estimates. But what is the utility of this proof for the more relevant problem of estimating parameters from a sample for an *unknown* population? The statistical reasoning starts again from a known population and the results of the former simulation experiment. Knowing that the sample estimates are normally distributed around the true population parameter, one can predict an interval in which one expects with a given probability the parameter estimates. This is illustrated in the upper panel of Fig. 7.4. For example, in our case, we expect 95 % of the estimates in the interval ranging from $ll = (0.1108 - z_{(1-0.975)} \cdot \sigma_{\hat{\beta}_j})$ to

---

[3] One way to do that quantitatively is to compute the *mean square error*, which is a function of both the bias and the variance of the estimates.

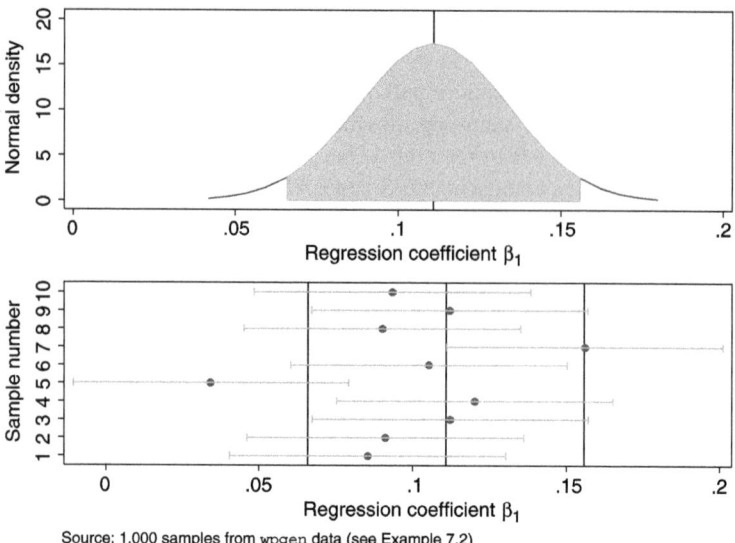

Fig. 7.4 Drawing inferences from a sample about an unknown population

$ul = (0.1108 + z_{(1-0.975)} \cdot \sigma_{\hat{\beta}_j})$ (see the shaded area under the normal distribution in the upper panel of Fig. 7.4). $z_{(1-0.975)} = 1.96$ is the 97.5th percentile value of the normal distribution.[4] This result can then be used for the reverse (and more realistic) case with a given sample and an unknown population. This is illustrated in the lower panel of Fig. 7.4. Instead of using the population parameter $\beta_1 = 0.1108$, one would insert the sample estimate $\hat{\beta}_1$ in the formulas and derive a *confidence interval* based on the sample estimate. However, depending on how much the sample estimate $\hat{\beta}_1$ itself is different from the true parameter β_1, this confidence interval does or does not include the population parameter (see the confidence intervals for sample 5 and 7 in the lower panel of Fig. 7.4. Only if the sample estimate $\hat{\beta}_1$ is one of the many estimates in the aforementioned interval $[ll, ul]$ (indicated by the vertical lines in the lower panel of Fig. 7.4, then the confidence interval based on the sample estimate will include the true population parameter and hence, will make a correct prediction. Since the interval $[ll, ul]$ includes 95 % of all possible sample estimates, this kind of reasoning will be correct in 95 % of the cases. Therefore, it is interpreted as a confidence interval that includes the true parameter with a probability of 95 % (for short, the 95 % confidence interval). However, social science applications never apply repeated sampling and with only one single sample at hand, one can never be sure whether its estimate of β_1 is within or without the interval $[ll, ul]$. Therefore,

---

[4]If one would be interested in 80 % of the estimates, one would have to use the 90th percentile value of the normal distribution. Generally speaking, one has to use $z_{(1-\alpha/2)}$ with α measuring the share of extreme z values either far below or far above the true population parameter (i.e., the z values in the white areas under the normal distribution in the upper panel of Fig. 7.4).

7.2.1.5 How to Test Parameters of an Unknown Population with a Sample of Data?

Similarly, one can *test hypotheses* about population parameters, say the hypothesis that an independent variable X_j has no effect and the corresponding regression coefficient equals zero ($H_0 : \beta_j^0 = 0$). The reasoning is pretty much the same as in the case of estimation. Since $\beta_j^0 = 0$ is a statement about the population, one computes first of all an interval $[ll, ul]$, in which one would expect with a certain probability sample estimates if the hypothesis would be true. One uses the same formulas as before, except that the true population parameter is exchanged for the value assumed in the null hypothesis: $ll = (\beta_j^0 - z_{(1-\alpha/2)} \cdot \sigma_{\hat{\beta}_j})$ and $ul = (\beta_j^0 + z_{(1-\alpha/2)} \cdot \sigma_{\hat{\beta}_j})$. If the sample estimate $\hat{\beta}_j$ falls outside the corresponding interval, the null hypothesis is rejected. α is now the *significance level* of the test. In case of a two-sided alternative hypothesis ($H_1 : \beta_j^1 \neq 0$), one has to use the $(1 - \alpha/2)$ quantile value $z_{(1-\alpha/2)}$ of the normal distribution to compute the interval. In case of a one-sided alternative hypothesis, one has to use the $(1 - \alpha)$ quantile value $z_{(1-\alpha)}$ of the normal distribution and the interval is either $[ll, +\infty]$ in case of $H_1 : \beta_j^1 < 0$ or $[-\infty, ul]$ in case of $H_1 : \beta_j^1 > 0$. Alternatively, one can ask how far—relative to the spread of the normal distribution (as measured by $\sigma_{\hat{\beta}_j}$)—the sample estimate $\hat{\beta}_j$ is away from the value β_j^0 assumed in the null hypothesis:

$$z = \frac{\hat{\beta}_j - \beta_j^0}{\sigma_{\hat{\beta}_j}} \quad (7.9)$$

If this so-called Z *statistic* falls outside the interval $[-z_{(1-\alpha/2)}, +z_{(1-\alpha/2)}]$ in case of a two-sided alternative hypothesis, the null hypothesis is rejected. In case of a one-sided alternative hypothesis, the corresponding interval is either $[-z_{(1-\alpha)}, +\infty]$ in case of $H_1 : \beta_j^1 < 0$ or $[-\infty, +z_{(1-\alpha)}]$ in case of $H_1 : \beta_j^1 > 0$.

In practice, estimation and testing is a little bit more complicated, because in order to compute the standard error of the estimated regression coefficient one needs the variance σ^2 of the error terms e_i, which are unknown by definition. Hence, they have to be estimated from the residuals $(y_i - \hat{y}_i)$ of the model and the formula for the standard error turns into:

$$\hat{\sigma}_{\hat{\beta}_j} = \sqrt{\frac{\hat{\sigma}^2}{SST_j \cdot (1 - R_j^2)}} \quad \text{with } SST_j = \sum_{i=1}^{n}(x_{ij} - \bar{x}_j)^2 \text{ and } \hat{\sigma}^2 = \frac{\sum_{i=1}^{n}(y_i - \hat{y}_i)}{n - k - 1}$$

(7.10)

$\hat{\sigma}^2$ is called *standard error of the regression* or *root mean squared error*. It is something like an average squared residual, where the sample size n has been corrected for the lost degrees of freedom due to k estimated regression coefficients and one estimated regression constant. The larger the sample size and the less complicated the model in terms of the number of parameters (i.e., the more degrees of freedom available), the smaller the root mean squared error and consequently, the standard error of the estimated regression coefficient $\hat{\beta}_j$.

Since one of the three components of the standard error has been estimated, (7.10) shows the *estimated* standard error of the regression coefficient (as opposed to the *theoretical* standard error in (7.8)). Similar to estimates of regression coefficients, the sample estimate of the error variance $\hat{\sigma}^2$ can be different from its "true" value σ^2 in the population and accordingly, estimation and testing have to be adapted to this additional imprecision. This is done by using Student's T distribution instead of the normal distribution. There is a statistical reason why exactly the T distribution is preferred over the normal distribution, but this is not of interest here. From a practical point of view, not much changes. In the former formulas, one only has to exchange the z values by the quantile values $t_{(1-\alpha/2,\, n-k-1)}$ resp. $t_{(1-\alpha,\, n-k-1)}$ of the T distribution. Hence, the confidence interval is defined as follows

$$ll = \hat{\beta}_j - t_{(1-\alpha/2,\, n-k-1)} \cdot \hat{\sigma}_{\hat{\beta}_j} \leq \beta_j \leq \hat{\beta}_j + t_{(1-\alpha/2,\, n-k-1)} \cdot \hat{\sigma}_{\hat{\beta}_j} = ul \qquad (7.11)$$

and (7.9) is turned into a so-called T *statistic*,

$$t = \frac{\hat{\beta}_j - \beta_j^0}{\hat{\sigma}_{\hat{\beta}_j}} \qquad (7.12)$$

which has to be compared to a T distribution to decide the test. The T distribution looks like the bell-shaped normal distribution, but in smaller samples it provides larger confidence intervals (i.e., less precise estimates) and more conservative test statistics. This is exactly what we would like to do to control for the additional imprecision due to estimating σ^2.

Besides testing single regression coefficients, there are different methods for testing several regression coefficients and/or for testing (linear) relationships between regression coefficients. A standard test for the overall fit of a regression model is for example the hypothesis that none of the independent variables has an effect significantly different from zero. For a trivariate regression model with two independent variables this includes essentially two null hypotheses a and b: $H_0^a : \beta_1 = 0$ and $H_0^b : \beta_2 = 0$. Or think about an extended earnings model that includes besides schooling X_1 a measure of labor market experience X_2 (both variables measured in years). One may want to test the hypothesis that the return to experience is as large as the return to schooling. In this example the null hypothesis assumes $H_0^a : \beta_1 = \beta_2$ or equivalently $\beta_1 - \beta_2 = 0$, which implies again one null hypothesis a, but now including a relationship between two parameters of the model. As a final example, let us assume that a researcher uses in his regression model instead years of education three different *levels* of education. Medium and high educational level are measured by two dummy variables, while low educational level is used as the reference category. In a model that includes only the two dummies, the regression constant estimates average earnings for employees with low educational level, while the regression coefficients of the two dummies estimate how average earnings for

7.2 Estimation and Testing

employees with medium (β_1) and high (β_2) educational level differ from average earnings in the reference group. Now the researcher wants to test whether the increase in earnings over different educational levels is linear, which implies that the regression coefficient of the "high" dummy is twice as large as the coefficient of the "medium" dummy: $H_0^a : \beta_2 = 2\beta_1$ or equivalently, $\beta_2 - 2\beta_1 = 0$.

Before showing how to perform these tests, let us first see how to formalize these slightly more complicated null hypotheses. This is easily done by using a table that includes as many rows as there are hypotheses and as many columns as there are parameters in the model. For the three examples these tables look like the following:

Example 1	β_0	β_1	β_2		
H_0^a	0	1	0	$\Rightarrow$	$0 \cdot \beta_0 + 1 \cdot \beta_1 + 0 \cdot \beta_2 = 0$
H_0^b	0	0	1	$\Rightarrow$	$0 \cdot \beta_0 + 0 \cdot \beta_1 + 1 \cdot \beta_2 = 0$

Example 2	β_0	β_1	β_2		
H_0^a	0	1	-1	$\Rightarrow$	$0 \cdot \beta_0 + 1 \cdot \beta_1 - 1 \cdot \beta_2 = 0$

Example 3	β_0	β_1	β_2		
H_0^a	0	-2	1	$\Rightarrow$	$0 \cdot \beta_0 - 2 \cdot \beta_1 + 1 \cdot \beta_2 = 0$

As the arrows indicate, these tables should be read row-wise. In each cell, the number has to be multiplied with the corresponding (column) parameter and all products within a row have to be added. If one constrains each row sum to be equal to zero, one replicates the aforementioned hypotheses. Obviously, the numbers in the tables define a matrix of linear constraints (or restrictions) on the parameters of the model. Therefore, it is called a constraint matrix **C** and the multiplication, addition and setting equal to zero is easily specified with matrix algebra: $\mathbf{C}\beta = 0$. In this matrix equation, β is a column vector including all parameters of the model.

You do not need knowledge of matrix algebra to understand the following test procedures, but the concept of constraints resp. restrictions is useful as a general way of testing parameters including many classical test procedures as special cases (e.g., the former T test is based on a contrast matrix, which has only one row including only one 1). There are basically two approaches for testing the restrictions implied in **C**. One is based on comparing the fit of two hierarchically nested models M_u and M_r, in which the *restricted model* M_r includes only a subset of the parameters of the *unrestricted model* M_u. Or to put it differently: While in M_u all parameters have to be estimated, in model M_r only a subset of the parameters has to be estimated, because some of the parameters in M_r have to have specific values defined in the constraint matrix **C**. A good example for this kind of reasoning is the classical F test of the overall model fit. Its null hypothesis assumes that none of the independent variables has an effect (see the constraint matrix for the aforementioned example 1). It is equivalent to comparing a model M_u including all the independent variables with a model M_r that only estimates a regression constant and hence, implicitly assumes that all regression coefficients are restricted to be zero except the constant.[5] If SSR_u is the sum of squared residuals of the unrestricted model and SSR_r the corresponding sum for the restricted model, then the test statistic is defined as follows:

$$f = \frac{(SSR_r - SSR_u)/q}{SSR_u/(n-k-1)} \quad (7.13)$$

It is F distributed with $df_1 = q$ and $df_2 = n-k-1$ degrees of freedom. q equals the number of restrictions tested (i.e., the number of rows in **C**) and k equals the number of independent variables in the unrestricted model. If the f statistic is significant, at least one hypothesis specified in the constraint matrix has to be rejected. In case of example 1 that would mean that at least one independent variable has a significant effect.[6]

The second approach for testing linear restrictions is less intuitive. It is based on the estimation results for one single model and uses both the estimated parameters and their estimated variances and covariances. For instance, if you want to test the difference between two regression coefficients, say $(\hat{\beta}_1 - \hat{\beta}_2)$, you would need the standard error of the difference $\hat{\sigma}_{(\hat{\beta}_1 - \hat{\beta}_2)}$. As shown above, both regression coefficients are normally distributed random variables with variances $\hat{\sigma}^2_{\hat{\beta}_1}$ and $\hat{\sigma}^2_{\hat{\beta}_2}$. Since the difference is a linear function of both random variables, it is also a normally distributed random variable. Using the algebra of variances and covariances (see Sect. 7.1), its variance can be computed as follows: $\hat{\sigma}^2_{(\hat{\beta}_1-\hat{\beta}_2)} = \hat{\sigma}^2_{\hat{\beta}_1} + \hat{\sigma}^2_{\hat{\beta}_2} - 2 \cdot \hat{\sigma}_{\hat{\beta}_1, \hat{\beta}_2}$. The term $\hat{\sigma}_{\hat{\beta}_1, \hat{\beta}_2}$ is the estimated covariance of both parameter estimates.[7] Knowing the estimated parameter difference and its estimated standard error, one can easily test whether it is different from zero by using a standard T test,

$$t = \frac{(\hat{\beta}_1 - \hat{\beta}_2) - 0}{\hat{\sigma}^2_{(\hat{\beta}_1-\hat{\beta}_2)}} \quad (7.14)$$

It can also be shown that the square of the test statistic, t^2, is distributed as χ^2 with $df = 1$ degree of freedom. The same result can be achieved with the constraint matrix $\mathbf{C} = [0 + 1 - 1]$ from the aforementioned example 2 by solving the following matrix equation:

---

[5] How a restricted model can be specified in case of example 2 or 3 is not that obvious, but it can be done. We do not discuss this topic, because most statistical software supplies commands for testing linear restrictions and the user has not to bother about specifying a restricted model.

[6] The sum of squared residuals in the constant only model equals the total variation of Y and hence, $SSR_r = SST$ in the constant only model. Moreover, with the constant only as the restricted model, $SSR_r - SSR_u$ in (7.13) equals $SST - SSR_u = SSE$ and the number of restrictions (compared to the full model) equals the number of independent variables: $q = k$. Inserting all of this into (7.13) results in the well-known f statistic for the overall test of model fit: $f = (SSE/k)/[SSR/(n - k - 1)]$.

[7] Parameter estimates do not only vary between different samples, but also show some kind of covariation across samples. This is immediately evident from the former simulation study. For example, if the regression coefficient β_1 is slightly overestimated, the regression constant most probably is slightly underestimated (and vice versa).

7.2 Estimation and Testing

$$W^2_{r/u} = (C\hat{\beta} - 0)'(C\hat{V}_{\hat{\beta}}C')^{-1}(C\hat{\beta} - 0) = (C\hat{\beta})'(C\hat{V}_{\hat{\beta}}C')^{-1}C\hat{\beta} \qquad (7.15)$$

This so-called *Wald statistic* compares again an unrestricted and a restricted model; now by using a constraint matrix C, the vector of estimated parameters $\hat{\beta}$ and a matrix $\hat{V}_{\hat{\beta}}$ including all estimated variances and covariances of the parameter estimates. Again, you do not need to understand this matrix equation, because many statistical software packages have already implemented this test. $W^2_{r/u}$ is distributed as χ^2 with $df = q$ degrees of freedom, q being the number of restrictions tested (i.e., the number of rows of the constraint matrix C). In the simple example 1, $q = 1$ and $W^2_{r/u}$ will equal t^2 as computed from (7.14). But obviously, the advantage of the Wald statistic is that it can test much more general hypotheses.

In sum, there are different choices when testing model fit and parameter restrictions: T tests for testing single restrictions with the possibility of using one-sided alternative hypotheses, F tests based on fit measures of two different (restricted and unrestricted) models, and Wald tests that use the information of the estimated variances and covariances of the estimates in the current model. In the case of linear models, all three alternatives will provide identical results. This is different for non-linear models and ML estimation, as we will see in Sect. 7.2.2. Moreover, testing can be made robust against violations of the OLS assumptions of homoscedasticity and absence of autocorrelation by using so-called *empirical standard errors* (or more generally, empirical variances and covariances). While the former *theoretical* standard error (7.10) is derived analytically from these assumptions, robust standard errors are computed empirically from the distribution of residuals of the model. How this is done in detail, is beyond the scope of this section.

7.2.2 Maximum Likelihood

Now let us return to the sixcases data (see Example 7.1) to explain maximum likelihood (ML) estimation. Let us assume that instead of the original continuous values we observe two different categories $y_i = 1$ and $y_i = 0$. The first category is observed for units $i = 3, 4$ and 6 (i.e., when the original Y was larger or equal to 5) and the second category is observed for units $i = 1, 2$ and 5 (i.e., when the original Y was smaller than 5). In case of a categorical dependent variable, the idea of ML is easy to understand. ML estimates of the regression coefficients are defined as those estimates that—given the independent variables X—maximize the joint probability of observing the values of the dependent variable Y. Hence, if we want to know the probability of observing all the ones and zeros in the sample, we have to multiply all the probabilities of observing either $y_i = 1$ or $y_i = 0$ for each unit $i = 1, \ldots, n$. If the six observations in the sixcases data are coded 0, 0, 1, 1, 0, and 1, their overall likelihood would look like this:[8]

---

[8] Remember that the probability of a combined event that results from an *and*-combination of the single events equals the *product* of the corresponding probabilities.

$$L = \left(1 - \Pr(y_1 = 1)\right) \cdot \left(1 - \Pr(y_2 = 1)\right) \cdot \Pr(y_3 = 1)$$
$$\cdot \Pr(y_4 = 1) \cdot \left(1 - \Pr(y_5 = 1)\right) \cdot \Pr(y_6 = 1)$$

According to the logistic regression model, the probability of observing 1—given the value of X—equals the following expression:

$$\Pr(y_i = 1) = \frac{\exp(\beta_0 + \beta_1 x_i)}{1 + \exp(\beta_0 + \beta_1 x_i)} \tag{7.16}$$

For the whole sample of n units, the likelihood function equals the following expression:

$$L(\beta_0, \beta_1) = \prod_{i=1}^{n} \left[\Pr(y_i = 1)\right]^{y_i} \cdot \left[1 - \Pr(y_i = 1)\right]^{(1-y_i)}$$

$$= \prod_{i=1}^{n} \left[\frac{\exp(\beta_0 + \beta_1 x_i)}{1 + \exp(\beta_0 + \beta_1 x_i)}\right]^{y_i} \cdot \left[1 - \frac{(\beta_0 + \beta_1 x_i)}{1 + \exp(\beta_0 + \beta_1 x_i)}\right]^{(1-y_i)} \tag{7.17}$$

This likelihood function looks quite formidable, but think of the two exponents y_i and $(1 - y_i)$ like switches that switch on either $\Pr(y_i = 1)$ or $(1 - \Pr(y_i = 1))$ depending on whether the dependent variable for the corresponding observation equals $y_i = 1$ or $y_i = 0$ (and hence, $1 - y_i = 1$). If you now insert into (7.17) some arbitrary starting values β_0^0 and β_1^0 for the ML estimates, the values of the independent variable X and the values of the dependent binary variable Y for each unit $i = 1, \ldots, n$, (7.17) will provide the overall likelihood of the observed sample given the parameters β_0^0 and β_1^0: $L(\beta_0^0, \beta_1^0)$. Since the likelihood equals the product of many small numbers (probabilities are numbers in the unit interval $[0, 1]$), $L(\beta_0, \beta_1)$ itself is a very small number and not very practical to search for the estimates that maximize it. It is also much easier to maximize an additive function and therefore, one uses the natural logarithm of the likelihood, which—since it is a monotone transformation—provides exactly the same parameters that also maximize $L(\beta_0, \beta_1)$:

$$\ln L(\beta_0, \beta_1) = \sum_{i=1}^{n} y_i \cdot \ln\left(\frac{\exp(\beta_0 + \beta_1 x_i)}{1 + \exp(\beta_0 + \beta_1 x_i)}\right)$$
$$+ \sum_{i=1}^{n} (1 - y_i) \cdot \ln\left(1 - \frac{\exp(\beta_0 + \beta_1 x_i)}{1 + \exp(\beta_0 + \beta_1 x_i)}\right) \tag{7.18}$$

Since the likelihood function of the logistic regression model is "well-behaved" (i.e., has a maximum that is easy to find), one can choose any starting values for the regression coefficients that look plausible. For example, a good starting point is the assumption that the independent variable has no effect ($\beta_1^0 = 0$) and the probability of observing a 1 equals the proportion of ones, $\hat{\pi}$, in the sample. This assumption results in $\beta_0^0 = \ln(\hat{\pi}/(1 - \hat{\pi}))$. If you now use the starting values $\beta_1^0 = 0$

7.2 Estimation and Testing

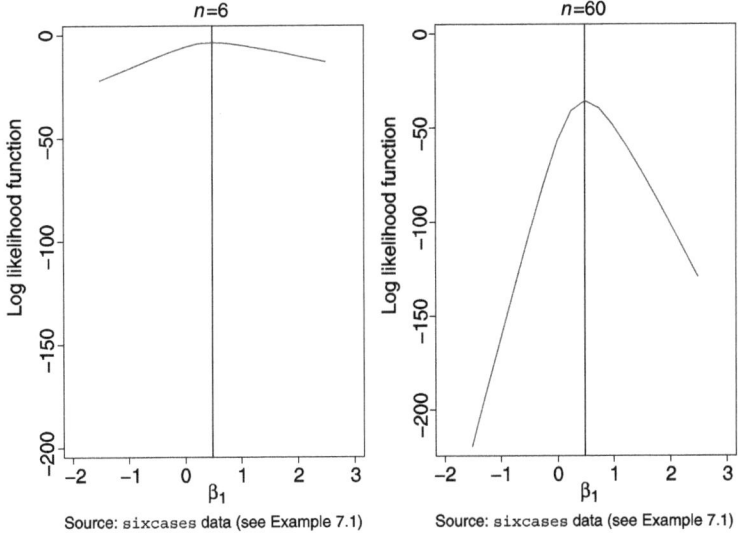

Fig. 7.5 Log likelihood function (β_1 varied, β_0 fixed at $\beta_0 = -1.4606$)

and $\beta_0^0 = 0$ ($\hat{\pi} = 0.5$ for the sixcases data), the values of the independent variable X and the values of the dependent binary variable Y for each observation, (7.18) provides the logarithm of the overall probability of the observed sample given the two chosen starting values. For the sixcases data, we arrive at a value of $\ln L(\beta_0^0, \beta_1^0) = -4.1589$.[9] Obviously, the result depends on the chosen starting values. You may try other values and in doing so, try to maximize $\ln L(\beta_0, \beta_1)$ (see the Excel file on the web site). This is a cumbersome task, especially in a multivariate model with many independent variables. Contrary to OLS estimation, there is also no analytical solution to find the parameters that maximize $\ln L(\beta_0, \beta_1)$. Therefore, you need numerical optimization algorithms that solve this problem efficiently for you. The parameters β_0 and β_1 that maximize the log likelihood are called *ML estimates*. In case of the sixcases data they amount to $\hat{\beta}_0 = -1.4606$ and $\hat{\beta}_1 = 0.4869$ and the maximum of the likelihood equals $L(\hat{\beta}_0, \hat{\beta}_1) = -3.5782$. The left panel of Fig. 7.5 shows the log likelihood function for different values of β_1, assuming that the ML estimate $\hat{\beta}_0 = -1.4606$ for the regression constant has already been found by the numerical optimization algorithm. The vertical line at 0.4869 shows that at this point the log likelihood is maximal and hence, $\hat{\beta}_1 = 0.4869$ is the ML estimate of the regression coefficient of X.

A more difficult question is how variances and covariances of the ML estimates can be estimated. Again, and contrary to the case of OLS estimation, there are no

---

[9] As already mentioned, the likelihood (7.17) is a number smaller than 1 and hence, the log likelihood (7.18) is a negative number. At this point you have to be very attentive to what your statistical software prints out. Since dealing with negative numbers may be confusing, some programs print the *negative* log likelihood, which is defined as $\mathcal{L} = -\ln L$.

analytical formulas for the (co)variances. However, they depend on similar quantities as in the case of OLS estimation, such as sample size (negatively), variance of the independent variable (negatively), and multicollinearity among the independent variables (positively). This is illustrated in the right panel of Fig. 7.5, where we have again used the `sixcases` data, however duplicating each observation ten times. In other words, the analysis is now based on a much larger sample, but still on the same non-linear (logistic) relationship between X and the categorized Y (because no new x and y values have been introduced). The likelihood function is now much steeper and consequently, it is much more easy to identify precisely the maximum, i.e., the ML estimates of β_0 and β_1. This result is exactly what we would expect in larger samples: They provide us with more precise estimates of the model parameters, because they include more information.

Obviously, the precision of the estimate (and hence, its standard error) is related to the curvature of the likelihood function. The smaller the sample size, the flatter the likelihood function and the larger the imprecision resp. the standard error. Mathematically, the maximum of a function is found by computing its first derivative with respect to the parameter in question, while the curvature at the maximum is found by computing the second derivative (note that in a multivariate model one computes partial derivatives). Therefore, variances and covariances of ML estimates are computed from the second partial derivatives of the log likelihood function with respect to the regression parameters. More specifically, the variance-covariance matrix of the ML estimates equals the inverse of the negative expected value of the matrix of second derivatives of the log likelihood function. The formulas for this computation are more difficult than in the linear case with OLS estimation, but as already mentioned, their main determinants are similar. Standard errors decrease with sample size, variation of the corresponding independent variable and its independence from the other explanatory variables in the model. Furthermore, a robust version of the variance-covariance matrix is also available.

As noted in Sect. 5.1.1.2, ML estimates are asymptotically normally distributed given the assumptions specified in Textbox 5.1. The fact they are only asymptotically normal implies that we need a sufficiently large sample ($n > 100$) for estimation and testing. The behavior of ML estimates in small samples is often unclear. Therefore, we have to use the normal (and not the T) distribution and z statistics to test the significance of single parameters

$$z = \frac{\hat{\beta}_j - \beta_j^0}{\hat{\sigma}_{\hat{\beta}_j}} \qquad (7.19)$$

and to compute confidence intervals

$$ll = \hat{\beta}_j - z_{(1-\alpha/2)} \cdot \hat{\sigma}_{\hat{\beta}_j} \leq \beta_j \leq \hat{\beta}_j + z_{(1-\alpha/2)} \cdot \hat{\sigma}_{\hat{\beta}_j} = ul \qquad (7.20)$$

For testing model fit and more general parameter restrictions the same kind of reasoning applies like in the linear case with OLS estimation. Either one compares the fit of an appropriately restricted model with the unrestricted model or one specifies

7.2 Estimation and Testing

a constraint matrix $\mathbf{C}$ and computes a Wald test. In the latter case, the Wald statistic is computed using (7.15). In the former case, an equivalent to the (linear) fit measure *SSR* is needed. This is again the maximum of the fit function, which in case of ML estimation is the value $\ln L$ of the maximized log likelihood function. Whether restricted and unrestricted model differ significantly is tested with the following *likelihood ratio test*:

$$L^2 = -2 \cdot \ln \frac{L_r}{L_u} = 2 \cdot (\ln L_u - \ln L_r) \qquad (7.21)$$

It is distributed as χ^2 with $df = q$ degrees of freedom, q being again the number of restrictions tested. For example, one can test whether the inclusion of X in case of the former sixcases data ($n = 6$) provides a significant improvement over a model that includes only the regression constant. The test statistic amounts to $L^2 = 2 \cdot (-3.5782 - -4.1589) = 1.1614$, which is not significant with $df = 1$ degree of freedom ($p = 0.2812$). Contrary to the linear case and OLS estimation, Z tests, Wald tests and likelihood ratio tests are only asymptotically equivalent in case of ML estimation and hence, may lead to different conclusions the smaller the sample size.

Apart from estimation and testing, there is also the question of how to describe the fit of a model estimated with ML. Unfortunately, there is no good equivalent to the coefficient of determination that can be interpreted as the amount of explained variance. There are various attempts to define similar measures for ML estimation, all of which only have similar mathematical properties (i.e., are numbers between 0 and 1), but should not be interpreted as measures of explained variance. The most prominent one is McFadden's Pseudo R^2, which is also used in this textbook:

$$\text{Pseudo } R^2 = \frac{\ln L_0 - \ln L_u}{\ln L_0} \qquad (7.22)$$

u refers to the model of interest, while 0 refers to a model that only includes a constant (the so-called *null model*). For example, in case of the sixcases data ($n = 6$), the Pseudo R^2 for the model including X amounts to Pseudo $R^2 = (-4.1589 - -3.5782)/(-4.1589) = 0.1396$. Note that Pseudo R^2 values are usually much smaller than R^2 values in linear regression (Pseudo $R^2 < 0.05$ indicates low fit, Pseudo $R^2 > 0.20$ indicates a very good fit, and Pseudo $R^2 > 0.40$ is hardly observed).[10]

Alternatively, one can compute Akaike's information criterion *AIC* or its modification *BIC*. Both fit measures are computed using the value $\ln L$ of the maximized log likelihood function:

$$\begin{aligned} AIC &= -2\ln L + 2 \cdot (k+1) \\ BIC &= -2\ln L + \ln n \cdot (k+1) \end{aligned} \qquad (7.23)$$

---

[10] Note also that the maximum value (Pseudo $R^2 = 1$) is only a theoretical value. It would imply that all predicted probabilities would be either 1 or 0; a distribution, which is impossible to fit with a logistic curve.

($k+1$) measures the number of estimated parameters (k regression coefficients plus one regression constant) and n the sample size. The smaller *AIC* or *BIC*, the better the fit of the model. *AIC* and *BIC* are often used when comparing models that are not hierarchically nested. In principle, *AIC* and *BIC* can also be computed for linear models, because the parameters of a linear model can be estimated with ML and the ML estimates are identical with the OLS estimates.

7.3 Web Site of the Textbook

All computations, estimations, and most of the figures for this textbook have been made with the statistical software package Stata. The book's web site (eswf.uni-koeln.de/panel) provides all necessary data sets and Stata syntax files to replicate our findings. For readers not being familiar with this software we also include the printed output, so that they can follow the computations without having to apply the software itself. In the future, the web site may also provide syntax files for other statistical software packages. We are also interested in your feedback and therefore, would like to encourage you to send comments to the email address mentioned on the web site. We have made every endeavor to keep this textbook as error-free as possible. However, if you think that you have encountered an error, please send us an email and we will include it in a list of errors that is also provided on the web site.

References

Abbott, A., & Tsay, A. (2000). Sequence analysis and optimal matching methods in sociology. *Sociological Methods & Research, 29*(1), 3–33.
Agresti, A. (2002). *Categorical data analysis*. New York: Wiley.
Agresti, A. (2010). *Analysis of ordinal categorical data*. New Jersey: Wiley.
Aldrich, J., & Nelson, F. (1984). *Linear probability, logit, and probit models*. Newbury Park: Sage.
Allison, P. (1982). Discrete-time methods for the analysis of event histories. *Sociological Methodology, 12*, 61–98.
Allison, P. (1984). *Event history analysis. Regression for longitudinal event data*. Beverly Hills: Sage.
Allison, P. (1995). *Survival analysis using the SAS system. A practical guide*. Cary: SAS Institute.
Allison, P. (2001). *Logistic regression using the SAS system. Theory and application*. Cary: SAS Institute.
Allison, P. (2002). *Missing data*. Thousand Oaks: Sage.
Allison, P. (2009). *Fixed effects regression models*. Thousand Oaks: Sage.
Allison, P. D. (1996). Fixed-effects partial likelihood for repeated events. *Sociological Methods & Research, 25*, 207–222.
Allison, P. D., & Christakis, N. A. (2006). Fixed-effects methods for the analysis of nonrepeated events. *Sociological Methodology, 36*(1), 155–172.
Amemiya, T. (1981). Qualitative response models: A survey. *Journal of Economic Literature, 19*, 1483–1536.
Andreß, H.-J., & Bröckel, M. (2007). Income and life satisfaction after marital disruption in Germany. *Journal of Marriage and Family, 69*(2), 500–512.
Annacker, D., & Hildebrandt, L. (2004). Unobservable effects in structural models of business performance. *Journal of Business Research, 57*(5), 507–517.
Arellano, M., & Carrasco, R. (2003). Binary choice panel data models with predetermined variables. *Journal of Econometrics, 115*, 125–157.
Atkinson, T., Cantillon, B., Marlier, E., & Nolan, B. (2002). *Social indicators: The EU and social inclusion*. Oxford: Oxford University Press.
Baltagi, B. (2008). *Econometric analysis of panel data*. Chichester: Wiley.
Beck, N. (2001). Time-series-cross-section-data: What have we learned in the past few years? *Annual Review of Political Science, 4*, 271–293.
Beck, N., & Katz, J. N. (1995). What to do (and not to do) with time-series cross-section data? *American Political Science Review, 89*(3), 634–647.
Beck, N., & Katz, J. N. (1996). Nuisance vs. substance: Specifying and estimating time-series-cross-section models. *Political Analysis, 6*(1), 1–36.
Beck, N., & Katz, J. N. (2011). Modeling dynamics in time-series-cross-section political economy data. *Annual Review of Political Science, 14*(1), 331–352.
Begg, C., & Gray, R. (1984). Calculation of polychotomous logistic regression parameters using individualized regressions. *Biometrika, 71*(1), 11–18.
Berelson, B. R., Lazarsfeld, P. F., & McPhee, W. N. (1954). *Voting. A study of opinion formation in a presidential campaign*. Chicago: University of Chicago Press.

Bergsma, W., Croon, M., & Hagenaars, J. (2010). *Marginal models: For dependent, clustered, and longitudinal categorical data*. Dordrecht: Springer.

Berk, R. (1983). An introduction to sample selection bias in sociological data. *American Sociological Review*, 48(3), 386–398.

Bethlehem, J. (2009). *Applied survey methods. A statistical perspective*. Hoboken: Wiley.

Blalock, H. M. J. (1970). Estimating measurement error using multiple indicators and several points in time. *American Sociological Review*, 35(1), 101–111.

Blossfeld, H.-P., Golsch, K., & Rohwer, G. (2007). *Event history analysis with Stata*. Mahwah: Lawrence Erlbaum Associates.

Blossfeld, H.-P., Hamerle, A., & Mayer, K. U. (1989). *Event history analysis*. Hillsdale: Lawrence Erlbaum Associates.

Blossfeld, H.-P., & Rohwer, G. (2002). *Techniques of event history modeling. New approaches to causal analysis*. Mahwah: Lawrence Erlbaum Associates.

Blundell, R., Griffith, R., & Windmeijer, F. (2002). Individual effects and dynamics in count data models. *Journal of Econometrics*, 108, 113–131.

Bollen, K. (1989). *Structural equations with latent variables*. New York: Wiley.

Bollen, K., & Curran, P. (2006). *Latent curve models: A structural equation perspective*. New York: Wiley-Interscience.

Bollen, K. A., & Brand, J. E. (2010). A general panel model with random and fixed effects: A structural equations approach. *Social Forces*, 89(1), 1–34.

Borooah, V. K. (2001). *Logit and probit: Ordered and multinomial models*. Thousand Oaks: Sage.

Bowles, S., Gintis, H., & Groves, M. (2008). *Unequal chances: Family background and economic success*. Princeton: Princeton University Press.

Box-Steffensmeier, J., De Boef, S., & Joyce, K. A. (2007). Event dependence and heterogeneity in duration models: The conditional frailty model. *Political Analysis*, 15, 237–256.

Box-Steffensmeier, J., & Zorn, C. (2002). Duration models for repeated events. *Journal of Politics*, 64, 1069–1094.

Breusch, T. S., & Pagan, A. R. (1980). The Lagrange multiplier test and its applications to model specification in econometrics. *The Review of Economic Studies*, 47(1), 239–253.

Brzinsky-Fay, C., & Kohler, U. (2010). New developments in sequence analysis. *Sociological Methods & Research*, 38(3), 359–364.

Cameron, A., & Trivedi, P. (1986). Econometric models based on count data: Comparisons and applications of some estimators and test. *Journal of Applied Econometrics*, 1, 29–53.

Cameron, A., & Trivedi, P. (1998). *Regression analysis of count data*. Cambridge: Cambridge University Press.

Cameron, A., & Trivedi, P. (2008). *Microeconometrics using Stata*. College Station: Stata Press.

Cameron, A. C., & Trivedi, P. K. (2005). *Microeconometrics: Methods and applications*. Cambridge: Cambridge University Press.

Chamberlain, G. (1980). Analysis of covariance with qualitative data. *Review of Economic Studies*, 47, 225–238.

Clark, T. S., & Linzer, D. A. (2012). *Should I use fixed or random effects?* Technical report, Emory University: Department of Political Science. http://polmeth.wustl.edu/mediaDetail.php?docId=1315 (accessed 18.4.2012).

Cleves, M., Gould, W., & Gutierrez, R. (2002). *An introduction to survival analysis using Stata*. College Station: Stata Press.

Cox, D., & Oakes, D. (1984). *Analysis of survival data*. London: Chapman Hall.

Das, M., Toepoel, V., & van Soest, A. (2011). Nonparametric tests of panel conditioning and attrition bias in panel surveys. *Sociological Methods & Research*, 40(1), 32–56.

Daykin, A., & Moffatt, P. (2002). Analyzing ordered responses: A review of the ordered probit model. *Understanding Statistics*, 1(1), 157–166.

De Leeuw, E. (2001). Reducing missing data in surveys: An overview of methods. *Quality & Quantity*, 35, 147–160.

Diggle, P., Heagerty, P., Liang, K., & Zeger, S. (2002). *Analysis of longitudinal data*. Cornwall: International Ltd.

DuMouchel, W., & Duncan, G. (1983). Using sample survey weights in multiple regression analyses of stratified samples. *Journal of the American Statistical Association, 78*, 535–543.

Duncan, G. J., Gustafsson, B., Hauser, R., Schmauss, G., Messinger, H., Muffels, R., Nolan, B., & Ray, J.-C. (1993). Poverty dynamics in eight countries. *Journal of Population Economics, 6*(3), 215–234.

Duncan, T., Duncan, S., & Strycker, L. (2006). *An introduction to latent variable growth curve modeling: Concepts, issues, and applications.* Mahwah: Lawrence Erlbaum Associates.

Ferrer-i-Carbonell, A. (2005). Income and well-being: An empirical analysis of the comparison income effect. *Journal of Public Economics, 89*, 997–1019.

Ferrer-i-Carbonell, A., & Frijters, P. (2004). How important is methodology for the estimates of the determinants of happiness? *The Economic Journal, 114*(497), 641–659.

Fitzmaurice, G., Laird, N., & Ware, J. (2011). *Applied longitudinal analysis.* Hoboken: Wiley.

Forthofer, R. N., & Lehnen, R. (1981). *Public program analysis: A new categorical data approach.* Belmont: Wadsworth.

Frees, E. W. (2004). *Longitudinal and panel data. Analysis and applications in the social sciences.* Cambridge: Cambridge University Press.

Frick, J. R., Goebel, J., Schechtman, E., Wagner, G. G., & Yitzhaki, S. (2006). Using analysis of Gini (ANOGI) for detecting whether two subsamples represent the same universe. *Sociological Methods & Research, 34*(4), 427–468.

Garrett, G., & Mitchell, D. (2001). Globalization, government spending and taxation in the OECD. *European Journal of Political Research, 39*, 145–177.

Goldstein, H. (2011). *Multilevel statistical models.* Hoboken: Wiley.

Goldstein, H., Pan, H., & Bynner, J. (2004). A flexible procedure for analysing longitudinal event histories using a multilevel model. *Understanding Statistics, 3*, 85–99.

Greene, W. (2004a). The behaviour of the maximum likelihood estimator of limited dependent variable models in the presence of fixed effects. *Econometrics Journal, 7*(1), 98–119.

Greene, W. (2004b). Fixed effects and bias due to the incidental parameters problem in the Tobit model. *Econometric Reviews, 23*(2), 125–147.

Greene, W. H. (2008). *Econometric analysis.* Upper Saddle River: Prentice Hall.

Grizzle, J. E., Starmer, C. F., & Koch, G. G. (1969). Analysis of categorical data by linear models. *Biometrics, 25*(3), 489–504.

Groves, R., Dillman, D. A., Eltinge, J. L., & Little, R. J. (2002). *Survey nonresponse.* New York: Wiley.

Groves, R., Fowler, F., Couper, M., Lepkowski, J., Singer, E., & Tourangeau, R. (2004). *Survey methodology.* Hoboken: Wiley.

Hachen, D. (1988). The competing risks model. A method for analyzing processes with multiple types of events. *Sociological Methods & Research, 17*, 21–54.

Hagenaars, J. (1990). *Categorical longitudinal data. Log-linear panel, trend, and cohort analysis.* London: Sage.

Hank, K. (2004). Effects of early life family events on women's late life labour market behaviour: An analysis of the relationship between childbearing and retirement in western Germany. *European Sociological Review, 20*(3), 189–198.

Hardin, J., & Hilbe, J. (2012). *Generalized estimating equations* (2nd ed.). London: Taylor & Francis.

Hausman, J. (1978). Specification tests in econometrics. *Econometrica, 46*, 1251–1271.

Hausman, J., Hall, B., & Griliches, Z. (1984). Econometric models for count data with an application to the patents-r&d relationship. *Econometrica, 52*, 909–938.

Hausman, J. A., & Taylor, W. E. (1981). Panel data and unobservable individual effects. *Econometrica, 49*(6), 1377–1398.

Heckman, J. (1979). Sample selection bias as a specification error. *Econometrica, 47*(4), 153–161.

Heckman, J. (1981a). Heterogeneity and state dependence. In S. Rosen (Ed.), *Studies in labor markets* (pp. 91–140). Cambridge: MIT Press.

Heckman, J. (1981b). Statistical models for discrete panel data. In C. Manski & D. McFadden (Eds.), *Structural analysis of discrete data* (Chapter 3). Cambridge: MIT Press.

Heckman, J., & Willis, R. (1976). Estimation of a stochastic model of reproduction: An econometric approach. In *Household production and consumption* (pp. 99–146). New York: National Bureau of Economic Research.

Heineck, G., & Schwarze, J. (2004). Fly me to the moon: The determinants of secondary jobholding in Germany and the UK. Technical report 1358, IZA discussion paper.

Heise, D. R. (1969). Separating reliability and stability in test-retest correlation. *American Sociological Review, 34*(1), 93–101.

Hill, D., Axinn, W., & Thornton, A. (1993). Competing hazards with shared unmeasured risk factors. *Sociological Methodology, 23*, 245–277.

Hill, M. (1997). *The panel study of income dynamics. A user's guide*. Newbury Park: Sage.

Hirano, K., Imbens, G., Ridder, G., & Rubin, D. (2001). Combining panel data sets with attrition and refreshment samples. *Econometrica, 69*(6), 1645–1659.

Honoré, B. (2002). Nonlinear models with panel data. *Portuguese Economic Journal, 1*, 163–179.

Honoré, B., & Kyriazidou, E. (2000). Panel data discrete choice models with lagged dependent variables. *Econometrica, 68*(4), 839–874.

Honoré, B., & Lewbel, A. (2002). Semiparametric binary choice panel data models without strictly exogeneous regressors. *Econometrica, 70*, 2053–2063.

Hosmer, D., & Lemeshow, S. (2000). *Applied logistic regression*. New York: Wiley.

Hox, J. (2010). *Multilevel analysis: Techniques and applications*. London: Routledge.

Hsiao, C. (2003). *Analysis of panel data (Econometric Society Monographs)*. Cambridge: Cambridge University Press.

Inglehart, R. (1971). The silent revolution in Europe: Intergenerational change in postindustrial societies. *American Political Science Review, 65*, 991–1017.

Johnson, D. R., & Wu, J. (2002). An empirical test of crisis, social selection, and role explanations of the relationship between marital disruption and psychological distress: A pooled time-series analysis of four-wave panel data. *Journal of Marriage and Family, 64*(1), 211–224.

Kalbfleisch, J. D., & Prentice, R. L. (1980). *The statistical analysis of failure time data*. New York: Wiley.

Kalton, G. (1986). Handling wave nonresponse in panel surveys. *Journal of Official Statistics, 2*(3), 303–314.

Kalton, G. (1987). *Introduction to survey sampling*. Beverly Hills: Sage.

Kasprzyk, D., Duncan, G., Kalton, G., & Singh, M. (Eds.) (1989). *Panel surveys*. New York: Wiley.

Kirk, R. (1995). *Experimental design: Procedures for the behavioral sciences*. Pacific Grove: Brooks/Cole.

Kish, L. (1965). *Survey sampling*. New York: Wiley.

Kittel, B., & Winner, H. (2005). How reliable is pooled analysis in political economy? The globalization-welfare state nexus revisited. *European Journal of Political Research, 44*(2), 269–293.

Klein, M., & Pötschke, M. (2004). Die intra-individuelle Stabilität gesellschaftlicher Wertorientierungen. *KZfSS Kölner Zeitschrift für Soziologie und Sozialpsychologie, 56*(3), 432–456.

Kline, R. (2010). *Principles and practice of structural equation modeling* (3rd ed.). New York: Guilford Press.

Kohler, U., & Kreuter, F. (2005). *Data analysis using Stata*. College Station: Stata Press.

Kroh, M. (2012). *Documentation of sample sizes and panel attrition in the German Socio Economic Panel (SOEP) (1984 until 2011)*, No. 66. Deutsches Institut für Wirtschaftsforschung, Berlin.

Lancaster, T. (1990). *The econometric analysis of transition data*. Cambridge: Cambridge University Press.

Lazarsfeld, P. F. (1940). Panel studies. *Public Opinion Quarterly, 4*, 122–128.

Lazarsfeld, P. F., Berelson, B., & Gaudet, H. (1944). *The peoples choice. How the voter makes up his mind in a presidential campaign*. New York: Duell, Sloan and Pearce.

Lazarsfeld, P. F., & Fiske, M. (1938). The panel as a new tool for measuring opinion. *Public Opinion Quarterly, 2*, 596–612.

Lee, L. (1982). Specification error in multinomial logit models: Analysis of the omitted variable bias. *Journal of Econometrics, 20*, 197–209.

Liang, K.-Y., & Zeger, S. L. (1986). Longitudinal data analysis using generalized linear models. *Biometrika, 73*, 13–22.

Little, R., & Rubin, D. (1987). *Statistical analysis with missing data*. New York: Wiley.

Little, T. D., Preacher, K. J., Selig, J. P., & Card, N. A. (2007). New developments in latent variable panel analyses of longitudinal data. *International Journal of Behavioral Development, 31*(4), 357–365.

Long, S. (1997). *Regression models for categorical and limited dependent variables*. Thousand Oaks: Sage.

Long, S., & Freese, J. (2006). *Regression models for categorical dependent variables using Stata* (2nd ed.). College Station: Stata Press.

Lynn, P. (2009). *Methodology of longitudinal surveys*. Chichester: Wiley.

Lynn, P., Buck, N., Burton, J., Jäckle, A., & Laurie, H. (2005). *A review of methodological research pertinent to longitudinal survey design and data collection*. Technical report 29, University of Essex, Institute for Social and Economic Research.

MacIndoe, H., & Abbott, A. (2004). Sequence analysis and optimal matching techniques for social science data. In M. Hardy & A. Bryman (Eds.), *Handbook of data analysis* (pp. 387–406). London: Sage.

Maclure, M. (1991). The case-crossover design: A method for studying transient effects on the risk of acute events. *American Journal of Epidemiology, 133*(2), 144–153.

Maddala, G. S. (1987). Limited dependent variable models using panel data. *The Journal of Human Resources, 22*(3), 307–338.

McKelvey, R., & Zavoina, W. (1975). A statistical model for the analysis of ordinal level dependent variables. *Journal of Mathematical Sociology, 4*, 103–120.

Menard, S. (2008). *Handbook of longitudinal research. Design, measurement, and analysis*. Amsterdam: Elsevier/Academic Press.

Mitchell, M. (2010). *Data management using Stata: A practical handbook*. College Station: Stata Press.

Mundlak, Y. (1978). On the pooling of time series and cross section data. *Econometrica, 46*(1), 69–85.

Oud, J. H. L., & Delsing, M. J. M. H. (2010). Continuous time modeling of panel data by means of SEM. In K. van Montfort, J. Oud, & A. Satorra (Eds.), *Longitudinal research with latent variables* (pp. 201–244). New York: Springer.

Pendergast, J., Gange, S., Newton, M., Lindstrom, M., Palta, M., & Fisher, M. (1996). A survey of methods for analyzing clustered binary response data. *International Statistical Review, 64*, 89–118.

Pfeffermann, D. (1993). The role of sampling weights when modeling survey data. *International Statistical Review, 61*, 317–337.

Pfeffermann, D., & Nathan, G. (2002). Imputation for wave nonresponse: Existing methods and a time series approach. In R. Groves, D. Dilman, J. Eltinge & R. Little (Eds.), *Survey nonresponse*, New York: Wiley.

Puhani, P. (2000). The Heckman correction for sample selection and its critique. *Journal of Economic Surveys, 14*(1), 53–68.

Rabe-Hesketh, S., & Skrondal, A. (2008). *Multilevel and longitudinal modeling using Stata*. College Station: Stata Press.

Rose, D. (Ed.) (2000). *Researching social and economic change: The uses of household panel studies*. London: Routledge.

Rubin, D. (1976). Inference with missing data. *Biometrika, 63*, 581–592.

Rubin, D. (1987). *Multiple imputation*. New York: Wiley.

Ruspini, E. (2002). *Introduction to longitudinal research*. London: Routledge.

Schafer, J., & Olsen, M. (1998). Multiple imputation for multivariate missing-data problems: A data analyst's perspective. *Multivariate Behavioral Research, 33*(4), 545–571.

Schumacker, R., & Lomax, R. (2004). *A beginner's guide to structural equation modeling*. New Jersey: Lawrence Erlbaum Associates.
Schupp, J., & Wagner, G. G. (1995). Die Zuwanderer-Stichprobe des Sozio-ökonomischen Panels (SOEP). *Vierteljahreshefte zur Wirtschaftsforschung, 64*, 16–25.
Shrout, P. E., & Fleiss, J. L. (1979). Intraclass correlations: Uses in assessing rater reliability. *Psychological Bulletin, 86*(2), 420–428.
Singer, J., & Willett, J. (2003). *Applied longitudinal data analysis. Modeling change and event occurrence*. Oxford: Oxford University Press.
Snijders, T., & Bosker, R. (2011). *Multilevel analysis: An introduction to basic and advanced multilevel modeling*. Thousand Oaks: Sage.
Steele, F., Goldstein, H., & Browne, W. (2004). A general multistate competing risks model for event history data, with an application to a study of contraceptive use dynamics. *Statistical Modelling, 4*, 145–159.
Sturgis, P., Allum, N., & Brunton-Smith, I. (2009). Attitudes over time: The psychology of panel conditioning. In P. Lynn (Ed.), *Methodology of longitudinal surveys* (pp. 113–126). Hoboken: Wiley.
Suissa, S. (1995). The case-time-control design. *Epidemiology, 6*(3), 248–253.
Taris, T. (2000). *A primer in longitudinal analysis*. London: Sage.
van de Pol, F., & Langeheine, R. (1990). Mixed Markov latent class models. *Sociological Methodology, 20*, 213–247.
Van der Zouwen, J., & Van Tilburg, T. (2001). Reactivity in panel studies and its consequences for testing causal hypotheses. *Sociological Methods & Research, 30*(1), 35–56.
Vella, F., & Verbeek, M. (1998). Whose wages do unions raise? A dynamic model of unionism and wage rate determination for young men. *Journal of Applied Econometrics, 13*, 163–183.
Vermunt, J. (1997). *Log linear models for event histories*. Thousand Oaks: Sage.
Vermunt, J., Tran, B., & Magidson, J. (2008). Latent class models in longitudinal research. In S. Menard (Ed.), *Handbook of longitudinal research: Design, measurement, and analysis* (pp. 373–385). Amsterdam: Elsevier.
White, H. (1980). A heteroskedasticity-consistent covariance matrix estimator and a direct test for heteroskedasticity. *Econometrica, 48*(4), 817–830.
Wiley, D. E., & Wiley, J. A. (1970). The estimation of measurement error in panel data. *American Sociological Review, 35*(1), 112–117.
Winkelmann, L., & Winkelmann, R. (1998). Why are the unemployed so unhappy? Evidence from panel data. *Economica, 65*(1), 1–15.
Winkelmann, R. (2003). *Econometric analysis of count data* (3rd ed.). Berlin: Springer.
Winkelmann, R., & Zimmermann, K. (1986). Recent developments in count data modelling: Theory and application. *Journal of Economic Surveys, 9*(1), 1–24.
Winship, C., & Radbill, L. (1994). Sampling weights and regression analysis. *Sociological Methods & Research, 23*(2), 230–257.
Wooldridge, J. M. (2009). *Introductory econometrics. A modern approach*. Mason: South-Western.
Wooldridge, J. M. (2010). *Econometric analysis of cross section and panel data* (2nd ed.). Cambridge: MIT Press.
Yamaguchi, K. (1986). Alternative approaches to unobserved heterogeneity in the analysis of repeatable events. *Sociological Methodology, 16*, 213–249.
Yamaguchi, K. (1991). *Event history analysis*. Newbury Park: Sage.

Index

A
Age effects, 5, 80, 192
AIC, *see* Akaike's information criterion
Akaike's information criterion (AIC), 265, 311
Analysis of variance, 291
Appending data, 20, 30
Attrition, *see* Panel attrition
Autocorrelation, *see* Serial correlation
Autoregressive correlation structure, *see* Correlation structure

B
Balanced panel, *see* Panel data
Bayesian information criterion (BIC), 265, 311
BE, *see* Between effects
Best linear unbiased estimator (BLUE), 123, 136, 185, 297
Between effects (BE), 158, 163
Between-unit variance, 76, 158
BHPS, *see* British Household Panel Survey
Bias, 99, 142
    measurement attenuation, 113
    measurement error, 109, 113
    non-response, 9, 49, 56
    omitted variable, 5, 99, 108, 141, 198, 205, 266
BIC, *see* Baysian information criterion
BLUE, *see* Best linear unbiased estimator
Breusch–Pagan test, 151
British Household Panel Survey (BHPS), 205, 288

C
Case-crossover design, 274
Case-time-control design, 276
Categorical variables, *see* Variables
Causality, 6
Censoring, 92, 252
    interval-censoring, 278
    left censoring, 252
    right censoring, 252

Centered effects, 130
Change scores, 89, 180
CML, *see* Maximum likelihood, conditional
CNEF, *see* Cross National Equivalent File
Coefficient of determination, *see* Explained variance
Cohort
    cohort analysis, 80, 192
        identification problem, 194
    cohort effects, 5, 80, 192
    synthetic, 5
    true, 5
Complementary log–log hazard model, 278
Conditional effect plot, 215
Conditional maximum likelihood (CML), *see* Maximum likelihood
Confidence interval, 302, 304
Constraint matrix, 305
Continuous variables, *see* Variables
Continuous-time data, 251
Continuous-time process, 278
Cornered effects, 130
Correlation structure
    autoregressive, 86, 176
    exchangeable, 86, 176
    independent, 86
Costs, 11
Covariance structure, 86
Cross National Equivalent File (CNEF), 288

D
Demeaning, *see also* Variables, time-demeaned
    degrees of freedom, 137
    demeaning parameter θ, 154, 171
Dependent observations, 6, 40, 65
Destination state, 69
Differencing, 176
Discrete change, 215
Discrete response models, 88, 204
Discrete-time data, 251
Distributed lag models, 94

Dummy coding, 130
Duration effects, 260
Dynamic models, 94, 110, 247

E
ECHP, *see* European Community Household Panel
Effect coding, 130
Efficiency, 55, 99, 107, 123, 160, 297
Empty model, *see* Null model
Endogeneity, 120
Episodes, 251
    multiple, 283
Equivalized disposable household income, 34, 39
Error components model, 152
Error components models, *see* Variance components models
Error (term), 122, 127, 146, 175
    error variance, 146, 240
    heteroscedastic, 107, 174
    idiosyncratic, 102, 114, 127, 146, 227
    first-order autoregressive, 176
Errors of classification, 110
Estimation, 297
Estimation assumptions, 114, *see also* Textboxes
EU-SILC, *see* European Union Statistics of Income and Living Conditions
European Community Household Panel (ECHP), 288
European Union Statistics of Income and Living Conditions (EU-SILC), 288
Event, 72, 90, 251
    event history, 251
    multiple events, 251
    repeatable events, 251
    single event, 251
Event history analysis, 72, 92, 248, 291
Exchangeable correlation structure, *see* Correlation structure
Exogeneity
    contemporaneous, 115
    exogeneity assumption, 107, 114, 205
    strict, 115, 135, 153, 184
Experiment, 6, 104
Explained variance, 138, 156, 296
    adjusted R^2, 296
    McFadden's Pseudo R^2, 311
    within R^2, 138

F
F test, 139, 157, 306
    linear restrictions, 305

overall model fit, 305
Factor analysis, 113
FD, *see* First difference
FE, *see* Fixed effects
Feasible generalized least squares (FGLS), 155, 198
FGLS, *see* Feasible generalized least squares
First difference (FD), 89, 95, 103, 120, 180, 184
    FD versus FE, 189
Fixed effects event history model, 274
Fixed effects (FE), 104, 127, 128, 133, 135, 229
    FE versus FD, 189
    FE versus RE, 163, 170
    forecasts, 170
    regression constant, 138
    residuals, 144
    serial correlation, 144
    testing differences to RE, 166

G
GEE, *see* Generalized estimation equation
Generalized estimating equations (GEE), 86, 109
Generalized ordinary least squares (GLS), 152, 154
German Socio-economic Panel (SOEP), 8, 31, 181, 193, 249, 288
GLS, *see* Generalized ordinary least squares
Growth model, 75, 88, 291

H
Hausman test, 168, 243
Hazard rate, *see also* Event history analysis
    discrete-time, 73, 91, 253
    hazard function, 74
    hazard rate model, 92
    instantaneous, 280
    interval-specific, 281
    logistic discrete-time hazard model, 257
    proportional hazard rate model, 279
Heterogeneity
    correlated, 128, 229, 274
    estimate of unit-specific, 142
    uncorrelated, 127, 147, 236, 271
    unobserved, 99, 102, 125, 126, 142, 151, 227, 266
    correlation of, 143
Heteroscedastic consistent estimator, *see* Standard errors, robust
Heteroscedasticity, 98, 108, 109, 212, *see also* Error (term)
Hierarchical data, *see* Nested data

Index

Hierarchical modeling, *see* Multilevel modeling
HILDA, *see* Household, Income and Labor Dynamics in Australia Survey
Homoscedasticity, 107, 211
    homoscedasticity assumption, 108, 114, 123
Household, Income and Labor Dynamics in Australia (HILDA), 288
Hybrid models, 163, 164, 173, 245
    type 1, 164
    type 2, 167

I

IAB Establishment Panel (IAB-EP), 2, 288
IAB-EP, *see* IAB Establishment Panel
ICC, *see* Intra-class correlation coefficient
Idiosyncratic error, *see* Error (term)
Impact functions, 90, 185
    linear, 191
Imputation, 16
Incidental parameters problem, 230
Independent correlation structure, *see* Correlation structure
Independent observations, 108
Individual change, 2
Inference, 49, 301
Intra-class correlation coefficient, 77, 146, 153

K

KLIPS, *see* Korea Labor Income Panel Survey
Korea Labor Income Panel Survey (KLIPS), 288

L

LAD, *see* Least absolute difference estimation
Lagrange multiplier (LM) test, 151
Late entry, 16
Latent class, 110
Latent class analysis, 113
LCA, *see* Latent class analysis (LCA)
Least absolute difference estimation (LAD), 296
Least absolute difference (LAD), 300
Least squares dummy variables (LSDV), 128
Level, 3
Likelihood ratio test, 263, 311
Linear probability model
    pooled, 209
Linear regression model, 119
Linear restrictions, 139, 157, 190, 304, *see also* F test
LM, *see* Lagrange multiplier

Logistic discrete-time hazard rate model, *see* Hazard rate
Logistic distribution function, 254
Logistic regression, 88, 257, 308
    logistic regression function, 214
    multinomial, 248, 283
    ordinal, 248
    pooled, 213
Logit, 216
Long format, *see* Panel data
Longitudinal population, 9, 43, 52
LSDV, *see* Least squares dummy variables

M

Macro panels, 63, 129, 291
Marginal effect, 215
Markov modeling, 71, 91, 92, 292
Master data set, 23, 33
Maximum likelihood (ML), 197, 217, 307
    asymptotically normally distributed estimates, 310
    conditional (CML), 106, 231, 274
    full, 198
    likelihood function, 217, 307
    restricted, 198
    variance-covariance matrix of ML estimates, 310
McNemar's test, 105
Mean square error, 301
Measurement error, 7, 107, 115
Measurement models, 110
Merging data, 21, 30
    many-to-one merge, 22, 26
    one-to-many merge, 22, 37
    one-to-one merge, 38
Micro panels, 63
Missing data, 16, 49, 249
    missing at random, 56, 57
    missing completely at random, 56
    not missing at random, 56
Mixed models, 96
Multilevel modeling, 65, 96, 291
Multiple-risk model, 283

N

National Longitudinal Surveys (NLS), 288
Nested data, 18, 23, 64
NLS, *see* National Longitudinal Surveys
No autocorrelation assumption, *see* Serial correlation
Non-linearity, 214
Non-recursive models, 94
Non-response, 56
    item, 49

Non-response (cont.)
    temporary, 16, 53
    unit, 15, 49
Null model, 124, 311

O

Odds, 216
Odds ratios, 217
OLS, *see* Ordinary least squares
Omitted variable bias, *see* Bias
Ordinary least squares (OLS), 122, 295, 296
    pooled, 121, 170
        matrix weighted average of BE and FE, 158
        serial correlation, 123
Origin state, 69

P

Panel attrition, 8, 16, 53
Panel conditioning, 10
Panel data, 1
    balanced, 15, 53, 62
    long format, 16, 27, 42, 63
    raw, 19
    raw data, 33
    unbalanced, 15, 53, 62, 177, 218, 260
    wide format, 16, 27, 42, 47, 63, 177
Panel effect, 10
Panel mortality, *see* Panel attrition
Panel Study of Income Dynamics (PSID), 4, 288
Period effects, 80, 192
Poisson model, 248
Pooled time-series cross-section data (TSCS), 64, 291
Population parameters, 49, 98, 123, 198
Pre-test post-test design, 6, 104
Probit regression, 88
    ordinal, 248
    pooled, 223
    probit function, 224
Proportional hazard rate model, *see* Hazard rate
PSID, *see* Panel Study of Income Dynamics

Q

Quasi-demeaning, 154

R

R^2, *see* Explained variance
Random coefficients, 95, 174, 199
Random effects event history model, 271
Random effects (RE), 127, 147, 152, 236
    matrix weighted average of BE and FE, 158
    ML estimation of, 197
    RE versus FE, 163, 171
    serial correlation, 153
    testing differences to FE, 166
Random parameters, 95
Random slopes, *see* Random coefficients
Randomization, 6, 104
RE, *see* Random effects
Recursive models, 94
Replicates, 298
Research examples
    Andress and Bröckel 2007 (genderdiff data), 181
    cancer data, 267
    data management (mypanel data), 19
    efficiency data, 160
    Garrett and Mitchell 2001 (garmit data), 128
    Hank 2004 (hank data), 249
    Heineck and Schwarze 2004 (heineck-schwarze data), 205
    heterogeneity bias (hetbias and nohetbias data), 100
    Johnson and Wu 2002 (johnson-wu data), 147, 191, 207
    Klein and Potschke 2004 (postmat data), 193
    sixcases data, 295
    SOEP data management (SOEP data), 31
    SOEP wage panel (wpgen data), 297
    Vella and Verbeek 1998 (wagepan data), 65, 120, 140
Restricted maximum likelihood, *see* Maximum likelihood
Restricted model, 305
Risk period, 251
Risk set, 251
Robust standard errors, *see* Standard errors
Root mean squared error, 303
Rotating panels, 10

S

Sample
    refreshment, 10
    small sample size, 6
Sampling design, 49
Sampling distribution
    empirical, 298
    theoretical, 298
Sampling errors, 50, 298
Seam effects, 11
Selection effect, 269

Index 323

Selection probabilities, 49
SEM, *see* Structural equation modeling (SEM)
Semi-parametric model, 266
Sequence, 71
Sequence analysis, 292
Serial correlation, 83, 107, 112, 144, 211, *see also* Correlation structure
    autocorrelation, 67, 175, 211
    equal correlation assumption, 147
    no autocorrelation assumption, 107, 115, 123, 151
    serial correlation coefficient, 67, 124
SHP, *see* Swiss Household Panel
Simple random sample, 49, 55, 122, 197
Simulation, 298
Social Expenditure Data Base (SOCX), 2, 288
SOCX, *see* Social Expenditure Data Base
SOEP, *see* German Socio-economic Panel
Spell, 72, 251
SRS, *see* Simple random sample
Standard errors, 57, 70, 99, 299, 303
    empirical, *see* Robust
    estimated, 304
    robust, 57, 109, 125, 205, 212, 221, 307
        cluster, 126, 170, 212
        heteroscedastic consistent estimator, 57, 109, 212
    standard error of the regression, *see* Root mean squared error
    theoretical, 107, 304, 307
State dependence
    spurious, 83
    true, 83, 94, 110
State probability, 69
Static models, 94
Structural equation modeling (SEM), 113, 179, 291, 292
Survey population, 32
Survival analysis, *see* Event history analysis
Survival probability, 72, 253
Survivor function, 72, 253
Swiss Household Panel (SHP), 288

T

T test, 304
    for dependent observations, 103
Target population, 15, 33
Test–retest reliability, 7
Test-retest reliability, 113
Testing, 303
Textboxes
    BE estimation, 158
    degrees of freedom for time-demeaned data, 137
    FD assumptions, 184
    FE assumptions, 135
    FE versus FD estimation, 189
    hierarchical linear models, 96
    intra-class correlation coefficient, 77
    latent variable specification for discrete response model, 228
    ML assumptions, 219
    ML estimation of random effects models, 197
    OLS assumptions, 122
    RE assumptions, 152
Time trends, 192
    continuous, 88, 190
    discontinuous, 82, 88
    linear, 190
Time-constant variables, *see* Variables
Time-series operators, 45
Time-varying variables, *see* Variables
Trajectories, 74
Transition analysis, *see* Event history analysis
Transition intensity, 280
Transition probability, 69
    conditional, 73, 91, 253, 278
Trend, 3
TSCS, *see* Pooled time-series cross-section data

U

Unbalanced panel, *see* Panel data
Unbiasedness, 99, 123, 297, *see also* Bias
Unit, 62
Unobserved heterogeneity, *see* Heterogeneity
Unrestricted model, 305

V

Variables
    categorical, 61, 88, 91, 105, 203
    continuous, 61, 87, 89, 119
    count, 248
    key, 17, 18, 30
    lagged, 44, 66, 83, 91, 94, 227, 247
    latent, 110
    lead, 66
    manifest, 110
    names, 17, 30
    quasi-demeaned, 154
    random, 152
    time-constant, 46, 79, 93, 147, 234, 236
    time-demeaned, 84, 133, 134, 229

Variables (*cont.*)
   time-varying, 46, 79, 93, 184, 257
Variance components, 155, 199
Variance components models, 98, 152, 194, 291

W

Wald test, 157, 168, 244, 307
Web site, 312
Weighting, 16, 49
   cross-sectional weights, 50
   design weights, 49
   longitudinal weights, 53
   population weights, 49
   redressment weights, *see* Population weights
Wide format, *see* Panel data
Within estimator, 136
Within-unit variance, 76, 89, 136
Working data set, 15, 33

Z

Z test, 156, 303, 310

Author Index

A
Abbott and Tsay (2000), 292
Agresti (2002), 247, 285
Agresti (2010), 248
Aldrich and Nelson (1984), 285
Allison (1982), 283, 285
Allison (1984), 285
Allison (1995), 283, 285
Allison (1996), 284
Allison (2001), 285
Allison (2002), 58
Allison (2009), 202, 285
Allison and Christakis (2006), 274
Amemiya (1981), 285
Andreß and Bröckel (2007), 181, 185–187
Annacker and Hildebrandt (2004), 180
Arellano and Carrasco (2003), 247
Atkinson et al. (2002), 2

B
Baltagi (2008), 12, 158, 201, 291
Beck (2001), 291
Beck and Katz (1995), 291
Beck and Katz (1996), 291
Beck and Katz (2011), 291
Begg and Gray (1984), 283
Berelson et al. (1954), 1
Bergsma et al. (2010), 292
Berk (1983), 58
Bethlehem (2009), 59
Blalock (1970), 292
Blossfeld and Rohwer (2002), 285
Blossfeld et al. (1989), 285
Blossfeld et al. (2007), 285, 292
Blundell et al. (2002), 248
Bollen (1989), 292
Bollen and Brand (2010), 180, 202
Bollen and Curran (2006), 202, 291
Borooah (2001), 248
Bowles et al. (2008), 50
Box-Steffensmeier and Zorn (2002), 284
Box-Steffensmeier et al. (2007), 284
Breusch and Pagan (1980), 151
Brzinsky-Fay and Kohler (2010), 292

C
Cameron and Trivedi (1986), 248
Cameron and Trivedi (1998), 248
Cameron and Trivedi (2005), 12, 202, 291
Cameron and Trivedi (2008), 12
Chamberlain (1980), 248, 285
Clark and Linzer (2012), 173
Cleves et al. (2002), 285
Cox and Oakes (1984), 285

D
Das et al. (2011), 10
Daykin and Moffatt (2002), 248
De Leeuw (2001), 59
Diggle et al. (2002), 291
DuMouchel and Duncan (1983), 57, 58
Duncan et al. (1993), 3, 4
Duncan et al. (2006), 202, 291

F
Ferrer-i-Carbonell (2005), 248
Ferrer-i-Carbonell and Frijters (2004), 248
Fitzmaurice et al. (2011), 291
Forthofer and Lehnen (1981), 217
Frees (2004), 291
Frick et al. (2006), 10

G
Garrett and Mitchell (2001), 63, 128
Goldstein (2011), 291
Goldstein et al. (2004), 283
Greene (2004a), 230
Greene (2004b), 230
Greene (2008), 154, 158, 177, 202, 291, 293
Grizzle et al. (1969), 217
Groves et al. (2002), 58
Groves et al. (2004), 59

H
Hachen (1988), 283
Hagenaars (1990), 292
Hank (2004), 249, 253, 255, 263, 266
Hardin and Hilbe (2012), 86
Hausman (1978), 168, 169, 243
Hausman and Taylor (1981), 169
Hausman et al. (1984), 248
Heckman (1979), 57, 58
Heckman (1981a), 247
Heckman (1981b), 247, 285
Heckman and Willis (1976), 285
Heineck and Schwarze (2004), 205, 206, 209, 223, 236
Heise (1969), 292
Hill (1997), 287
Hill et al. (1993), 283
Hirano et al. (2001), 58
Honoré (2002), 247
Honoré and Kyriazidou (2000), 247
Honoré and Lewbel (2002), 247
Hosmer and Lemeshow (2000), 247, 248, 285
Hox (2010), 202, 291
Hsiao (2003), 12, 202, 285, 291

I
Inglehart (1971), 193

J
Johnson and Wu (2002), 147, 156, 164, 191

K
Kalbfleisch and Prentice (1980), 285
Kalton (1986), 59
Kalton (1987), 59
Kasprzyk et al. (1989), 287
Kirk (1995), 291
Kish (1965), 59
Kittel and Winner (2005), 129
Klein and Pötschke (2004), 193, 195, 201
Kline (2010), 292
Kohler and Kreuter (2005), 58
Kroh (2012), 8

L
Lancaster (1990), 285
Lazarsfeld (1940), 1
Lazarsfeld and Fiske (1938), 1
Lazarsfeld et al. (1944), 1
Lee (1982), 248
Liang and Zeger (1986), 285
Little and Rubin (1987), 58
Little et al. (2007), 292

Long (1997), 247
Long and Freese (2006), 247, 285
Lynn (2009), 287
Lynn et al. (2005), 287

M
MacIndoe and Abbott (2004), 292
Maclure (1991), 274
Maddala (1987), 247, 285
McKelvey and Zavoina (1975), 248
Menard (2008), 287
Mitchell (2010), 58
Mundlak (1978), 164

O
Oud and Delsing (2010), 291

P
Pendergast et al. (1996), 285
Pfeffermann (1993), 58
Pfeffermann and Nathan (2002), 59
Puhani (2000), 58

R
Rabe-Hesketh and Skrondal (2008), 12
Rose (2000), 287
Rubin (1976), 58
Rubin (1987), 59
Ruspini (2002), 287

S
Schafer and Olsen (1998), 59
Schumacker and Lomax (2004), 292
Schupp and Wagner (1995), 10
Shrout and Fleiss (1979), 78
Singer and Willett (2003), 285, 291, 292
Snijders and Bosker (2011), 202, 291
Steele et al. (2004), 283
Sturgis et al. (2009), 10
Suissa (1995), 274

T
Taris (2000), 287, 291

V
van de Pol and Langeheine (1990), 292
Van der Zouwen and Van Tilburg (2001), 10
Vella and Verbeek (1998), 63, 65, 120, 121
Vermunt (1997), 283
Vermunt et al. (2008), 292

W
White (1980), 57, 58
Wiley and Wiley (1970), 292
Winkelmann (2003), 248
Winkelmann and Winkelmann (1998), 248

Winkelmann and Zimmermann (1986), 248
Winship and Radbill (1994), 58
Wooldridge (2009), 12, 189, 201
Wooldridge (2010), 12, 136, 153, 185, 202, 291

Y
Yamaguchi (1986), 284
Yamaguchi (1991), 285, 292

GPSR Compliance

The European Union's (EU) General Product Safety Regulation (GPSR) is a set of rules that requires consumer products to be safe and our obligations to ensure this.

If you have any concerns about our products, you can contact us on

ProductSafety@springernature.com

In case Publisher is established outside the EU, the EU authorized representative is:

Springer Nature Customer Service Center GmbH
Europaplatz 3
69115 Heidelberg, Germany

www.ingramcontent.com/pod-product-compliance
Ingram Content Group UK Ltd.
Pitfield, Milton Keynes, MK11 3LW, UK
UKHW022132230426
470314UK00005BA/34